Data Science

Data Science: Measuring Uncertainties

Editors

Carlos Alberto De Bragança Pereira
Adriano Polpo
Agatha Rodrigues

MDPI • Basel • Beijing • Wuhan • Barcelona • Belgrade • Manchester • Tokyo • Cluj • Tianjin

Editors
Carlos Alberto De Bragança Pereira
Brazil and University of Sao Paulo
Brazil

Adriano Polpo
University of Western Australia
Australia

Agatha Rodrigues
Federal University of Espirito Santo
Brazil

Editorial Office
MDPI
St. Alban-Anlage 66
4052 Basel, Switzerland

This is a reprint of articles from the Special Issue published online in the open access journal *Entropy* (ISSN 1099-4300) (available at: https://www.mdpi.com/journal/entropy/special_issues/data_science_uncertainties).

For citation purposes, cite each article independently as indicated on the article page online and as indicated below:

LastName, A.A.; LastName, B.B.; LastName, C.C. Article Title. *Journal Name* **Year**, *Volume Number*, Page Range.

ISBN 978-3-0365-0792-7 (Hbk)
ISBN 978-3-0365-0793-4 (PDF)

© 2021 by the authors. Articles in this book are Open Access and distributed under the Creative Commons Attribution (CC BY) license, which allows users to download, copy and build upon published articles, as long as the author and publisher are properly credited, which ensures maximum dissemination and a wider impact of our publications.

The book as a whole is distributed by MDPI under the terms and conditions of the Creative Commons license CC BY-NC-ND.

Contents

About the Editors vii

Carlos Alberto de Braganca Pereira, Adriano Polpo and Agatha Sacramento Rodrigues
Data Science: Measuring Uncertainties
Reprinted from: *Entropy* **2020**, *22*, 1438, doi:10.3390/e22121438 1

Erlandson Saraiva, Adriano Suzuki, Luis Milan and Carlos Pereira
An Integrated Approach for Making Inference on the Number of Clusters in a Mixture Model
Reprinted from: *Entropy* **2019**, *21*, 1063, doi:10.3390/e21111063 7

Jenny Farmer, Zach Merino, Alexander Gray and Donald Jacobs
Universal Sample Size Invariant Measures for Uncertainty Quantification in Density Estimation
Reprinted from: *Entropy* **2019**, *21*, 1120, doi:10.3390/e21111120 25

Graziadei, Lijoi, Lopes, Marques and Prünster
Prior Sensitivity Analysis in a Semi-Parametric Integer-Valued Time Series Model
Reprinted from: *Entropy* **2020**, *22*, 69, doi:10.3390/e22010069 47

Paulo Canas Rodrigues, Jonatha Pimentel, Patrick Messala, Mohammad Kazemi
The Decomposition and Forecasting of Mutual Investment Funds Using Singular Spectrum Analysis
Reprinted from: *Entropy* **2020**, *22*, 83, doi:10.3390/e22010083 59

Chenguang Lu
Channels' Confirmation and Predictions' Confirmation: From the Medical Test to the Raven Paradox
Reprinted from: *Entropy* **2020**, *22*, 384, doi:10.3390/e22040384 81

Enrique Hernández-Lemus
On a Class of Tensor Markov Fields
Reprinted from: *Entropy* **2020**, *22*, 451, doi:10.3390/e22040451 109

Ang Li, Luis Pericchi and Kun Wang
Objective Bayesian Inference in Probit Models with Intrinsic Priors Using Variational Approximations
Reprinted from: *Entropy* **2020**, *22*, 513, doi:10.3390/e22050513 123

Ping Li, Ying Ji, Zhong Wu and Shao-Jian Qu
A New Multi-Attribute Emergency Decision-Making Algorithm Based on Intuitionistic Fuzzy Cross-Entropy and Comprehensive Grey Correlation Analysis
Reprinted from: *Entropy* **2020**, *22*, 768, doi:10.3390/e22070768 143

Marcio A. Diniz, Carlos A. B. Pereira and Julio M. Stern
Cointegration and Unit Root Tests: A Fully Bayesian Approach
Reprinted from: *Entropy* **2020**, *22*, 968, doi:10.3390/e22090968 165

Yarong Luo, Chi Guo, Shengyong You and Jingnan Liu
A Novel Perspective of the Kalman Filter from the Rényi Entropy
Reprinted from: *Entropy* **2020**, *22*, 982, doi:10.3390/e22090982 189

Oday A. Hassen, Saad M. Darwish, Nur A. Abu and Zaheera Z. Abidin
Application of Cloud Model in Qualitative Forecasting for Stock Market Trends
Reprinted from: *Entropy* **2020**, *22*, 991, doi:10.3390/e22090991 205

Hossein Bonakdari, Azadeh Gholami, Amir Mosavi, Amin Kazemian-Kale-Kale, Isa Ebtehaj and Amir Hossein Azimi
A Novel Comprehensive Evaluation Method for Estimating the Bank Profile Shape and Dimensions of Stable Channels Using the Maximum Entropy Principle
Reprinted from: *Entropy* **2020**, *22*, 1218, doi:10.3390/e22111218 225

About the Editors

Carlos Alberto De Bragança Pereira received his B.Sc. degree in statistics from the National School of Statistical Science, Brazil, in 1968, M.Sc. degree in statistics from the University of Sao Paulo (USP), Brazil, in 1971, and Ph.D. degree in statistics from Florida State University, USA, in 1980. He was Director of the Institute of Mathematics and Statistics, USP, from 1994 to 1998, Head of the Statistical Department in three separate appointments, and Director of the Bioinformatics Scientific Center, USP, from 2006 to 2009. He is currently Senior Professor with the Department of Statistics, USP, and has served as Visiting Professor with the Federal University of Mato Grosso do Sul from 2018 to 2020. He has authored or coauthored more than 200 papers and seven books. He was President of the Brazilian Statistical Society from 1998 to 1990 and an elected member of the International Statistical Institute. He has edited three Special Issues of *Entropy* in 2018–2020.

Adriano Polpo is a statistician who likes to develop new methods and to study the statistical theory underlying practical approaches, and to do so properly, he considers that it is necessary to work with real challenges. In the search of such challenges, he was introduced to many different problems in statistics and collaborated with many researchers in other fields. Adriano is a graduate of the State University of Campinas, Brazil, in 2001, obtained his Ph.D. from the University of Sao Paulo, Brazil, in 2005, and has taken a postdoctoral sabbatical at Florida State University, USA, from 2011 to 2012. He was Associate Professor at Federal University of Sao Carlos, from 2006 to 2018, and Head of the Department of Statistics (from 2013 to 2015). He was Elected Secretary (2011–2012), Elected President (2013–2014), and Member of Director Board (2015–2016) of ISBrA (Brazilian Chapter of ISBA, the International Society for Bayesian Analysis). Since 2019, Adriano has been Associate Professor at the University of Western Australia and Chief Investigator of the Australian Research Council, Training Centre for Transforming Maintenance through Data Science. He is an investigator of the Brazilian Obstetric Observatory project since 2021, funded by the Bill & Melinda Gates Foundation. He has mainly been working with reliability/survival analysis, regression models for counting data, and Bayesian nonparametric methods. He has been also working with statistical hypothesis testing. His research agenda aims at providing solutions to many different problems, from functional analysis of human gait, studies of the impact of copper nanoparticles in algae (carbon farming), and latency to treatment in OCD patients to methods for the proper use of statistical hypothesis tests. He has served as chair of several scientific events and authored or coauthored more than 40 papers and three books. He has edited four Special Issues of *Entropy* in 2018–2020.

Agatha Rodrigues received her B.Sc. degree in statistics from the Federal University of Sao Carlos, Brazil, in 2010, and M.Sc. and Ph.D. degrees in statistics from the University of Sao Paulo, Brazil, in 2013 and 2018, respectively. Currently, she works as Professor in the Statistics Department at the Federal University of Espírito Santo, Brazil. She has authored or coauthored more than 20 papers. Her current research interests include biostatistics, reliability, and survival analysis. She has edited one Special Issue of Entropy (2020). Dr. Rodrigues is also the principal investigator of the Brazilian Obstetric Observatory project, funded by the Bill & Melinda Gates Foundation.

Editorial

Data Science: Measuring Uncertainties

Carlos Alberto de Braganca Pereira [1,*,†], **Adriano Polpo** [2,†] **and Agatha Sacramento Rodrigues** [3,†]

1. Department of Statistics, Institute of Mathematics and Statistics, University of Sao Paulo, Sao Paulo 05508-090, Brazil
2. Department of Mathematics and Statistics, The University of Western Australia, Perth, WA 6009, Australia; adriano.polpo@uwa.edu.au
3. Department of Statistics, Federal University of Espirito Santo, Vitoria 29075-910, Brazil; agatha.rodrigues@ufes.br

* Correspondence: cpereira@ime.usp.br
† These authors contributed equally to this work.

Received: 26 November 2020; Accepted: 15 December 2020; Published: 20 December 2020

With the increase in data processing and storage capacity, a large amount of data is available. Data without analysis does not have much value. Thus, the demand for data analysis is increasing daily, and the consequence is the appearance of a large number of jobs and published articles in this area.

Data science has emerged as a multidisciplinary field to support data-driven activities, integrating and developing ideas, methods and processes to extract information from data. There are methods built from different areas of knowledge: Statistics, Computer Science, Mathematics, Physics, Information Science and Engineering, among others. This mixture of areas gave rise to what we call Data Science.

New solutions to new problems have been proposed rapidly for large volumes of data. Current and future challenges require greater care in creating new solutions that satisfy the rationality of each type of problem. Labels such as Big Data, Data Science, Machine Learning, Statistical Learning and Artificial Intelligence are demanding more sophistication in their foundations and in the way they are being applied. This point highlights the importance of building the foundations of Data Science.

This Special Issue is dedicated to solutions and discussions of measuring uncertainties in data analysis problems. The twelve articles in this edition discuss data science problems. The articles consider the reasoning behind their proposed solutions and illustrate how to apply them either in a real dataset or simulated dataset.

As stated earlier, multidisciplinarity is an important feature of data science, and this is clearly presented in this Special Issue. Ref. [1] proposes a new method for modelling problems and a data-clustering framework, and ref. [2] considers the estimation of the probability density function. In terms of the stochastic process, ref. [3] considers the fundamental properties of Tensor Markov Fields. Under a Bayesian paradigm of Statistical Inference, ref. [4] proposes a solution to classification problems.

Time series is one of the most prominent areas in data science, and some of the articles published here propose solutions with practical motivations in this area [5–8]. As mentioned before, this Special Issue encouraged articles on the foundations of measuring uncertainty [9–12].

The first article of this Special Issue was published on 30 October 2019, and the last on 26 October 2020. The articles are briefly discussed below, in order of the date of submission.

Due to its flexibility for treating heterogeneous populations, mixture models have been increasingly considered in modelling problems, and it provides a better cluster interpretation under a data-clustering framework [13].

In the traditional literature solutions, the results of the mixture model fit are highly dependent on the choice of the number of components fixed a priori. Thus, selecting an incorrect number of mixture components may cause the non-convergence of the algorithm and/or a short exploration of the clusterings [1].

Ref. [1] is the first published article in this issue. The authors propose an integrated approach that jointly selects the number of clusters and estimates the parameters of interest, without needing to specify (fix) the number of components. The authors developed the ISEM (integrated stochastic expectation maximisation) algorithm where the allocation probabilities depend on the number of clusters, and they are independent of the number of components of the mixture model.

In addition to theoretical development and evaluation of the proposed algorithm through simulation studies, the authors analyse two datasets. The first one refers to velocity in km/s of 82 galaxies from 6 well-separated conic sections of an unfilled survey of the Corona Borealis region; this is well-known Galaxy data in the literature. The second dataset refers to an acidity index measured in a sample of 155 lakes in central-north Wisconsin.

By considering the estimation of the probability density function (pdf), ref. [2] presented a wide range of applications for pdf estimation are provided, exemplifying its ubiquitous importance in data analysis. They discuss the need for developing universal measures to quantify error and uncertainties to enable comparisons across distribution classes, by establishing a robust distribution-free method to make estimates rapidly while quantifying the error of an estimate.

The authors consider a high-throughput, non-parametric maximum entropy method that employs a log-likelihood scoring function to characterise uncertainty in trial probability density estimates through a scaled quantile residual (SQR). This work is based on [14]. The SQR for the true probability density has universal sample size invariant properties equivalent to the sampled uniform random data (SURD).

Several alternative scoring functions that use SQR were considered, and they compared the sensitivity in quantifying the quality of a pdf estimate. The scoring function must exhibit distribution-free and sample size invariant properties so that it can be applied to any random sample of a continuous random variable. It is worth noting that all the scoring functions presented in the article exhibit desirable properties with similar or greater efficacy than the Anderson Darling scoring function and all are useful for assessing the quality of density estimates.

They present a numerical study to explore different types of measures for SQR quality. The initial emphasis was on constructing sensitive quality measures that are universal and sample size invariant. These scoring functions based on SQR properties can be applied to quantifying the "goodness of fit" of a pdf estimate created by any methodology, without knowledge of the true pdf.

The scoring function effectiveness is evaluated using receiver operator characteristics (ROC) to identify the most discriminating scoring function, by comparing overall performance characteristics during density estimation across a diverse test set of known probability distributions.

Integer-valued time series are relevant to many fields of knowledge, and an extensive number of models has been proposed, such as the first-order integer-valued autoregressive (INAR(1)) model. Ref. [5] considered a hierarchical Bayesian version of the INAR(p) model with variable innovation rates clustered according to a Pitman–Yor process placed at the top of the model hierarchy.

Using the full conditional distributions of the innovation rates, they inspected the behaviour of the model as concentrating or spreading the mass of the Pitman–Yor base measure. Then, they presented a graphical criterion that identified an elbow in the posterior expectation of the number of clusters as varying the hyperparameters of the base measure. The authors investigated the prior sensitivity and found ways to control the hyperparameters in order to achieve robust results. A significant contribution is a graphical criterion, which guides the specification of the hyperparameters of the Pitman–Yor process base measure.

Besides the theoretical development, the proposed graphical criterion was evaluated in simulated data. Considering a time series of yearly worldwide earthquakes events of substantial magnitude (equal or greater than 7 points on the Richter scale) from 1900 to 2018, they compared the forecasting performance of their model against the original INAR(p) model. Ref. [6] considered the problem of model fit and model forecasting in time series. For that, the authors considered the singular spectrum analysis (SSA), that is a powerful non-parametric technique to decompose the original time series into a set of components that can be interpreted, such as trend components, seasonal components, and noise components. They proposed a robust SSA algorithm by replacing the standard least-squares singular value decomposition (SVD) by a robust SVD algorithm based on the L1 norm and a robust SSA algorithm. The robust SVD was based on the Huber function. Then, a forecasting strategy was presented for the robust SSA algorithms, based on the linear recurrent SSA forecasting algorithm.

Considering a simulation example and time-series data from investment funds, the algorithms were compared to other versions of the SSA algorithm and classical ARIMA. The comparisons considered the computational time and the accuracy for model fit and model forecast. Ref. [9] presented a discussion about hypothetical judgment and measures to evaluate that, and exemplified it using a diagnostic of the infection of the Coronavirus Disease (COVID-19). Their purposes are (1) to distinguish channel confirmation measures that are compatible with the likelihood ratio and prediction confirmation measures that can be used to assess probability predictions, and (2) to use a prediction confirmation measure to eliminate the Raven Paradox and to explain that confirmation and falsification may be compatible.

They consider the measure F, that is one of few confirmation measures which possess the desirable properties as identified by many authors: symmetries and asymmetries, normalisation, and monotonicity. Also, the measure $b*$, the degree of belief, was considered and optimised with a sampling distribution seen as a confirmation measure, which is similar to the measure F and also possesses the above-mentioned desirable properties.

From the diagnosis of the infection of the COVID-19, they show that only measures that are functions of the likelihood ratio, such as F and $b*$, can help to diagnose the infection or choose a better result that can be accepted by the medical society. However, measures F and $b*$ do not reflect the probability of the infection. Furthermore, using F or $b*$ is still difficult to eliminate the Raven Paradox.

The measures F and $b*$ indicate how good a hypothesis test of means is compared to the probability predictions. Hence, the authors proposed a measure $c*$ that can indicate how good a probability prediction is. $c*$ is called the prediction confirmation measure and $b*$ is the channel confirmation measure. The measure $c*$ accords to the Nicod criterion and undermines the Equivalence Condition, and hence can be used to eliminate the Raven Paradox. Ref. [3] presented the definitions and properties of Tensor Markov Fields (random Markov fields over tensor spaces). The author shows that tensor Markov fields are indeed Gibbs fields whenever strictly positive probability measures are considered. It is also proved how this class of Markov fields can be built based on statistical dependency structures inferred on information-theoretical grounds over empirical data. The author discusses how the Tensor Markov Fields described in the article can be useful for mathematical modelling and data analysis due to their intrinsic simplicity and generality. Ref. [4] proposed a variational approximation on probit regression models with intrinsic priors to deal with a classification problem. Some of the authors' motivations to combine intrinsic prior methodology and variational inference are to automatically generate a family of non-informative priors; to apply intrinsic priors on inference problems; intrinsic priors have flat tails that prevent finite sample inconsistency; for inference problems with a large dataset, variational approximation methods are much faster than MCMC-based methods.

The proposed method is applied to the LendingClub dataset (https://www.lendingclub.com). The LendingClub is a peer-to-peer lending platform that enables borrowers to create unsecured personal loans between $1000 and $40,000. Investors can search and browse the loan listings on the LendingClub website and select loans that they want to invest in. In addition, the information about the borrower, amount of loan, loan grade, and loan purpose was provided to them. The variable loan status (paid-off or charged-off) is the target variable, and [4] considers a set of predictive covariates, as loan term in months, employment length in years, annual income, among others. [10] constructed a decision-making model based on intuitionistic fuzzy cross-entropy and a comprehensive grey correlation analysis algorithm. Their motivation is the fact that despite the fact that intuitionistic fuzzy distance measurement is an effective method to study multi-attribute emergency decision-making (MAEDM) problems, the traditional intuitionistic fuzzy distance measurement method cannot accurately reflect the difference between membership and non-membership data, where it is easy to cause information confusion.

The intuitionistic fuzzy cross-entropy distance measurement method was introduced, which can not only retain the integrity of decision information but also directly reflect the differences between intuitionistic fuzzy data. Focusing on the weight problem in MAEDM, the authors analysed and compared the known and unknown attribute weights, which significantly improved the reliability and stability of decision-making results. The intuitionistic fuzzy cross-entropy and grey correlation analysis algorithm were introduced into the emergency decision-making problems such as the location ranking of shelters in earthquake disaster areas, which significantly reduced the risk of decision-making. The validity of the proposed method was verified by comparing the traditional intuitionistic fuzzy distance to the intuitionistic fuzzy cross-entropy.

The authors highlight that the proposed method applies to emergency decision-making problems with certain subjective preference. In addition to the theoretical approach and highlighting the importance to deal with disasters motivations, the authors took the Wenchuan Earthquake on May 12th 2008 as a case of study, constructing and solving the ranking problem of shelters.

Motivated by time series problems, ref. [7] reviewed the shortcomings of unit root and cointegration tests. They proposed a Bayesian approach based on the Full Bayesian Significance Test (FBST), a procedure designed to test a sharp or precise hypothesis.

The importance of studying this is justified by the fact that one should be able to assess if a time series present deterministic or stochastic trends to perform statistical inference. For univariate analysis, one way to detect stochastic trends is to test if the series has unit-roots (unit root tests). For multivariate studies, determining stationary linear relationships between the series, or if they cointegrate (cointegration tests) are important.

The Augmented Dickey–Fuller test is one of the most popular tests used to assess if a time series has a stochastic trend or if they have a unit root for series described by auto-regressive models. When one is searching for long-term relationships between multiple series, it is crucial to know if there are stationary linear combinations of these series, i.e., if the series are cointegrated. One of the most used tests is the maximum eigenvalue test.

Besides proposing the method considering FBST, the authors also compared its performance with the most used frequentist alternatives. They have shown that the FBST works considerably well even when one uses improper priors, a choice that may preclude the derivation of Bayes Factors, a standard Bayesian procedure in hypotheses testing. Ref. [11] considered a Kalman filter and a Rényi entropy. The Rényi entropy was employed to measure the uncertainty of the multivariate Gaussian probability density function. The authors proposed calculation of the temporal derivative of the Rényi entropy of the Kalman filter's mean square error matrix, which provided the optimal recursive solution mathematically and was minimised to obtain the Kalman filter gain.

One of the findings of this manuscript was that, from the physical point of view, the continuous Kalman filter approached a steady state when the temporal derivative of the Rényi entropy was equal to zero, which means that the Rényi entropy remained stable.

A numerical experiment of falling body tracking in noisy conditions with radar using the unscented Kalman filter, and a practical experiment of loosely-coupled integration, are provided to demonstrate the effectiveness of the above statements and to show the Rényi entropy truly stays stable when the system becomes steady.

The knowledge about future values and the stock market trend has attracted the attention of researchers, investors, financial experts, and brokers. Ref. [8] proposed a stock trend prediction model by utilising a combination of the cloud model, Heikin–Ashi candlesticks, and fuzzy time series in a unified model.

By incorporating probability and fuzzy set theories, the cloud model can aid the required transformation between the qualitative concepts and quantitative data. The degree of certainty associated with candlestick patterns can be calculated through repeated assessments by employing the normal cloud model. The hybrid weighting method comprising the fuzzy time series, and Heikin–Ashi candlestick was employed for determining the weights of the indicators in the multi-criteria decision-making process. The cloud model constructs fuzzy membership functions to deal effectively with uncertainty and vagueness of the historical stock data to predict the next open, high, low, and close prices for the stock.

The objective of the proposed model is to handle qualitative forecasting and not quantitative only. The experimental results prove the feasibility and high forecasting accuracy of the constructed model. Ref. [12] uses the maximum entropy principle to provide an equation to calculate the Lagrange multipliers. Accordingly, an equation was developed to predict the bank profile shape of threshold channels.

The relation between ratio with the entropy parameter and the hydraulic and geometric characteristics of channels was evaluated. The Entropy-based Design Model of Threshold Channels (EDMTC) for estimating the shape of banks profiles and the channel dimensions was designed based on the maximum entropy principle in combination with the Gene Expression Programming regression model.

The results indicate that the entropy model is capable of predicting the bank profile shape trend with acceptable error. The proposed EDMTC can be used in threshold channel design and for cases when the channel characteristics are unknown.

It is our understanding that this Special Issue contributes to increasing knowledge in the data science field, by fostering discussions of measuring uncertainties in data analysis problems. The discussion of foundations/theoretical aspects of the methods is essential to avoid the use of black-box procedures, as well as the presentation of the methods in real problem data. Theory and application are both important to the development of data science.

Funding: This research received no external funding.

Conflicts of Interest: The authors declare no conflict of interest.

References

1. Saraiva, E.F.; Suzuki, A.K.; Milan, L.A.; Pereira, C.A.B. An Integrated Approach for Making Inference on the Number of Clusters in a Mixture Model. *Entropy* **2019**, *21*, 1063. [CrossRef]
2. Farmer, J.; Merino, Z.; Gray, A.; Jacobs, D. Universal Sample Size Invariant Measures for Uncertainty Quantification in Density Estimation. *Entropy* **2019**, *21*, 1120. [CrossRef]
3. Hernández-Lemus, E. On a Class of Tensor Markov Fields. *Entropy* **2020**, *22*, 451. [CrossRef] [PubMed]
4. Li, A.; Pericchi, L.; Wang, K. Objective Bayesian Inference in Probit Models with Intrinsic Priors Using Variational Approximations. *Entropy* **2020**, *22*, 513. [CrossRef] [PubMed]
5. Graziadei, H.; Lijoi, A.; Lopes, H.F.; Marques F, P.C.; Prünster, I. Prior Sensitivity Analysis in a Semi-Parametric Integer-Valued Time Series Model. *Entropy* **2020**, *22*, 69. [CrossRef] [PubMed]
6. Rodrigues, P.C.; Pimentel, J.; Messala, P.; Kazemi, M. The Decomposition and Forecasting of Mutual Investment Funds Using Singular Spectrum Analysis. *Entropy* **2020**, *22*, 83. [CrossRef] [PubMed]
7. Diniz, M.A.; Pereira, C.A.B.; Stern, J.M. Cointegration and unit root tests: A fully Bayesian approach. *Entropy* **2020**, *22*, 968. [CrossRef] [PubMed]
8. Hassen, O.A.; Darwish, S.M.; Abu, N.A.; Abidin, Z.Z. Application of Cloud Model in Qualitative Forecasting for Stock Market Trends. *Entropy* **2020**, *22*, 991. [CrossRef] [PubMed]
9. Lu, C. Channels' Confirmation and Predictions' Confirmation: From the Medical Test to the Raven Paradox. *Entropy* **2020**, *22*, 384. [CrossRef] [PubMed]
10. Li, P.; Ji, Y.; Wu, Z.; Qu, S.J. A new multi-attribute emergency decision-making algorithm based on intuitionistic fuzzy cross-entropy and comprehensive grey correlation analysis. *Entropy* **2020**, *22*, 768. [CrossRef] [PubMed]

11. Luo, Y.; Guo, C.; You, S.; Liu, J. A Novel Perspective of the Kalman Filter from the Rényi Entropy. *Entropy* **2020**, *22*, 982. [CrossRef] [PubMed]
12. Bonakdari, H.; Gholami, A.; Mosavi, A.; Kazemian-Kale-Kale, A.; Ebtehaj, I.; Azimi, A.H. A Novel Comprehensive Evaluation Method for Estimating the Bank Profile Shape and Dimensions of Stable Channels Using the Maximum Entropy Principle. *Entropy* **2020**, *22*, 1218. [CrossRef] [PubMed]
13. Bouveyron, C.; Brunet-Saumard, C. Model-based clustering of high-dimensional data: A review. *Comput. Stat. Data Anal.* **2014**, *71*, 52–78. [CrossRef]
14. Farmer, J.; Jacobs, D. High throughput nonparametric probability density estimation. *PLoS ONE* **2018**, *13*, e0196937. [CrossRef] [PubMed]

Publisher's Note: MDPI stays neutral with regard to jurisdictional claims in published maps and institutional affiliations.

© 2020 by the authors. Licensee MDPI, Basel, Switzerland. This article is an open access article distributed under the terms and conditions of the Creative Commons Attribution (CC BY) license (http://creativecommons.org/licenses/by/4.0/).

Article

An Integrated Approach for Making Inference on the Number of Clusters in a Mixture Model

Erlandson Ferreira Saraiva [1,*], Adriano Kamimura Suzuki [2], Luis Aparecido Milan [3] and Carlos Alberto de Bragança Pereira [1,4]

1. Instituto de Matemática, Universidade Federal de Mato Grosso do Sul, Campo Grande 79070-900, Brazil; cpereira@ime.usp.br
2. Departamento de Matemática Aplicada e Estatística, Universidade de São Paulo, São Carlos 13566-590, Brazil; suzuki@icmc.usp.br
3. Departamento de Estatística, Universidade Federal de São Carlos, São Carlos 13565-905, Brazil; dlam@ufscar.br
4. Instituto de Matemática e Estatística, Universidade de São Paulo, São Paulo 05508-090, Brazil
* Correspondence: erlandson.saraiva@ufms.br; Tel.: +55-67-3345-7511

Received: 23 September 2019; Accepted: 26 October 2019; Published: 30 October 2019

Abstract: This paper presents an integrated approach for the estimation of the parameters of a mixture model in the context of data clustering. The method is designed to estimate the unknown number of clusters from observed data. For this, we marginalize out the weights for getting allocation probabilities that depend on the number of clusters but not on the number of components of the mixture model. As an alternative to the stochastic expectation maximization (SEM) algorithm, we propose the integrated stochastic expectation maximization (ISEM) algorithm, which in contrast to SEM, does not need the specification, a priori, of the number of components of the mixture. Using this algorithm, one estimates the parameters associated with the clusters, with at least two observations, via local maximization of the likelihood function. In addition, at each iteration of the algorithm, there exists a positive probability of a new cluster being created by a single observation. Using simulated datasets, we compare the performance of the ISEM algorithm against both SEM and reversible jump (RJ) algorithms. The obtained results show that ISEM outperforms SEM and RJ algorithms. We also provide the performance of the three algorithms in two real datasets.

Keywords: model-based clustering; mixture model; EM algorithm; integrated approach

1. Introduction

Recently, there has been increasing interest in modeling using mixture models. This is mainly due to the flexibility for treating heterogeneous populations. Under a data-clustering framework, this model has the advantage of being probabilistic, and then the obtained clusters can have a better interpretation from a statistical point of view [1]. This contrasts with usual methods, such as k-means or hierarchical clustering, in which clusters are not statistically based, as discussed by [2].

From a frequentist viewpoint, the standard method to get the maximum likelihood estimates for the parameters of a mixture model is based on the use of the Expectation Maximization (EM) algorithm [3]. However, for the use of this algorithm, the number of components k of the mixture model needs to be known a priori. As the resulting model is highly dependent on the choice of this value, the main question is how to set the k value. Selecting an erroneous k value may cause the non-convergence of the algorithm and/or a low exploration of the clusterings. In addition, depending on the k value chosen we may have empty components, and therefore, there are no maximum likelihood estimates for these components.

An approach frequently used to determine the best k value among a fixed set of values is the use of the stochastic version of the EM algorithm (SEM) with some model selection criterion, such as the Akaike information criterion (AIC) [4,5] or the BIC [6]. In this approach, models are fitted for a set of predefined k values, and the best model is the one that has the smallest AIC or BIC value.

However, as discussed by [7], to adjust several models for a predefined set of values for the number of the cluster and compare them using some model selection criterion is not a practical and efficient procedure. Therefore, it is desirable to have an efficient algorithm to calculate the optimal number of clusters together with the estimation of the parameters of each mixture component. In this scenario, the Bayesian approach was successfully performed considering the Markov chain Monte Carlo (MCMC) algorithm with reversible jumps, described by [8] in the context of univariate normal mixture models. On the other hand, a difficulty often encountered in implementing a reversible jump algorithm (RJ) is the construction of efficient transition proposals that lead to a reasonable acceptance rate.

Following in the line of MCMC algorithms, [9] proposes a split–merge MCMC procedure for the conjugated Dirichlet process mixture model using a restricted Gibbs sampling scan to determine a split proposal, where the number of scans (tuning parameter) must be previously fixed by the user, and [10] extend their method to a nonconjugated Dirichlet process mixture model. [11] proposes a data-driven split-and-merge approach. In this proposal, the number of clusters is updated according to the creation of a new component based on a single observation and using a split–merge strategy, developed based on the use of the Kullback–Leibler divergence. A difficulty encountered for implementing this algorithm is the obtaining of the mathematical expression for the Kullback–Leibler divergence, which does not always have known analytical expression. In addition, the sequential allocation used in the split–merge strategy of these three works may make the algorithm slow when the sample size is great, and the computation implementation of these methods is not so simple.

The present work proposes an integrated approach that, in a joint way, selects the number of clusters and estimates the parameters of interest. With this approach, the mixture weights are integrated out to obtain allocation probabilities that depend on the number of clusters (nonempty components) but do not depend on the number of components k. In addition, considering k tending to infinity, this procedure introduces a positive probability of a new cluster being created by a single observation. When a new cluster is created, the parameters associated with it are generated from its posterior distribution. We then developed the ISEM (integrated stochastic expectation maximization) algorithm to estimate the parameters of interest. This algorithm configures a setting for latent allocation variables **c** according to allocation probabilities, and then the cluster parameters are updated conditionally on **c** as follows: for clusters with at least two observations, the parameter values are the maximum likelihood estimates; for the clusters with only one observation, the parameter values are generated from their posterior distribution.

In order to illustrate the computation implementation of the method and verify its performance, we have considered a specific model in which data are generated from mixtures of univariate normal distributions. This model allows us to avoid the label switching problem by considering the labeling of the components according to the increasing order of the component averages, as done by [8,11–13], among others. But we emphasize that our algorithm is not restricted to this particular model. For instance, for the multivariate case, we may consider the labeling of the components according to the eigenvalues of the current covariance matrix, as done by [14]. However, a detailed discussion of the multivariate case will be done in a future paper.

We also compare the performance of the ISEM with both SEM and RJ algorithms. The criteria used to compare the methods are the estimated probability of the number of clusters, convergence of the sampled values, mixing, autocorrelation, and computation time. We also applied the three algorithms to two real datasets. The first is the well-known Galaxy data, and the second is a dataset on Acidity.

The remainder of the paper is as follows. Section 2 describes the mixture model and the estimation process based on the SEM algorithm. Section 3 develops the integrated approach and describes the ISEM algorithm. Section 4 shows how we applied the algorithm to simulated datasets in order to assess its performance. Section 5 describes the application of the three algorithms to two real datasets. Section 6 is about our final remarks. Additional details are in the Supplementary Material, which is referred to as "SM" in this paper. Table 1 presents the main notations used throughout the article.

Table 1. Main mathematical notation used throughout the paper.

Notation	Description
k	Number of components
k_c	Number of clusters
θ_j	Parameter of the j-th component, for $j = 1, \ldots, k$
$\boldsymbol{\theta}_k = (\theta_1, \ldots, \theta_k)$	The whole vector of parameters
w_j	Weight of the j-th component, for $j = 1, \ldots, k$
Y_i	The i-th sampled value, for $i = 1, \ldots, n$
c_i	The i-th indicator variable, for $i = 1, \ldots, n$
$\mathbf{y} = (y_1, \ldots, y_n)$	The vector of independent observations
$\mathbf{c} = (c_1, \ldots, c_n)$	The vector of latent indicator variables
$k_{\mathbf{c}_{-i}}$	Number of clusters excluding the i-th observation
$n_{j,-i}$	Number of observations assigned to the j-th component, excluding the i-th observation

2. Mixture Model and SEM Algorithm

Let $\mathbf{y} = (y_1, \ldots, y_n)$ be a vector of independent observations from a mixture model with k components, i.e.,

$$f(y_i | \mathbf{w}, \boldsymbol{\theta}_k, k) = \sum_{j=1}^{k} w_j f(y_i | \theta_j), \tag{1}$$

where $f(y_i | \theta_j)$ is the density of a family of parametric distributions with parameters θ_j (scalar or vector), $\boldsymbol{\theta}_k = (\theta_1, \ldots, \theta_k)$ are the parameters of the components, and $\mathbf{w} = (w_1, \ldots, w_k)$, $w_j > 0$ and $\sum_{j=1}^{k} w_j = 1$ are component weights.

The log-likelihood function for $(\boldsymbol{\theta}_k, \mathbf{w})$ is given by

$$l(\boldsymbol{\theta}_k, \mathbf{w} | \mathbf{y}, k) = \log \left\{ \prod_{i=1}^{n} \left[\sum_{j=1}^{k} w_j f(y_i | \theta_j) \right] \right\} = \sum_{i=1}^{n} \log \left\{ \left[\sum_{j=1}^{k} w_j f(y_i | \theta_j) \right] \right\}.$$

The mathematical notation $l(\boldsymbol{\theta}_k, \mathbf{w} | \mathbf{y}, k)$ is given as in the book of Casella and Berger (2002).

The usual procedure to obtain the maximum likelihood estimators consists of getting partial derivatives of $l(\boldsymbol{\theta}_k, \mathbf{w} | \mathbf{y})$ in relation to θ_j and then equalizing the result to zero, i.e.,

$$\frac{d l(\boldsymbol{\theta}_k, \mathbf{w} | \mathbf{y})}{d \theta_j} = \sum_{i=1}^{n} \frac{w_j f(y_i | \theta_j)}{\sum_{j=1}^{k} w_j f(y_i | \theta_j)} \frac{d \log \left[f(y_i | \theta_j) \right]}{d \theta_j} = 0, \tag{2}$$

for $j = 1, \ldots, k$.

But, note that in (2), the maximization procedure consists of a weighted maximization process of the log-likelihood function with each observation y_i having a weight associated to component j given by

$$w_{ij}^* = \frac{w_j f(y_i|\theta_j)}{\sum_{j=1}^{k} w_j f(y_i|\theta_j)}, \qquad (3)$$

for $i = 1, \ldots, n$ and $j = 1, \ldots, k$. However, these weights depend on the parameters that we are trying to estimate. In this way, we cannot obtain a "closed" mathematical expression that allows the direct maximization of the log-likelihood function. Due to this, the mixture problem is reformulated as a complete-data problem [12,15].

Complete-Data Formulation

Consider associated to each observation y_i a latent indicator variable c_i not known, so that if $c_i = j$, then y_i is from component j, for $i = 1, \ldots, n$ and $j = 1, \ldots, k$. The probability of $c_i = j$ is w_j, $P(c_i = j|\mathbf{w}, k) = w_j$, for $i = 1, \ldots, n$ and $j = 1, \ldots, k$. Letting n_j be the number of observations from component j (i.e., the number of c_is equals to j), the joint probability for $\mathbf{c} = (c_1, \ldots, c_n)$ given $\mathbf{w}$ and k is

$$\pi(\mathbf{c}|\mathbf{w}, k) = \prod_{j=1}^{k} w_j^{n_j}. \qquad (4)$$

The distribution of the number of observations assigned to each component, $n_1, \ldots, n_k$, called the occupation number, is multinomial, $(n_1, \ldots, n_k|n, \mathbf{w}) \sim Multinomial(n, \mathbf{w})$, where $n = n_1 + \ldots + n_k$.

Thus, under this augmented framework, we have that

(1) the probability of $c_i = j$, conditional on observation y_i and on component parameters θ_k, is w_{ij}^*, i.e., $P(c_i = j|y_i, \theta_k, k) = w_{ij}^*$, for w_{ij}^* given in Equation (3), for $i = 1, \ldots, n$ and $j = 1, \ldots, k$. That is, although the indicator variables are nonobservable, they are implicitly present in the estimation procedure given in Equation (2).

(2) the log-likelihood function for $(\theta_k, \mathbf{w})$, conditional on complete data $(\mathbf{y}, \mathbf{c})$, is given by

$$l(\theta_k, \mathbf{w}|\mathbf{y}, \mathbf{c}) = \log\left\{\prod_{j=1}^{k} w_j^{n_j} L(\theta_j|\mathbf{y})\right\} = \sum_{j=1}^{k} \left[n_j \log(w_j) + l(\theta_j|\mathbf{y})\right],$$

where $l(\theta_j|\mathbf{y}) = \log\left[L(\theta_j|\mathbf{y})\right]$ is the log-likelihood function for component j, for $j = 1, \ldots, k$. Thus, the estimation procedure of the k component parameters reduce to k independent problems of estimation. For example, for a normal mixture model, the maximum likelihood estimates for component parameters $\theta_j = (\mu_j, \sigma_j^2)$ is $\hat{\theta}_j = (\hat{\mu}_j, \hat{\sigma}_j^2) = (\bar{y}_j, s_j^2)$, where $\bar{y}_j$ and s_j^2 are, respectively, the average and variance of the observations allocated to component j, for $j = 1, \ldots, k$.

From this complete-data formulation, the estimation procedure is given by an iterative process with two steps. In the first one, the allocation indicator variables are updated conditional on component parameters, and in the subsequent step, the component parameters are updated conditional on configuration of the allocation indicator variables.

The usual algorithm used to implement these two steps is the EM algorithm [3]. The stochastic version of the EM algorithm (SEM) can be implemented according to Algorithm 1.

Algorithm 1 SEM Algorithm

1: Initialize the algorithm with a configuration $\mathbf{c}^{(0)} = \left(c_1^{(0)}, \ldots, c_n^{(0)}\right)$ for allocation indicator variables.
2: **procedure** For the s-th iteration of the algorithm, $s = 1, \ldots$
3:     get the maximum likelihood estimates $\hat{\boldsymbol{\theta}}_k^{(s)} = \left(\hat{\theta}_1^{(s)}, \ldots, \hat{\theta}_k^{(s)}\right)$ and $\hat{\mathbf{w}}^{(s)} = \left(\hat{w}_1^{(s)}, \ldots, \hat{w}_k^{(s)}\right)$ conditional on configuration $\mathbf{c}^{(s-1)}$;
4:     **if** $\left| \frac{l\left(\hat{\theta}_k^{(s)}, \hat{\mathbf{w}}^{(s)} | \mathbf{y}\right) - l\left(\hat{\theta}_k^{(s-1)}, \hat{\mathbf{w}}^{(s-1)} | \mathbf{y}\right)}{l\left(\hat{\theta}_k^{(s-1)}, \hat{\mathbf{w}}^{(s-1)} | \mathbf{y}\right)} \right| < \epsilon$, where ϵ is a threshold value previously fixed, **then** stop the algorithm. Otherwise, go to item (iii);
5:     conditional on $\hat{\boldsymbol{\theta}}_k^{(s)}$ and $\hat{\mathbf{w}}^{(s)}$, update $\mathbf{c} = (c_1, \ldots, c_n)$ as follows. For $i = 1, \ldots, n$ and $j = 1, \ldots, k$ do the following:
6:     Let $\mathbf{z}_i = (z_{i1}, \ldots, z_{ik})$ be a indicator vector, so that $z_{ij} = 0$ or $z_{ij} = 1$;
7:     Generate $\mathbf{z}_i \sim Multinomial(1, \mathbf{w}_i^*)$, where $\mathbf{w}_i^* = (w_{i1}^*, \ldots, w_{ik}^*)$ and w_{ij}^* is obtained from Equation (3) doing $\theta_j = \hat{\theta}_j$ and $w_j = \hat{w}_j$. If $z_{ij} = 1$, then do $c_i = j$;
8:     Do $s = s + 1$ and return to step (3).

Although it is simple to implement computationally, the SEM algorithm may present some practical problems. As discussed by [16], the algorithm may present a slow convergence. Due to this, some authors, such as [17,18], discuss how to set up the start values in order to increase the convergence. In addition, [15] discusses the non-existence of the global maximum estimator.

Moreover, in this algorithm, the k value must be known previously. For the cases in which k is an unknown quantity, the best k value is chosen by fitting a set of models associated with a set of predefined k values and comparing them according to AIC [4,5] or BIC [6] criteria. Furthermore, given a sample of size n and fixed a k value, there exists a positive probability, given by $(1 - w_j)^n \neq 0$, of the j-th component not having observations allocated in an iteration of the algorithm. In this case, we have an empty component, and the maximum likelihood estimates cannot be calculated for these components. Thus, in order to avoid the practical problems presented by the EM algorithm, we propose an integrated approach.

3. Integrated Approach

We start our integrated approach linking data clustering to a mixture model. For this, consider a sampling process from a heterogeneous population that is subdivided into k sub-populations. Thus, it is natural to assume that the sampling process consists of the realization of the following steps:

(i) choose a sub-population j with probability w_j, where w_j is the relative frequency of the j-th sub-population in relation to the overall population;
(ii) sample a Y_i value of this sub-population,

for $i = 1, \ldots, n$ and $j = 1, \ldots, k$, where n is the sample size.

Let (Y_i, c_i) be a sample unit, where c_i is an indicator allocation variable that assumes a value of the set $\{1, \ldots, k\}$ with probabilities $\{w_1, \ldots, w_k\}$, respectively. Thus, assuming that subpopulation j is modeled by a probability distribution $F(\theta_j)$ indexed by parameter θ_j (scalar or vector), we have that

$$(Y_i | c_i = j, \theta_j) \sim F(\theta_j) \quad \text{and} \quad P(c_i = j | \mathbf{w}) = w_j,$$

for $i = 1, \ldots, n$ and $j = 1, \ldots, k$.

However, in clustering problems, the c_i's values are non-observable. Thus, the probability of $c_i = j$ is w_j, and the marginal probability density function for $Y_i = y_i$ is given by Equation (1).

In addition, as the model in Equation (1) is a population model; so there exists a non-null probability $(1 - w_j)^n$ that the j-th component is an empty component. Thus, the number of clusters (i.e., non-empty components) is smaller than the number of components k. As viewed in the description of the EM algorithm, the number of clusters is defined by the configuration of the latent allocation variables **c**; thus hereafter, we will denote the number of clusters by $k_\mathbf{c}$, for $k_\mathbf{c} \leq k$.

Since the interest lies in the configuration of **c**, let us marginalize out the weights of the mixture model. Thus, integrating density (4) with respect to the prior $Dirichlet\left(\frac{\gamma}{k}, \ldots, \frac{\gamma}{k}\right)$ distribution of the weights, denoted by $(w_1, \ldots, w_k)|k, \gamma \sim Dirichlet\left(\frac{\gamma}{k}, \ldots, \frac{\gamma}{k}\right)$, the joint probability for **c** is given by (see Appendix 3 of the SM)

$$\pi(\mathbf{c}|\gamma, k) = \frac{\Gamma(\gamma)}{\Gamma(n+\gamma)} \prod_{j=1}^{k} \frac{\Gamma\left(n_j + \frac{\gamma}{k}\right)}{\Gamma\left(\frac{\gamma}{k}\right)}. \tag{5}$$

Similarly, the conditional probability for $c_i = j$ given $\mathbf{c}_{-i} = (c_1, \ldots, c_{i-1}, c_{i+1}, \ldots, c_n)$, is given by

$$\pi(c_i = j|\mathbf{c}_{-i}, \gamma, k) = \frac{n_{j,-i} + \frac{\gamma}{k}}{n + \gamma - 1}, \tag{6}$$

where $n_{j,-i}$ is the number of observations allocated to the j-th component, excluding the i-th observation, for $i = 1, \ldots, n$ and $j = 1, \ldots, k$.

As the main interest is in $k_\mathbf{c}$ and not k, we remove k from Equation (6) by letting k tend to infinity. Under this assumption, the probability reaches the following limit:

$$\pi(c_i = j|\mathbf{c}_{-i}, \gamma) = \frac{n_{j,-i}}{n + \gamma - 1}, \tag{7}$$

when $n_{j,-i} > 0$, for $i = 1, \ldots, n$ and $j = 1, \ldots, k_\mathbf{c}$, where $k_\mathbf{c}$ is the number of clusters defined by configuration **c**. In addition, we now have a probability of the i-th observation being allocated to one of the other infinite components, which is given by

$$\pi(c_i = j^*|\mathbf{c}_{-i}, \gamma) = \frac{\gamma}{n + \gamma - 1}, \tag{8}$$

for $j^* \notin \{1, \ldots, k_\mathbf{c}\}$. This is the probability of the observation y_i creating a new cluster, for $i = 1, \ldots, n$. The probabilities in (7) and (8) are equivalent to the update probabilities of a Dirichlet process mixture model. See, for example, [19–21].

Given y_i, the conditional probability for $c_i = j$, such that $n_{j,-i} > 0$, is

$$\pi_{ij} = \pi(c_i = j|y_i, \theta_j, \mathbf{c}_{-i}, \gamma) = \frac{n_{j,-i}}{n + \gamma - 1} f(y_i|\theta_j), \tag{9}$$

for $i = 1, \ldots, n$ and $j = 1, \ldots, k_{\mathbf{c}_{-i}}$, where $k_{\mathbf{c}_{-i}}$ is the number of clusters excluding the i-th observation. At this point, it is important to note that if an observation y_i is allocated to a component j, $c_i = j$, and $n_j > 1$, then $n_{j,-i} \geq 1$ and $k_{\mathbf{c}_{-i}} = k_\mathbf{c}$. But if $c_i = j$ and $n_j = 1$, then $n_{j,-i} = 0$ and $k_{\mathbf{c}_{-i}} = k_\mathbf{c} - 1$.

In order to define the conditional probability of the i-th observation creating a new cluster j^*, we integrate parameters out for this case, for $j^* = k_{\mathbf{c}_{-i}} + 1$. This was done because that probability does not depend on the parameter value θ_{j^*}. Thus, the conditional posterior probability for $C_i = j^*$ is

$$\pi_{ij^*} = \pi(c_i = j^*|y_i, \mathbf{c}_{-i}, \gamma) = \frac{\gamma}{n + \gamma - 1} \mathbf{I}(y_i), \tag{10}$$

where $\mathbf{I}(y_i) = \int f(y_i|\theta_{j^*}) \pi(\theta_{j^*}) d\theta_{j^*}$ and $\pi(\theta_{j^*})$ is the density of the prior distribution for θ_{j^*}, for $i = 1, \ldots, n$.

As is known from the literature, the likelihood function for a mixture model is non-identifiable, i.e., any permutation of the components' labels lead to the same likelihood function (see, for example, [8,11,22–24]). Thus, in order to get identifiability, we assume that $\mu_1, \ldots, \mu_{k_\mathbf{c}}$ are the component means for clusters and that $\mu_1 < \ldots < \mu_{k_\mathbf{c}}$. However, it does not prevent the algorithm described in the next Section from being applicable to another labeling criterion. Additional discussion about label switching can be found in [22,23].

3.1. Integrated SEM Algorithm

Using probabilities given in Equations (9) and (10), we update the allocation indicator variables according to Algorithm 2.

Conditional on a configuration **c**, we have k_c clusters. So, we update parameters of interest according to Algorithm 3. We then join Algorithms 2 and 3 to get the Algorithm 4.

After the S iterations, we discard the first B iterations as a burn-in. In the following, we also consider "jumps" of size h, i.e., only one draw every h is extracted from the original sequence in order to obtain a sub-sequence of size $H = (S - B)/h$ to make inferences. Denote this sub-sequence by $\mathbb{S}(H)$.

Consider $\mathbb{N}_{k_c}(j)$ to be the number of times that $k_c = j$ in $\mathbb{S}(H)$, for $j \in \{1, \ldots, K_{max}\}$, where K_{max} is the maximum k_c value sampled in the course of iterations. Thus, $\tilde{P}(k_c = j) = \frac{\mathbb{N}_{k_c}(j)}{H}$ is the estimated probability for $k_c = j$. We then consider

$$\tilde{k}_c = \underset{1 \leq j \leq K_{max}}{argmax} \left(\tilde{P}(k_c = j) \right)$$

as being the estimates for the number of components k_c.

Appendix 1 of the SM presents the mathematical expression used to determine a configuration for **c** and estimates for the parameters of the clusters, conditional on the estimate $\tilde{k}_c$.

Algorithm 2 Updating **c**

1: Let $\mathbf{c} = (c_1, \ldots, c_n)$ be the current configuration for latent allocation variables. Then, update **c** as follows.
2: **procedure** For $i = 1, \ldots, n$:
3: Remove c_i from the current state **c**, obtaining $\mathbf{c}_{-i}$ and $k_{\mathbf{c}_{-i}}$;
4: Generate a variable $\mathbf{Z}_i = (Z_{i1}, \ldots, Z_{ik_c}) \sim Multinomial(1, \mathbf{P}_i)$, where $\mathbf{P}_i = (\pi_{i1}, \ldots, \pi_{ik_{\mathbf{c}_{-i}}}, \pi_{ij^*})$ for π_{ij} given in (9) and π_{ij^*} given in (10), for $j = 1, \ldots, k_{\mathbf{c}_{-i}}$ and $j^* = k_{\mathbf{c}_{-i}} + 1$;
5: If $Z_{ij} = 1$, for $j \in \{1, \ldots, k_{\mathbf{c}_{-i}}\}$, set up $c_i = j$ and do $n_j = n_{j,-i} + 1$;
6: If $Z_{ij^*} = 1$, do $n_{j^*} = 1$ and $k_c = k_{\mathbf{c}_{-i}} + 1$. As this new cluster has just one observation allocated, set as component parameter $\theta_{j^*} = \theta_j^g$, where θ_j^g is a value generated from posterior distribution $\pi(\theta_{j^*}|y_i) \propto f(y_i|\theta_{j^*})\pi(\theta_{j^*})$, where $f(y_i|\theta_{j^*})$ is the probability density function for y_i conditional on θ_{j^*} and $\pi(\theta_{j^*})$ is the density of the prior distribution for θ_{j^*}. Relabel the k_c clusters in order to maintain the adjacency condition. If the component mean μ_{j^*} of the new cluster is such that:
7: $\mu_{j^*} = \underset{1 \leq j \leq k_c}{min}\ \mu_j$, then do $j^* = 1$ and relabel all other clusters doing $j + 1$;
8: $\mu_{j^*} = \underset{1 \leq j \leq k_c}{max}\ \mu_j$, then do $j^* = k_c$ and keep all other clusters labels;
9: $\mu_j < \mu_{j^*} < \mu_{j+1}$, for $j \neq \{1, k_c\}$, then do $j^* = j + 1$ and relabel all other clusters $j' \geq j + 1$ doing $j' = j' + 1$.

Algorithm 3 Updating cluster parameters

1: Let $\boldsymbol{\theta}_{k_c} = (\theta_1, \ldots, \theta_{k_c})$ be the current parameter values of the clusters. Conditional on configuration $\mathbf{c}$, get $\boldsymbol{\theta}_{k_c}^{updated} = (\theta_1^{updated}, \ldots, \theta_{k_c}^{updated})$ as follows:
2: **if** cluster j is such that $n_j > 1$, **then** do $\theta_j^{updated} = \hat{\theta}_j$, where $\hat{\theta}_j$ are the maximum likelihood estimates of the j-th cluster;
3: **if** cluster j is such that $n_j = 1$, **then** generate θ_j^g from conditional posterior distribution $\pi(\theta|y_i)$ and set $\theta_j^{updated} = \theta_j^g$;
4: Do $\boldsymbol{\theta}_k = \boldsymbol{\theta}_k^{updated}$ only if the adjacency condition $\mu_1^{updated} < \ldots < \mu_{k_c}^{updated}$ is met. Otherwise, keep $\boldsymbol{\theta}_{k_c}$ as the current value.

Algorithm 4 ISEM Algorithm

1: Initialize the algorithm with a configuration $\mathbf{c}^{(0)} = \left(c_1^{(0)}, \ldots, c_n^{(0)} \right)$ for allocation indicator variables.
2: **procedure** For the s-th iteration of the algorithm, $s = 1, \ldots, S$, do the following.
3:     Conditional on $\mathbf{c}^{(s-1)}$, update the parameters of the clusters according to algorithm 3 ;
4:     Obtain a new configuration $\mathbf{c}^{(s)}$ for the allocation of indicator variables using algorithm 2.

4. Simulation Study

In this section, we describe the results from a simulation study carried out to verify the performance of the proposed algorithm. To generate the artificial datasets, we considered univariate normal mixture models. We set up the number of clusters and parameter values according to the specified values in Table 2. We also fixed the sample size equal to $n = 200$.

Table 2. Number of clusters and parameter values used for simulating the datasets.

Artificial Dataset	Number of Clusters	Parameter Values
A_1	$k_c = 2$	$\mu_1 = 0$, $\mu_2 = 3$, $\sigma_1^2 = 1$, $\sigma_2^2 = 1$, $w_1 = 0.80$, $w_2 = 0.20$,
A_2	$k_c = 3$	$\mu_1 = -6$, $\mu_2 = 0$, $\mu_3 = 4$ $\sigma_1^2 = 3$, $\sigma_2^2 = 2$, $\sigma_3^2 = 1$ $w_1 = 0.50$, $w_2 = 0.30$, $w_3 = 0.20$
A_3	$k_c = 4$	$\mu_1 = -6$, $\mu_2 = 0$, $\mu_3 = 7$, $\mu_4 = 14$ $\sigma_1^2 = 1$, $\sigma_2^2 = 2$, $\sigma_3^2 = 2$, $\sigma_4^2 = 1$ $w_1 = 0.10$, $w_2 = 0.40$, $w_3 = 0.40$, $w_4 = 0.10$
A_4	$k_c = 5$	$\mu_1 = -13$, $\mu_2 = 7$, $\mu_3 = 0$, $\mu_4 = 6$, $\mu_5 = 11$ $\sigma_1 = 1$, $\sigma_2 = 2$, $\sigma_3 = 3$, $\sigma_4 = 2$, $\sigma_5 = 1$ $w_1 = 0.15$, $w_2 = 0.20$, $w_3 = 0.30$, $w_4 = 0.20$, $w_5 = 0.15$,

The procedure for generating the datasets is given by the following steps:

(i) For $i = 1, \ldots, n$, generate $U_i \sim \mathcal{U}(0,1)$; if $\sum_{j'=1}^{j-1} w_{j'} < u_i \leq \sum_{j'=1}^{j} w_{j'}$, generate $Y_i \sim \mathcal{N}\left(\mu_j, \sigma_j^2\right)$, with fixed parameter values according to Table 2, for $w_0 = 0$ and $j = 1, \ldots, k_c$.

(ii) In order to record from which component each observation is generated, we define $G = (G_1, \ldots, G_n)$ such that $G_i = j$ if $Y_i \sim \mathcal{N}\left(\mu_j, \sigma_j^2\right)$, for $i = 1, \ldots, n$ and $j = 1, \ldots, k_c$.

Having generated the datasets, we need to define the the probability of creating a new cluster and the posterior distribution for $\theta_{j^*} = \left(\mu_{j^*}, \sigma_{j^*}^2\right)$ given y_i, for $i = 1, \ldots, n$. For this, consider the following conjugated prior distributions for component parameters $\theta_j = \left(\mu_j, \sigma_j^2\right)$:

$$\mu_j | \sigma_j^2, \mu_0, \lambda \sim \mathcal{N}\left(\mu_0, \frac{\sigma_j^2}{\lambda}\right) \quad \text{and} \quad \sigma_j^{-2} | \alpha, \beta \sim \Gamma(\alpha, \beta),$$

where μ_0, λ, α, and β are hyperparameters. The parametrization of the gamma distribution is such that the mean is α/β and the variance is α/β^2.

Following [11,24], we consider the following procedure to define the values for the hyperparameters. Let R be the observed variation interval of the data and ε its midpoint. Then, we set up $\mu_0 = \varepsilon$ and $E(\sigma_j^{-2}) = R^{-2}$. Thus, we obtain $\beta = \alpha R^2$, and we fix $\alpha = 1$. In addition, to obtain a prior distribution with a large variance, we fixed $\lambda = 10^{-2}$, and for the hyperparameter γ, we consider the value 0.1, $\gamma = 0.1$.

Thus, the probability of creating a new cluster is given by Equation (10), in which

$$\mathbf{I}(y_i) = \left[\frac{\lambda}{2\beta\pi(1+\lambda)}\right]^{\frac{1}{2}} \frac{\Gamma(\alpha+1)}{\Gamma(\alpha)} \left[1 + \frac{y_i^2 + \lambda\mu_0^2}{2\beta} - \frac{(y_i + \lambda\mu_0)^2}{2\beta(1+\lambda)}\right]^{-(\alpha+\frac{1}{2})}, \quad (11)$$

and $j^* = k_c + 1$, for $i = 1, \ldots, n$.

When a new cluster is created, the new parameter values $\theta_{j^*} = (\mu_{j^*}, \sigma_{j^*}^2)$ are generated from the following conditional posterior distributions,

$$\mu_{j^*} | \sigma_{j^*}^2, y_i, \mathbf{c}, \mu_0, \lambda \sim \mathcal{N}\left(\frac{y_i + \lambda\mu_0}{1+\lambda}, \frac{\sigma_j^2}{1+\lambda}\right) \quad (12)$$

and

$$\sigma_{j^*}^{-2} | y_i, \mathbf{c}, \tau, \beta \sim \Gamma\left(\alpha + 1, \beta + \frac{y_i^2 + \lambda\mu_0^2}{2} - \frac{(y_i + \lambda\mu_0)^2}{2(1+\lambda)}\right), \quad (13)$$

for $j^* = k_{\mathbf{c}_{-i}} + 1$.

We run the ISEM algorithm for $S = 55{,}000$, $B = 5000$, and $h = 10$. From these values, we got a sub-sequence $\mathbb{S}(H)$ of size 5000 to make inferences. The algorithm was initialized with $k_c = 1$ and parameter values $\mu_1 = \bar{y}$ and $\sigma_1^2 = s^2$, the sample mean and variance of the generated dataset, respectively.

We also apply to the generated datasets the SEM algorithm, as describe in Section 2, and the RJ algorithm as proposed by [8]. In order to choose the number of clusters using the SEM algorithm, we consider the AIC and BIC model selection criteria. In addition, the algorithm was initialized using a configuration $\mathbf{c}^{(0)}$ obtained via the k-means algorithm [25]. As stop criterion, we set up the threshold $\varepsilon = 0.001$. For the RJ algorithm, we consider the same number of iterations, burn-in, and thin value used in the ISEM algorithm.

In order to compare the three algorithms in terms of the estimation of the number of clusters, we consider $M = 500$ simulated datasets. Table 3 shows the proportion of times that the ISEM and RJ algorithms put the highest estimated probability on the k_c values presented. This table also show the proportion of times that the AIC and BIC indicated the k_c value as the best among the tested values. The values highlighted in bold are the proportions on the k_c true value. As one can note, the ISEM shows a better performance, i.e., higher proportion of the k_c true value than the other two algorithms, especially in relation to the SEM algorithm with the selection of k_c via the AIC and BIC. The results also show that the AIC and BIC model selection criteria have a low success ratio, with a proportion of the k_c true value smaller than 0.50.

Table 3. Proportion of times the algorithms chose the k_c values as the number of clusters.

Data Set	k_c^{true}	k_c	$\tilde{P}(k_c = j\cdot)$ ISEM RJ	AIC	BIC	Data Set	k_c^{true}	k_c	$\tilde{P}(k_c = j\cdot)$ ISEM RJ	AIC	BIC
A_1	2	1	0.014 0.002	0.050	0.210	A_2	3	1	0.000 0.000	0.000	0.004
		2	**0.976 0.972**	0.294	**0.448**			2	0.276 0.094	0.104	0.438
		3	0.010 0.026	0.238	0.224			3	**0.720 0.672**	**0.304**	**0.384**
		4	0.000 0.000	0.152	0.082			4	0.004 0.232	0.262	0.138
		5	0.000 0.000	0.148	0.028			5	0.000 0.002	0.184	0.028
		6	0.000 0.000	0.118	0.008			6	0.000 0.000	0.146	0.008
A_3	4	1	0.000 0.000	0.000	0.000	A_4	5	1	0.000 0.000	0.000	0.000
		2	0.000 0.004	0.000	0.000			2	0.006 0.000	0.000	0.006
		3	0.000 0.000	0.010	0.066			3	0.006 0.000	0.000	0.018
		4	**0.956** 0.476	**0.226**	**0.450**			4	0.218 0.010	0.038	0.210
		5	0.044 **0.474**	0.252	0.296			5	**0.682 0.509**	**0.322**	**0.446**
		6	0.000 0.044	0.214	0.122			6	0.028 0.442	0.246	0.222
		7	0.000 0.000	0.184	0.056			7	0.000 0.039	0.210	0.072
		8	0.000 0.002	0.114	0.010			8	0.000 0.000	0.184	0.026

4.1. Results from a Single Simulated Data Set

We also analyze the results from a single dataset selected at random from the $M = 500$ generated datasets in each situation A_1 to A_4. Then, we discuss the convergence of the ISEM and RJ algorithms based on the sample generated across iterations, using graphical tools. In general, the graphical tools show whether the simulated chain stabilizes in some sense and provide useful feedback about the convergence [26].

Table 4 shows the estimated probabilities of k_c obtained with ISEM and RJ and the AIC and BIC values from the SEM algorithm for the selected dataset. In this table, the values highlighted in bold are the highest estimated probabilities and the smallest AIC and BIC values. As we can note, the ISEM algorithm set up a maximum probability for the k_c true value for the four simulated cases.

The RJ algorithm puts a higher probability on the k_c true value for datasets A_1 and A_2. However, the probability on the k_c true value is smaller than that estimated by ISEM. This indicates a higher precision for the ISEM algorithm. For datasets A_3 and A_4, the RJ attributes maximum probability to the wrong values, $k_c = 5$ and $k_c = 6$, respectively. Moreover, the probabilities estimated by RJ do not evidence a single value for k_c as being the best value since there are different values for k_c with similar probabilities. For example, for dataset A_2, the maximum is at $k_c = 3$ with $P(k_c = 3|\cdot) = 0.3836$, but one can argue that the estimated probabilities favor $k_c = 3$ or $k_c = 4$. For dataset A_3, there is similar support for k_c between 4 and 7, and for A_4 between 5 and 7.

Analogously to ISEM and RJ, the AIC and BIC model selection criteria indicate the k_c true value as the best value for datasets A_1 and A_2. For dataset A_3, similar to the RJ, the AIC indicates the wrong value $k_c = 5$ as the best value, while the BIC indicates the k_c true value as the best value. For dataset A_4, the AIC and BIC indicate the wrong value $k_c = 6$ as the best model.

4.2. An Empirical Check of the Convergence

We now empirically check the convergence of the sequence of the probability for k_c across iterations, the capacity to move for different values of k_c in the course of the iterations, and the estimated autocorrelation function the (acf) for the ISEM and RJ algorithms.

Table 4. Estimated probability for k_c.

Data Set	k_c^{true}	k_c	$\tilde{P}(k_c = j \cdot)$ ISEM	$\tilde{P}(k_c = j \cdot)$ RJ	AIC	BIC	Data Set	k_c^{true}	k_c	$\tilde{P}(k_c = j \cdot)$ ISEM	$\tilde{P}(k_c = j \cdot)$ RJ	AIC	BIC
A_1	2	1	0.0000	0.0000	786.7166	793.3133	A_2	3	1	0.0000	0.0004	1160.758	1167.355
		2	**0.9006**	**0.5252**	**762.5204**	**779.0120**			2	0.0122	0.0136	1129.981	1146.472
		3	0.0962	0.2862	764.1440	790.5305			3	**0.8694**	**0.3836**	**1114.024**	**1140.411**
		4	0.0032	0.1138	769.2648	805.5463			4	0.1124	0.3140	1118.789	1155.070
		5	0.0000	0.0466	768.0492	814.2256			5	0.0058	0.1716	1120.108	1166.284
		6	0.0000	0.0160	775.1082	831.1796			6	0.0002	0.0744	1130.558	1186.630
		≥7	0.0000	0.0122	-	-			≥7	0.0000	0.0424	-	-
A_3	4	1	0.0000	0.0000	1273.886	1280.482	A_4	5	1	0.0000	0.0002	1416.124	1422.721
		2	0.0000	0.0000	1276.281	1292.773			2	0.0000	0.0004	1388.738	1405.230
		3	0.0000	0.0002	1251.357	1277.743			3	0.0000	0.0028	1358.474	1384.861
		4	**0.8412**	0.1696	**1188.470**	**1224.751**			4	0.0014	0.0114	1357.037	1393.318
		5	0.1500	**0.3014**	1186.075	1232.252			5	**0.8340**	0.2788	**1355.922**	**1402.098**
		6	0.0088	0.2400	1191.747	1247.818			6	0.1520	**0.3004**	1325.927	1381.998
		7	0.0000	0.1632	1197.028	1262.995			7	0.0124	0.2224	1331.940	1397.907
		8	0.0000	0.0816	1200.337	1276.199			8	0.0002	0.0186	1331.352	1407.213
		≥9	0.0000	0.0440	-	-			≥9	0.0000	0.0750	-	-

Figure 1a,d,g,j presents the graphics of the probability for k_c in the course of the iterations, for the four simulated datasets. To maintain a better visualization, we plot in these graphics only the three higher $P(k_c|\cdot)$ estimates. Observing at these figures, it can be seen that the L iterations and the burn-in value B used were adequate to achieve stability for $P(k_c|\cdot)$. In addition, Figure 1b,e,h,k shows that the ISEM algorithm mixes well over k_c, i.e., "visits" mixture models with different values of k_c across iterations. As shown by Figure 1c,f,i,l, the sampled k_c values also do not have significant autocorrelation function (ACF). Thus, based on these graphical tools, there is no evidence against the convergence of the generated values by the ISEM algorithm.

Figure 2 shows the performance of the RJ algorithm. The probabilities of k_c present a satisfactory stability. The sampled k_c values have a satisfactory mix, and the estimated autocorrelation is non-significant. In addition, as can be noted in Figure 2, probabilities for the number of clusters do not differentiate a value of k_c in order to be chosen as the better value, as done by ISEM. This may happen due the fact that the performance of the RJ depends on the choice of the transition functions to do "good" jumping, meaning that a transition function that is adequate for one dataset may be not for another one. As the ISEM algorithm does not need the specification of transition functions to propose a change of the k_c value, these results shows us that ISEM may be an effective alternative in relation to RJ and SEM algorithms for the joint estimation of k_c and the cluster parameters of a mixture model.

Figure 1 in Appendix 2 of the SM shows the generated values for datasets A_1 to A_4. This Figure also shows the clusters identified by the ISEM algorithm. As can be seen, clusters are satisfactorily identified by the proposed algorithm.

We also compare ISEM and RJ algorithms in terms of CPU computation time. The simulations were realized on a MacBook Pro, 2.5 GHz Intel Core i5 dual core, 4 Gb MHz DDR3. Table 5 shows a summary of the times of iterations for the ISEM and RJ algorithms. The column denoted by s.d. presents the standard deviation values. For dataset A_1, the average time that RJ takes to run one iteration is 1.8491 times greater than the average time that ISEM takes to run an iteration. For datasets A_2, A_3, and A_4, the average time that RJ needs to run one iteration is 1.8175, 2.3239, and 1.8932 times greater than the average time that ISEM takes to run an iteration, respectively. These results show a better performance of the ISEM algorithm. The higher iteration times of the RJ algorithm are mainly due to the split–merge step used to increase the mixing of the Markov chain in relation to the number of clusters.

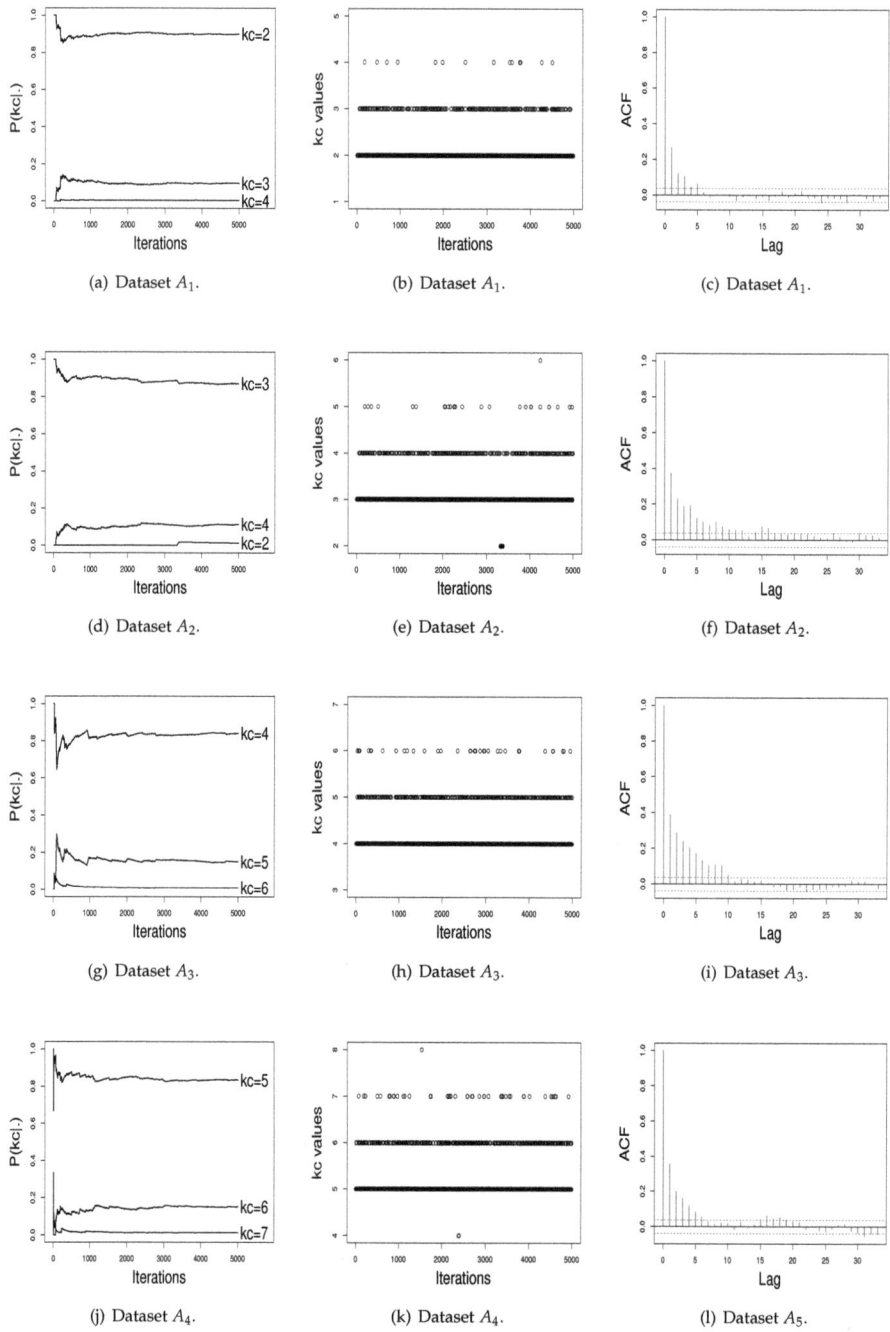

Figure 1. Performance of the ISEM algorithm across iterations.

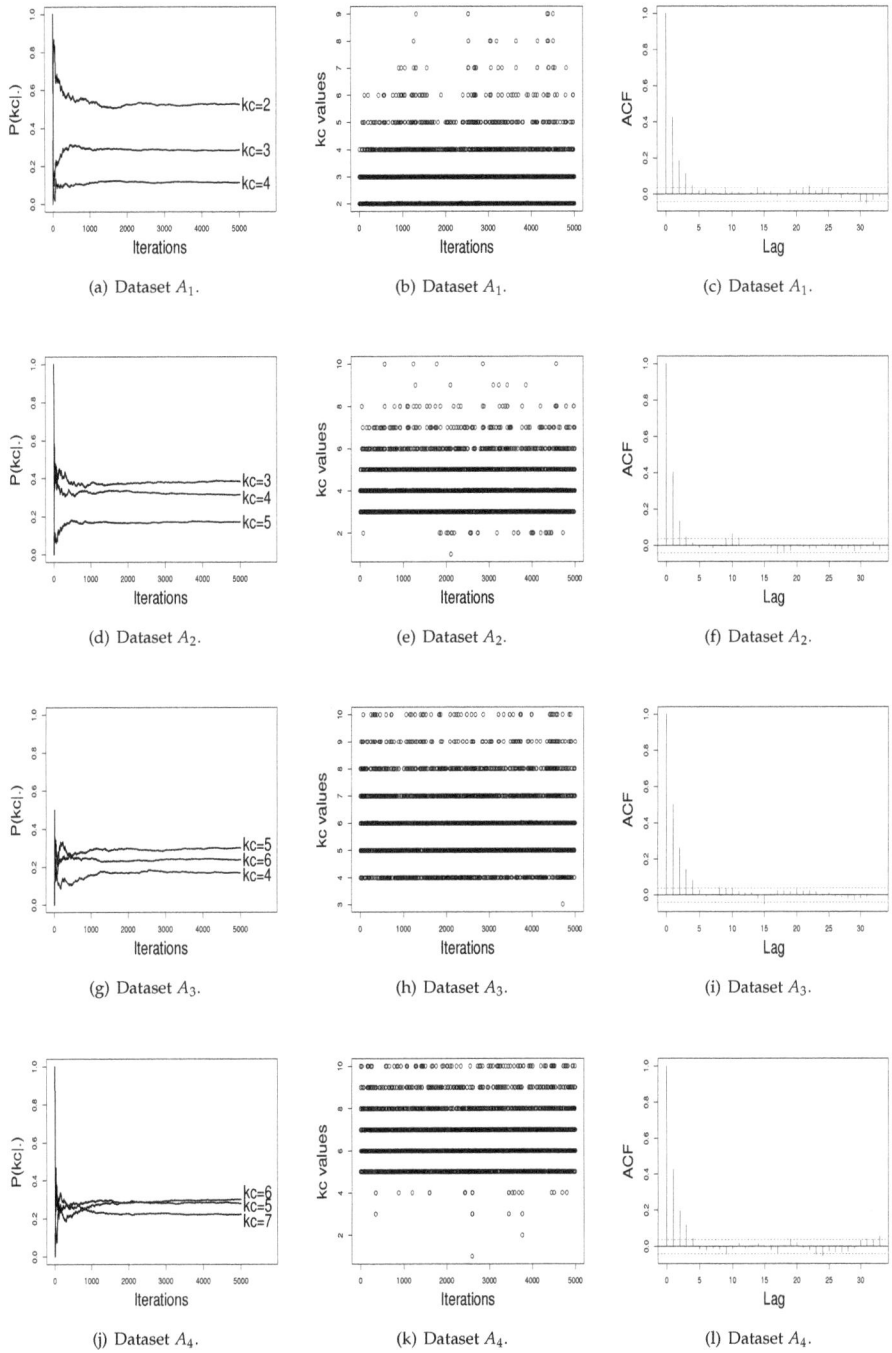

Figure 2. Performance of the RJ algorithm across iterations.

The results from these simulated datasets show that the ISEM algorithm may be an effective alternative to the RJ and SEM algorithms for data clustering in situations where the number of clusters is a unknown quantity.

Table 5. Times of the iterations, in seconds.

Artificial Dataset	Algorithm	Summary						
		Min	1º Q.	Med.	Mean	3º Q.	Max.	s.d.
A_1	ISEM	0.0064	0.0082	0.0091	0.0109	0.0105	0.4987	0.0107
	RJ	0.0032	0.0137	0.0158	0.0208	0.0202	0.3855	0.0174
A_2	ISEM	0.0055	0.0100	0.0114	0.0137	0.0146	0.3806	0.0108
	RJ	0.0032	0.0169	0.0196	0.0249	0.0243	0.7709	0.0181
A_3	ISEM	0.0059	0.0112	0.0123	0.0142	0.0139	0.4951	0.0100
	RJ	0.0020	0.0218	0.0255	0.0330	0.0320	0.4785	0.0239
A_4	ISEM	0.0059	0.0130	0.0146	0.0179	0.0187	0.5149	0.0108
	RJ	0.0026	0.0232	0.0266	0.0339	0.0323	0.5490	0.0231

5. Application

The three algorithms are now applied to two real datasets. The first real dataset refers to velocity in km/s of $n = 82$ galaxies from 6 well-separated conic sections of an unfilled survey of the Corona Borealis region. This dataset is known in the literature as the Galaxy data and has already been analyzed by [8,13,22,27], among others. This dataset is available in the R software. The second real dataset refers to an acidity index measured in a sample of $n = 155$ lakes in central-north Wisconsin. This dataset was downloaded from the website https://people.maths.bris.ac.uk/$\sim$mapjg/mixdata.

For application of ISEM and RJ algorithms, we consider the same number $L = 5500$, $B = 5000$, and $h = 10$. Table 6 shows the estimated probabilities for k_c obtained with ISEM and RJ and the AIC and BIC values from EM algorithm for each dataset. The maximum probability from ISEM and RJ and the minimum AIC and BIC values are highlighted in bold.

Table 6. Estimated probabilities for k_c, real datasets.

Data Set	k_c	$\tilde{P}(k_c = j \cdot)$		AIC	BIC	Data Set	k_c	$\tilde{P}(k_c = j \cdot)$		AIC	BIC
		ISEM	RJ					ISEM	RJ		
Galaxy	1	0.0000	0.0000	484.6819	489.4954	Acidity	1	0.0000	0.0000	455.5740	461.6608
	2	0.0000	0.0008	451.0018	463.0354		2	**0.7194**	0.0502	**380.3449**	**395.5620**
	3	**0.7024**	0.1200	426.7421	445.09959		3	0.2638	**0.3164**	382.7395	407.0869
	4	0.2748	0.2530	427.4915	453.9654		4	0.0152	0.3040	382.3660	415.8437
	5	0.0222	**0.2592**	**410.3666**	**444.0607**		5	0.0016	0.1724	391.7630	434.3709
	6	0.0006	0.1848	413.7755	454.6897		6	0.0000	0.0832	386.1420	437.8802
	7	0.0000	0.1084	422.1793	470.3137		7	0.0000	0.0452	388.1296	448.9981
	8	0.0000	0.0472	423.5542	478.9088		8	0.0000	0.0186	395.3957	465.3945
	≥9	0.0000	00226	-	-		≥9	0.0000	0.0010	-	-

For the Galaxy dataset, the ISEM and RJ algorithms put highest probability on $k_c = 3$ and $k_c = 5$, respectively. However, analogously to the simulation study, the probabilities estimated by RJ do not evidence a single value for k_c as being the best value. For this dataset, the estimated probabilities indicate a k_c value between 3 and 7. The AIC and BIC also indicate $k_c = 5$ as the best value. For the Acidity dataset, ISEM, AIC, and BIC indicate $k_c = 2$ as the best value. The probabilities estimated by RJ attribute similar values for $k_c = 3$ and $k_c = 4$.

Figures 3 and 4 show the performance of the ISEM and RJ algorithms across iterations for the Galaxy and Acidity datasets. The values sampled by the ISEM algorithm present satisfactory stability for estimated probability across iterations, mix well among different k_c values, and present no significant autocorrelation. That is, we do not have evidence against the convergence of the generated chain by the ISEM algorithm. In relation to the RJ, the sampled values mix well and do not present significant autocorrelation. However, although the values sampled by RJ present stability for $P(k_c)$, the estimated probabilities do not differentiate a value of k_c in order to be chosen as the better value, as done by ISEM. This result shows the need to run RJ for a greater number of iterations. With this, we have that for both real datasets, ISEM presents faster convergence than the RJ algorithm.

Table 7 shows a summary of the iteration times for the ISEM and RJ algorithms. For the Galaxy data, the average time that ISEM takes to run an iteration is 0.0053 s; while the average time for RJ is 0.0098 s. That is, the average time that RJ takes to run one iteration is 1.8491 times greater than the average time that ISEM takes to run an iteration. For the Acidity data, the average times that the ISEM and RJ algorithms take to run an iteration are 0.0085 and 0.0180 s, respectively. For this dataset, the average time that RJ needs to run an iteration is 2.2118 times greater than the average time that ISEM runs. Similarly to results from the simulation study, ISEM presents better results, i.e., a shorter time to run the iterations.

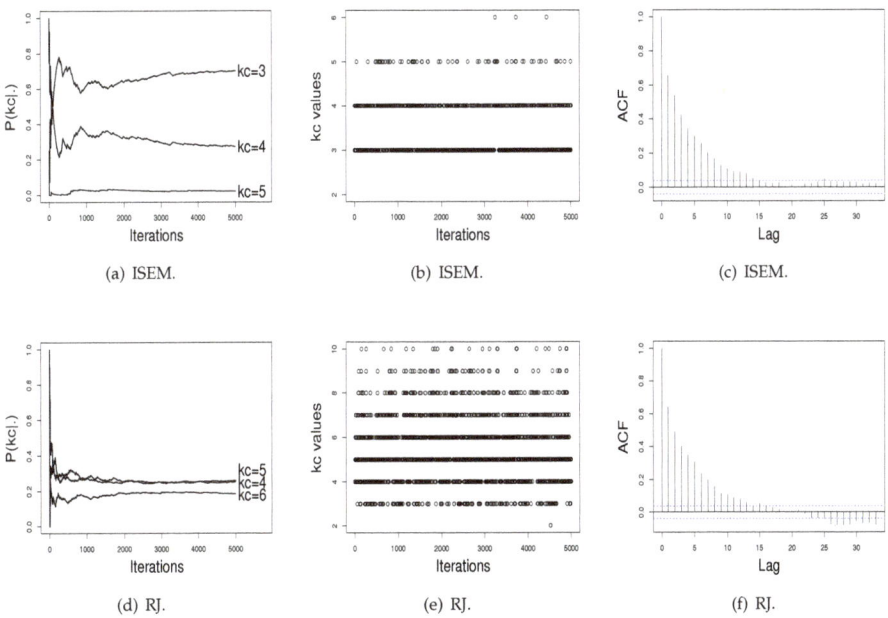

Figure 3. Performance of the ISEM and RJ algorithms for the Galaxy data.

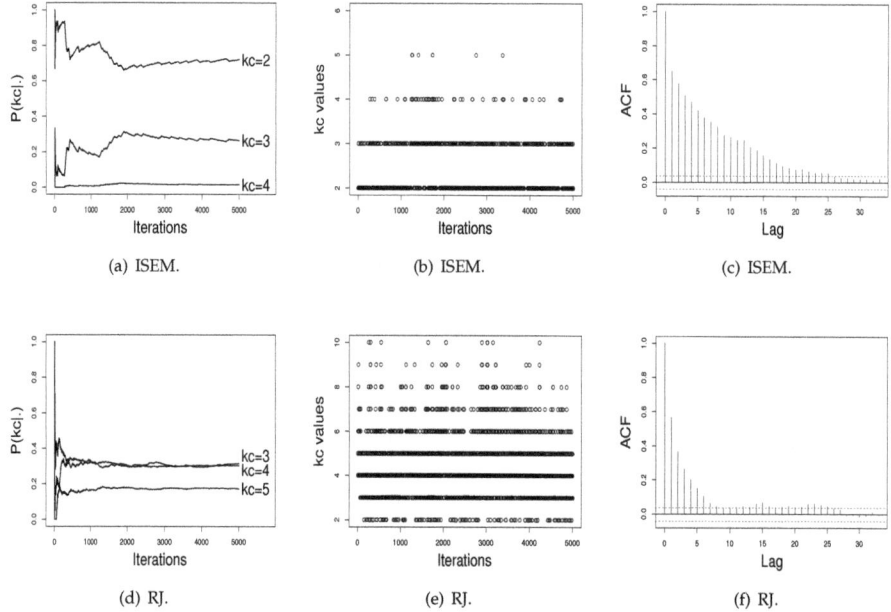

Figure 4. Performance of the RJ algorithm across iterations for the Acidity data.

Table 7. Iteration times in seconds.

Artificial Dataset	Algorithm	Summary						
		Min	1° Q.	Med.	Mean	3° Q.	Max.	s.d.
Galaxy	ISEM	0.0023	0.0038	0.0045	0.0053	0.0054	0.2468	0.0062
	RJ	0.0000	0.0000	0.0000	0.0000	0.0000	0.0000	0.0000
Acidity	ISEM	0.0055	0.0100	0.0114	0.0137	0.0146	0.3806	0.0108
	RJ	0.0046	0.0128	0.0149	0.0188	0.0180	0.4588	0.0160

6. Final Remarks

This article presents a discussion of how to estimate the parameters of a mixture model in the context of data clustering. We propose an alternative algorithm to the EM algorithm called ISEM. This algorithm was developed through an integrated approach in order to allow k_c to be estimated jointly with the other parameters of interest. In the ISEM algorithm, the allocation probabilities depend on the number of clusters k_c and are independent of the number of components k of the mixture model.

In addition, there exists a positive probability of a new cluster being created by a single observation. This is an advantage of the algorithm because it creates a new cluster without the need to specify transition functions. In addition, the cluster parameters are updated according to the number of allocated observations. For the clusters with at least two of these observations, the values of the parameters are taken by the maximum likelihood estimates. For a cluster with just one observation, the parameter values are generated from the posterior distribution.

In order to illustrate the performance of the ISEM algorithm, we developed a simulation study. In this simulation study, we considered four scenarios with artificial data generated from Gaussian mixture models. In addition, each one of the four scenarios was replicated $M = 500$ times, and the proportion of times that ISEM put a higher probability on the k_c true value was recorded. We applied this same procedure to the EM algorithm, choosing the number of clusters k_c via the AIC and BIC, and to the RJ algorithm. Then, the three algorithms were compared in terms of proportion of times that the k_c true value was selected as the best value. The results obtained show that the ISEM algorithm outperforms the RJ and SEM algorithms. Moreover, the results also show that the AIC and BIC model selection criteria should not be used to determine the number of clusters in a mixture model due to a low success rate.

We also compared the performance of ISEM and RJ in terms of empirical convergence of the sequence of values generated using graphical tools. For this, we selected at random an artificial dataset from each scenery, and then we plotted the probability estimates for k_c across iterations, the generated k_c values, and the estimated autocorrelation of the sampled values (see Figures 1 and 2). Again, the results show a better performance for the ISEM algorithm. While ISEM presents satisfactory stability for the probability of k_c and differentiates the true k_c as the best value, the probabilities estimated by RJ do not differentiate a value of k_c in order to be chosen as the better value.

In order to illustrate the practical use of the proposed algorithm and compare its performance with the SEM and RJ algorithms, we applied the three algorithms to two real datasets: the Galaxy and Acidity datasets. For the Galaxy dataset, ISEM indicates $k_c = 3$ with probability $P(k_c = 3|\cdot) = 0.7024$, while the RJ algorithm, the AIC, and the BIC indicate $k_c = 5$. However, as shown in Figure 3d, the RJ algorithm again does not differentiate a value of k_c, while ISEM differentiates the $k_c = 3$ value, and the generated values across iterations present satisfactory stability. For the Acidity dataset, the ISEM, AIC, and BIC indicate $k_c = 2$ as the best value, while RJ attributes similar probabilities for $k_c = 3$ and $k_c = 4$.

As mentioned in the Introduction, the generalization of the proposed algorithm for the multivariate case is the next step of our research. The simulation study and the application were done in R software, and the computational codes can be obtained by emailing the authors.

Supplementary Materials: The following are available online at http://www.mdpi.com/1099-4300/21/11/1063/s1.

Author Contributions: E.F.S. and C.A.d.B.P. developed the whole theoretical part of the research. E.F.S., A.K.S. and L.A.M. developed the simulation studies and real data application.

Funding: This research was funded by Conselho Nacional de Desenvolvimento Científico e Tecnológico, CNPq, grant number 308776/2014-3.

Conflicts of Interest: The authors declare no conflict of interest.

References

1. Bouveyron, C.; Brunet, C. Model-based clustering of high-dimensional data: A review. *Comput. Stat. Data Anal.* **2014**, *71*, 52–78. [CrossRef]
2. Oh, M.S.; Raftery, A.E. Model-based clustering with dissimilarities: A bayesian approach. *J. Comput. Graph. Stat.* **2007**, *16*, 559–585. [CrossRef]
3. Dempster, A.; Laird, N.; Rubin, D. Maximum likelihood from incomplete data via EM algorithm. *J. Roy. Statist. Soc. B* **1977**, *39*, 1–38. [CrossRef]
4. Akaike, H.A. New look at the statistical model identification. *IEEE Trans. Autom. Control* **1974**, *19*, 716–723. [CrossRef]
5. Bozdogan, H. Model selection and Akaike's information criterion (AIC): The general theory and its analytical extensions. *Psychometrica* **1987**, *52*, 345–370. [CrossRef]
6. Schwarz, G.E. Estimating the dimension of a model. *Ann. Stat.* **1978**, *6*, 461–464. [CrossRef]
7. Saraiva, E.F.; Suzuki, A.K.; Milan, L.A. A Bayesian sparse finite mixture model for clustering data from a heterogeneous population. *Braz. J. Probab. Stat.* **2019**.
8. Richardson, S.; Green, P.J. On Bayesian analysis of mixtures with an unknown number of components. *J. R. Stat. Soc. Ser. B* **1997**, *59*, 731–792; errata, *J. R. Stat. Soc. Ser. B* **1998**, *60*, U3. [CrossRef]
9. Jain, S.; Neal, R. A split–merge markov chain monte carlo procedure for the Dirichlet process mixture models. *J. Comput. Graph. Stat.* **2004**, *13*, 158–182. [CrossRef]

10. Jain, S.; Neal, R. Splitting and merging components of a nonconjugated Dirichlet process mixture model. *Bayesian Anal.* **2007**, *3*, 445–472. [CrossRef]
11. Saraiva, E.F.; Pereira, C.A.B.; Suzuki, A.K. A data-driven selection of the number of clusters in the Dirichlet allocation model via Bayesian mixture modelling. *J. Stat. Comput. Simul.* **2019**, *89*, 2848–2870. [CrossRef]
12. Diebolt, J.; Robert, C. Estimation of finite mixture distributions by Bayesian sampling. *J. R. Stat. Soc. B* **1994**, *56*, 363–375. [CrossRef]
13. Escobar, M.D.; West, M. Bayesian Density Estimation and Inference Using Mixtures. *J. Am. Stat. Assoc.* **1995**, *90*, 577–588. [CrossRef]
14. Dellapotas, P.; Papgeorgiou, I. Multivariate mixtures of normals with unknown number of components. *Stat. Comput.* **2006**, *16*, 57–68. [CrossRef]
15. McLachlan, G.; Peel, D. *Finite Mixture Models*; Wiley Interscience: New York, NY, USA, 2000.
16. Finch, S.; Mendell, N.; Thode, H. Probabilistic measures of adequacy of a numerical search for a global maximum. *J. Am. Stat. Assoc.* **1989**, *84*, 1020–1023. [CrossRef]
17. Karlis, D.; Xekalaki, W.D. Choosing initial values for the EM algorithm for Finite mixtures. *Comput. Stat. Data Anal.* **2003**, *41*, 577–590. [CrossRef]
18. Biernacki, C.; Celeux, G.; Govaert, G. Choosing starting values for the EM algorithm for getting the highest likelihood in multivariate gaussian mixture models. *Comput. Stat. Data Anal.* **2003**, *41*, 561–575. [CrossRef]
19. Ferguson, S.T. A bayesian analysis of some nonparametric problems. *Ann. Stat.* **1973**, *2*, 209–230. [CrossRef]
20. Blackwell, D.; MacQueen, J.B. Ferguson distributions via Polya urn scheme. *Ann. Stat.* **1973**, *1*, 353–355. [CrossRef]
21. Antoniak, C.E. Mixture of processes dirichlet with applications to bayesian nonparametric problems. *Ann. Stat.* **1974**, *2*, 1142–1174. [CrossRef]
22. Stephens, M. Dealing with label switching in mixture models. *J. R. Stat. Soc. B* **2000**, *62*, 795–809. [CrossRef]
23. Jasra, A.; Holmes, C.C.; Stephens, D.A. Markov Chain Monte Carlo methods and the label switching problem in Bayesian mixture modeling. *Stat. Sci.* **2005**, *20*, 50–67. [CrossRef]
24. Saraiva, E.F.; Suzuki, A.K.; Louzada, F.; Milan, L.A. Partitioning gene expression data by data-driven Markov chain Monte Carlo. *J. Appl. Stat.* **2016**, *43*, 1155–1173. [CrossRef]
25. MacQueen, J. Some methods for classification and analysis of multivariate observations. In *Proceedings of the Fifth Berkeley Symposium on Mathematical Statistics and Probability, Volume 1: Statistics*; University of California Press: Berkeley, CA, USA, 1967; pp. 281–297.
26. Sinharay, S. Assessing Convergence of the Markov Chain Monte Carlo Algorithms: A Review. *ETS Res. Rep. Ser.* **2003**, *2003*, i-52. Available online: http://www.ets.org/Media/Research/pdf/RR-03-07-Sinharay.pdf (accessed on 19 August 2019). [CrossRef]
27. Roeder, K.; Wasserman, L. Practical Bayesian Density Estimation Using Mixture of Normals. *J. Am. Stat. Assoc.* **1997**, *92*, 894–902. [CrossRef]

© 2019 by the authors. Licensee MDPI, Basel, Switzerland. This article is an open access article distributed under the terms and conditions of the Creative Commons Attribution (CC BY) license (http://creativecommons.org/licenses/by/4.0/).

Article

Universal Sample Size Invariant Measures for Uncertainty Quantification in Density Estimation

Jenny Farmer [1], Zach Merino [1], Alexander Gray [1] and Donald Jacobs [1,2,*]

[1] Department of Physics and Optical Science, University of North Carolina at Charlotte, Charlotte, NC 28223, USA; jfarmer6@uncc.edu (J.F.); zmerino@uncc.edu (Z.M.); agray36@uncc.edu (A.G.)
[2] Center for Biomedical Engineering and Science, University of North Carolina at Charlotte, Charlotte, NC 28223, USA
* Correspondence: djacobs1@uncc.edu

Received: 7 October 2019; Accepted: 8 November 2019; Published: 15 November 2019

Abstract: Previously, we developed a high throughput non-parametric maximum entropy method (PLOS ONE, 13(5): e0196937, 2018) that employs a log-likelihood scoring function to characterize uncertainty in trial probability density estimates through a scaled quantile residual (SQR). The SQR for the true probability density has universal sample size invariant properties equivalent to sampled uniform random data (SURD). Alternative scoring functions are considered that include the Anderson-Darling test. Scoring function effectiveness is evaluated using receiver operator characteristics to quantify efficacy in discriminating SURD from decoy-SURD, and by comparing overall performance characteristics during density estimation across a diverse test set of known probability distributions.

Keywords: density estimation; distribution free; non-parametric statistical test; decoy distributions; size invariance; scaled quantile residual; maximum entropy method; scoring function; outlier detection; overfitting detection

1. Introduction

The rapid and accurate estimate of the probability density function (pdf) for a random variable is important in many different fields and areas of research [1–6]. For example, accurate high throughput pdf estimation is sought in bioinformatics screening applications and in high frequency trading to evaluate profit/loss risks. In the era of big data, data analytics and machine learning, it has never been more important to strive for automated high-quality pdf estimation. Of course, there are numerous other traditional areas of low throughput applications where pdf estimation is also of great importance, such as damage detection in engineering [7], isotope analysis in archaeology [8], econometric data analysis in economics [9], and particle discrimination in high energy physics [10]. The wide range of applications for pdf estimation exemplifies its ubiquitous importance in data analysis. However, a continuing objective regarding pdf estimation is to establish a robust distribution free method to make estimates rapidly while quantifying error in an estimate. To this end, it is necessary to develop universal measures to quantify error and uncertainties to enable comparisons across distribution classes. To illustrate the need for universality, the pdf and cumulative distribution function (cdf) for four distinctly different distributions are shown in Figure 1a,b. Comparing the four cases of pdf and cdf over the same sample range, it is apparent that the data are distributed very differently.

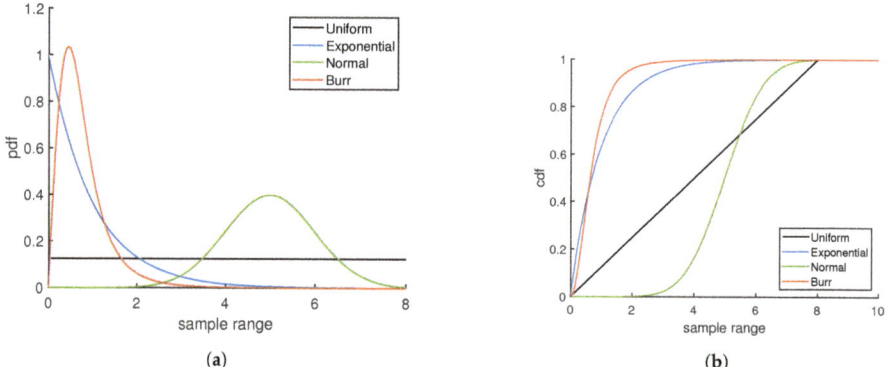

Figure 1. Examples of four distribution types in the form of (a) pdf and corresponding (b) cdf.

The process of estimating the pdf for a given sample of data is an inverse problem. Due to fluctuations in a sample of random data, many pdf estimates will be able to model the data sample well. If additional smoothness criteria are imposed, many proposed pdf estimates can be filtered out. Nevertheless, a pdf estimate will carry intrinsic uncertainty along with it. The development of a scoring function to measure uncertainty in a pdf estimate without knowing the form of the true pdf is indispensable in high throughput applications where human domain expertise cannot be applied to inspect every proposed solution for validity. Moreover, it is desirable to remove subjective bias from human (or artificial intelligence) intervention. Automation can be achieved by employing a scoring function that measures over-fitting and under-fitting quantitatively based solely on mathematical properties. The ultimate limit is set by statistical resolution, which depends on sample size.

Solving the inverse problem becomes a matter of optimizing a scoring function, which breaks down into two parts—first, developing a suitable measure that resists under- and over-fitting to the sampled data, which is the focus of this paper. Second, developing an efficient algorithm to optimize the score while adaptively constructing a non-parametric pdf. The second part will be accomplished by an algorithm involving a non-parametric maximum entropy method (NMEM) that was recently developed by JF and DJ [11] and implemented as the "PDFestimator." Similar to a traditional parametric maximum entropy method (MEM), NMEM employs Lagrange multipliers as coefficients to orthogonal functions within a generalized Fourier series. The non-parametric aspect of the process derives from employing a data driven scoring function to select an appropriate number of orthogonal functions, as their Lagrange multipliers are optimized to accurately represent the complexity of the data sample that ultimately determines the features of the pdf. The resolution of features that can be uncovered without over-fitting naturally depends on the sample size.

Some important results in statistics [12] that are critical to obtain universality in a scoring function are summarized here. For a univariate continuous random variable, X, the cdf is given by $F_X(x)$, which is a monotonically increasing function of x and, irrespective of the domain, the range of $F_X(x)$ is on the interval $(0, 1)$. A new random variable, R, that spans the interval $(0, 1)$ is obtained through the mapping $r = F_X(x)$. The cdf for the random variable R can be determined as follows,

$$F(r) = P(R \leq r) = P(F_X(x) \leq r) = P(X \leq F_X^{-1}(r)) = F_X(F_X^{-1}(r)) = r \quad (1)$$

Since the pdf for the random variable R is given as $f(r) = \frac{dF(r)}{dr} = 1$ it follows that R has a uniform pdf on the interval $(0, 1)$. Furthermore, due to the monotonically increasing property of $F_X(x)$ it follows that a sort ordered set of N random numbers $\{x_k\}_N$ maps to the transformed set of random numbers $\{r_k\}_N$ in a 1 to 1 fashion, where k is a labeling index that runs from 1 to N. In particular, for an index $k' > k$, it is the case that $r_{k'} \geq r_k$. The 1 to 1 mapping that takes $X \to R$ has important implications for assessing the quality of a pdf estimate. The universal nature of this approach is that, for a given sample of random data and no a priori knowledge of the underlying functional form of the true pdf, an evaluation can be made of the transformed data.

Given a high-quality pdf estimate from an estimation method, $\hat{f}_X(x)$, the corresponding estimated cdf, $\hat{F}_X(x)$, will exhibit sampled uniform random data (SURD). Conversely, for a given sample from the true pdf, a poor trial estimate, $\hat{f}_X(x)$, will yield transformed random variables that deviate from SURD. The objective of this work is to consider a variety of measures that can be used as a scoring function to quantify the uncertainty in how close the estimate $\hat{f}_X(x)$ is to the true pdf based on how closely the sort order statistics of $\hat{F}_X(\{x_k\})$ matches with the sort order statistics of SURD. The powerful concept of using sort order statistics to quantify the quality of density estimates [13] will be leveraged to construct universal scoring functions that are sample size invariant.

The strategy employed in the NMEM is to iteratively perturb a trial cdf and evaluate it with a scoring function. By means of a random search using adaptive perturbations, the trial cdf with the best score is tracked until the score reaches a threshold where optimization terminates. At this point, the trial cdf is within an acceptable tolerance to the true cdf and constitutes the pdf estimate. Different outcomes are possible since the method is based on a random fitness-selection process to solve an inverse problem. The role of the scoring function in the NMEM includes defining the objective target for optimizing the Lagrange multipliers, providing stopping criteria for adding orthogonal functions in the generalized Fourier series expansion and marking a point of diminishing returns where further optimizing the Lagrange multipliers results in over-fitting to the data. Simply put, the scoring function provides a means to quantify the quality of the NMEM density estimate. Optimizing the scoring function in NMEM differs from traditional MEM approaches that minimize error in estimates based on moments of the sampled data. Note that the universality of the scoring function eliminates problems with heavy tailed distributions that have divergent moments. Nevertheless, Lagrange multipliers are determined based on solving a well defined extremum problem in both cases.

Before tackling how to evaluate the efficacy of scoring functions, a brief description is given here on how the quality of a pdf estimate can be assessed without knowing the true pdf. Visualizing a quantile-quantile plot (QQ-plot) is a common approach in determining if two random samples come from the same pdf. Given a set of N sort ordered random variables $\{x_k\}_N$ that are monotonically increasing, along with a cdf estimate, the corresponding empirical quantiles are determined by the mapping $\{r_k\}_N = \hat{F}_X(\{x_k\}_N)$ as described above. It is not necessary to have a second data set to compare. As described previously [11], the empirical quantile can be plotted on the y-axis versus the theoretical average quantile for the true pdf plotted on the x-axis. From single order statistics (SOS) the expectation value of r_k is given by $\mu_k = k/(N+1)$ for $k = 1, 2, ...N$, which gives the mean quantile. Figure 2a illustrates the QQ plot for the distributions shown in Figure 1. The benefit of the QQ plot is that it is a universal measure. Unfortunately, for large sample sizes, the plot is no longer informative because all curves approach a perfect straight line as random fluctuations decrease with increasing sample size. A quantile residual (QR) allows deviations from the mean quantile to be readily visualized when one sample size is considered. However, as illustrated in Figure 2b, the residuals in a QR-plot decrease as sample size increases. Hence, the quantile residual is not sample size invariant.

The QR-plot is scaled [11] in such a way as to make the scaled quantile residual (SQR) sample size invariant. From SOS, the standard deviation for the empirical quantile to deviate from the mean quantile is well-known to be $\sigma_k = \sqrt{\mu_k(1-\mu_k)}/\sqrt{N+2}$ where k is the sort order index. Interestingly, all fluctuations regardless of the value for the mean quantile scale with sample size as $1/\sqrt{N+2}$. Sample size invariance is achieved by defining SQR as $\sqrt{N+2}(r_k - \mu_k)$ and, when plotted against μ_k, one obtains a SQR-plot. Figure 2c shows an SQR-plot for three different sample sizes for each of the four distributions considered in Figure 1. It is convenient to define contour lines using the formula $sf\sqrt{\mu(1-\mu)}$, where the scale factor, sf, can be adjusted to control how frequently points on the SQR plot will fall within a given contour. In particular, 99% of the time the SQR points will fall within the boundaries of the oval when bounded by $\pm 2.58\sqrt{\mu(1-\mu)}$. Scale factors of 1.65, 1.96, 2.58 and 3.40 lead to 90%, 95%, 99% and 99.9% of SQR points falling within the oval based on numerical simulation.

Interestingly, the scale factors of 1.65, 1.96, 2.58 and 3.40 respectively correspond to the z-values of a Gaussian distribution at the 90%, 95%, 99% and 99.9% confidence levels.

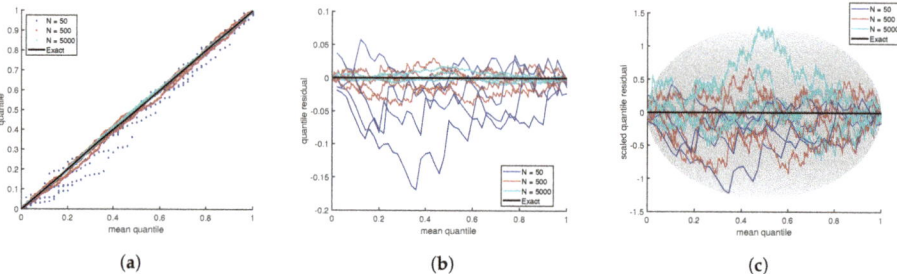

Figure 2. For each of the four distributions shown in Figure 1 and for sample sizes $N = 50, 500, 5000$ shown in all panels with same distinct colors, an empirical quantity is plotted as a function of the theoretical average quantile. The panels show (**a**) QQ-plot, (**b**) QR-plot and (**c**) SQR-plot. Only the SQR-plot is sample size invariant. As an illustration of universality in all panels, any of the colored lines could represent any one of the four distributions.

The SQR-plot provides a distribution free visualization tool to assess the quality of a cdf estimate in three ways. First, when the SQR falls appreciably within the oval that encloses 99% of the residual, it is not possible to reject the null hypothesis. Second, when the SQR exhibits non-random patterns, this is an indication of systematic error introduced by the estimator method. Finally, when the SQR has suppressed random fluctuations such that it is close to 0 for an extended interval, this indicates that the pdf estimate is over-fitting to the sample data. In general, over-fitting is hard to quantify [14]. As the graphical abstract shows, it is possible to plot the SQR against the original random variable x instead of the mean quantile. Doing this deforms the oval or "lemon drop" shape of the SQR-plot but it directly shows where problems in the estimate are locally occurring in relation to the pdf estimate. The aim of this paper is to quantify these salient features of an SQR-plot using a scoring function.

This work was motivated by the concern that different scoring functions will likely perform differently in terms of speed and accuracy in NMEM. The scoring function that was initially considered was constructed from the natural logarithm of the product of probabilities for each transformed random variable, given by $\hat{F}_X(\{x_k\})$. This log-likelihood scoring function provides one way to measure the quality of a proposed cdf. Interestingly, the log-likelihood scoring function has a mathematical structure similar to the commonly employed Anderson–Darling (AD) test [15,16]. As such, the current study considers several alternative scoring functions that use SQR and compares how sensitive they are in quantifying the quality of a pdf estimate. Other types of information measures that use cumulative relative entropy [17] or residual cumulative Kullback–Leibler information [18,19] are possible. However, these alternatives are outside the scope of this study, which focuses on leveraging SQR properties. The scoring function must exhibit distribution free and sample size invariant properties so that it can be applied to any sample of random data of a continuous variable and also to sub-partitions of the data when employed in the PDFestimator. It is worth noting that all the scoring functions presented in this paper exhibit desirable properties with similar or greater efficacy than the AD scoring function and all are useful for assessing the quality of density estimates.

In the remainder of this paper, a numerical study is presented to explore different types of measures for SQR quality. The initial emphasis is on constructing sensitive quality measures that are universal and sample size invariant. These scoring functions based on SQR properties can be applied to quantifying the accuracy (or "goodness of fit") of a pdf estimate created by any methodology, without knowledge of the true pdf. The SQR is readily calculated from the cdf which is obtained by integrating the pdf. To determine which scoring function best distinguishes between good and poor cdf estimates, the concept of decoy SURD is introduced. Once decoys are generated, Receiver

Operator Characteristics (ROC) are employed to identify the most discriminating scoring function [17]. In addition to ROC evaluation, performance of the PDFestimator for different plugged in scoring functions is evaluated. This benchmark is important because the scoring function is expected to affect the rate of convergence toward a satisfactory pdf estimate using the NMEM approach. After discussing the significance of the results, several conclusions are drawn from an extensive body of experiments.

2. Results

2.1. Sample Size Invariant Scoring Functions

Seven scoring functions are defined in Table 1. At the moment, the input to these scoring functions is SURD of sample size N. Specifically, N random numbers are independently and identically drawn uniformly on the interval (0,1) and then sort ordered to give SOS represented by the set $\{r_k\}_N$ where $0 < r_k \leq r_{k+1} < 1 \ \forall \ k = 1, 2, ...N$. For sample size, N, a scoring function of type t is evaluated as $S_t(\{r_k\}_N)$, which defines a new random variable that is simply denoted as $S_t(N)$. A scoring function is scale invariant if the probability density for $S_t(N)$ is independent of sample size, which typically holds only for large N. However, finite size corrections are made for each scoring function and are listed in Table 1. In all cases, the finite size corrections are empirically determined based on numerical simulation to achieve approximate scale invariance for $N \geq 9$. In all coefficients reported, there is a (3) error in the last significant figure, such as 0.406(3) or 11.32(3).

Table 1. Scoring function definitions and finite size corrections.

Anderson-Darling (AD)	Log-Likelihood (LL)
$S_{AD} = \frac{1}{N}\sum_{k=1}^{N}(1-2k)[log(r_k) + log(1-r_{n+1-k})]$ $S'_{AD} = S_{AD} - S^o_{AD}$ $\mu'_{AD} = 1 - 0.250/\sqrt{N} - 0.667/N$ $\sigma'_{AD} = 0.761 + 0.025/\sqrt{N}$	$S_{LL} = \frac{1}{N}\sum_{k=1}^{N} log\left[\frac{N!}{(k-1)!(N-k)!}(r_k)^{k-1}(1-r_k)^{N-k}\right]$ $S'_{LL} = S_{LL} - S^o_{LL}$ $\mu'_{AD} = 1 - 0.297/\sqrt{N} - 5.180/N + 7.56/N^{1.5}$ $\sigma'_{AD} = 0.761 - 0.120/\sqrt{N} - 0.351/N$
mean variance (VAR)	generalized moment ($S_{0.5}$)
$S_{VAR} = \frac{1}{N}\sum_{k=1}^{N} z_k^2$ $\mu_{VAR} = 1 - 0.003/\sqrt{N}$ $\sigma_{VAR} = 0.757 + 0.312/\sqrt{N} + 0.406/N$	$S_{0.5} = \left(\frac{1}{N}\sum_{k=1}^{N}|z_k|^{0.5}\right)^2$ $\mu_{0.5} = 0.704 - 0.008/\sqrt{N} + 0.009/N + 0.52/N^{1.5}$ $\sigma_{0.5} = 0.302 + 0.000/\sqrt{N} + 0.313/N$
root mean square of log-ratio (RMSLR)	generalized moment (S_4)
$RMSLR = \left[\frac{1}{R}\sum_{\forall(i,j)}\left(S_{LR}^{ij}\right)^2\right]^{\frac{1}{2}}$ where $R = N_b(N_b - 1)/2$ for distinct pairs. N_b = number of blocks.	$S_4 = \left(\frac{1}{N}\sum_{k=1}^{N}|z_k|^4\right)^{0.25}$ $\mu_4 = 1.153 + 0.129/\sqrt{N} - 1.630/N + 2.20/N^{1.5}$ $\sigma_4 = 0.345 + 0.303/\sqrt{N} + 0.762/N - 2.56/N^{1.5}$
mean log-ratio (SLR)	
$S_{LR}^{ij} = \frac{1}{m-1}\sum_{k=1}^{m-1} log\left(\frac{\delta_k^i}{\delta_k^j}\right)$ where m = block size for both the i-th and j-th blocks being compared. $\mu_{LR} = 0$ Let $x = (N_p - 1)/(N - 1)$ to account for the number of subsamples within partition, p. $\sigma_{LR} = \sqrt{\frac{2}{m}} \begin{cases} 1 + 0.1888x + 1.754x^2 - 13.71x^3 + 44.49x^4 - 47.01x^5, & x < 1/2 \\ -4.952 + 29.12x - 50.52x^2 + 38.95x^4 - 11.32x^4, & x \geq 1/2 \end{cases}$	

As defined in Table 1, the proposed scoring functions include the relevant part of the Anderson-Darling (AD) measure [15], denoted as S_{AD}, and the quasi log-likelihood formula [11], denoted as S_{LL}. Note that $S_{LL} = log[\prod_k p_k(r_k)]$ where $p_k(r_k)$ is the exact pdf corresponding to a beta distribution that describes the random variable r_k as derived from SOS [13]. The quasi log-likelihood is not an exact log-likelihood. Rather, S_{LL} corresponds to a mean field approximation where correlations between the random variables, $\{r_k\}_N$, are neglected. Another scoring function is defined as $S_{VAR} = \langle z_k^2 \rangle$, where $z_k = (r_k - \mu_k)/\sigma_k$. As mentioned in the Introduction, $\mu_k = \langle r_k \rangle = k/(N+1)$ is the mean quantile of the k-th random variable and $\sigma_k = \sqrt{\mu_k(\mu_k - 1)}/\sqrt{N+2}$ is the standard deviation of the k-th random variable about its mean. Essentially S_{VAR} is the mean variance of a "z-value" for SOS.

Despite sharing a similar mathematical form, the S_{AD} and S_{LL} scoring functions are not the same, even in the limit $N \to \infty$. At face value, these functions look very different. However, after shifting the origin of these functions to their natural reference points and scaling S_{LL} by a factor of -2, which was empirically determined to obtain data collapse, these two measures were remarkably similar. To demonstrate this, let $S'_{AD} \equiv S_{AD}(\{r_k\}) - S_{AD}(\{\mu_k\})$ and $S'_{LL} \equiv -2[S_{LL}(\{r_k\}) - S_{LL}(\{\mu_k\})]$. The natural reference points S^o_{AD} and S^o_{LL} are respectively defined as S_{AD} and S_{LL}, evaluated at the mean quantiles. Figure 3a,d show the pdf for S'_{AD} and the pdf for S'_{LL} are approximately sample size invariant and markedly similar. Interestingly, S'_{AD} has superior sample size invariance because it reaches its asymptotic limit extremely fast, as reported almost 60 years ago [20].

To improve or create a scale invariant scoring function, finite size corrections are incorporated by transforming $S_t(N)$ to a z-value. For all score types, $Z_t = (S_t - \mu_t)/\sigma_t$ where μ_t is the average of S_t and σ_t is the standard deviation of S_t about its mean. All shifts and scale factors used to transform $S_t(N) \to Z_t(N)$ are given in Table 1. Figure 3a,d,g show that, after finite size corrections, the pdf for the three scoring functions Z_{AD}, Z_{LL} and Z_{VAR} exhibit excellent scale invariance. Furthermore, the pdf for these scoring functions fall on top of one another in a massive data collapse (data not shown) indicating they share the same pdf for all practical purposes. It is worth noting that because this is a numerical study, there is uncertainty in the formulas that define the corrections to finite sample size. As can be clearly seen in Figure 3a, the AD measures before finite size corrections are applied display the most impressive data collapse. Indeed, the observed data collapse from numerical simulation are tighter than the intrinsic uncertainties in the correction to finite sample sizes. In contrast, the log likelihood measure has the most dispersion in its data collapse before finite size corrections are applied. In this case, the finite sample size corrections greatly improved the data collapse.

The most surprising result is that this numerical study demonstrates that Z_{VAR} shares the same pdf as Z_{AD}. This result is surprising because both Z_{AD} and Z_{LL} involve linear combinations of logarithms, while Z_{VAR} has no logarithms. However, it is not surprising that Z_{VAR} has good scaling properties because the function is defined in terms of the scaled variable, otherwise called the z-value. The transformation to the z-value naively sets the mean to the origin and normalizes the variance. As such, it would be somewhat surprising if Z_{VAR} did not exhibit data collapse as a function of the z-value. Given that Z_{VAR} scales, it is expected that generalized moments of the z-value variable will exhibit data collapse and also exhibit sample size invariance.

From a practical standpoint, it is computationally faster to work with Z_{VAR}. Therefore, additional scoring functions defined as $S_p = \langle |z_k|^p \rangle^{1/p}$ for $p = \frac{1}{2}$, 1, 2, 3, 4 were considered. Note that S_2 is the standard deviation of z_k and, after finite size corrections are applied, $S_p \to Z_p$. The cases $p = \frac{1}{2}$ and $p = 4$ are listed in Table 1 and exhibit scale invariance as shown in Figure 3b,e respectively. The $p = 1, 2, 3$ cases (data not shown) are similar and straddle the limiting cases smoothly. It is worth mentioning that the natural reference at the mean quantile is zero for S_{VAR} and S_p.

By exploring SURD for additional patterns, it was observed that two disjoint blocks of the same size can be compared using double order statistics (DOS). Among all random variables, $\{r_k\}_N$, the indices that span from k_o^1 to k_f^1 define block 1 and the indices that span from k_o^2 to k_f^2 define block 2. Without loss of generality, block 2 is taken to be to the right of block 1, such that $k_o^1 < k_f^1 < k_o^2 < k_f^2$. With m random variables in both blocks, $m - 1$ differences given by $\delta_k = r_{k+1} - r_k$ are used in the scoring function $S_{LR}^{2,1} = \langle log(\delta_k^2/\delta_k^1) \rangle$, which simplifies to $S_{LR}^{2,1} = \langle log(\delta_k^2) \rangle - \langle log(\delta_k^1) \rangle$. Importantly, $\langle log(\delta_k^j) \rangle$ is calculated for all disjoint blocks at once. By partitioning all random variables into equal blocks of indices, the mean log-ratio is calculated rapidly for all pairs of blocks. For any size block and for any pair of blocks, $S_{LR}^{(i,j)}$ exhibits strong scale invariance as shown in Figure 3h. Over a hundred diverse cases are shown as gray lines. Interestingly, the pdf for $S_{LR}^{(i,j)}$ is essentially a normal distribution shown as a red line.

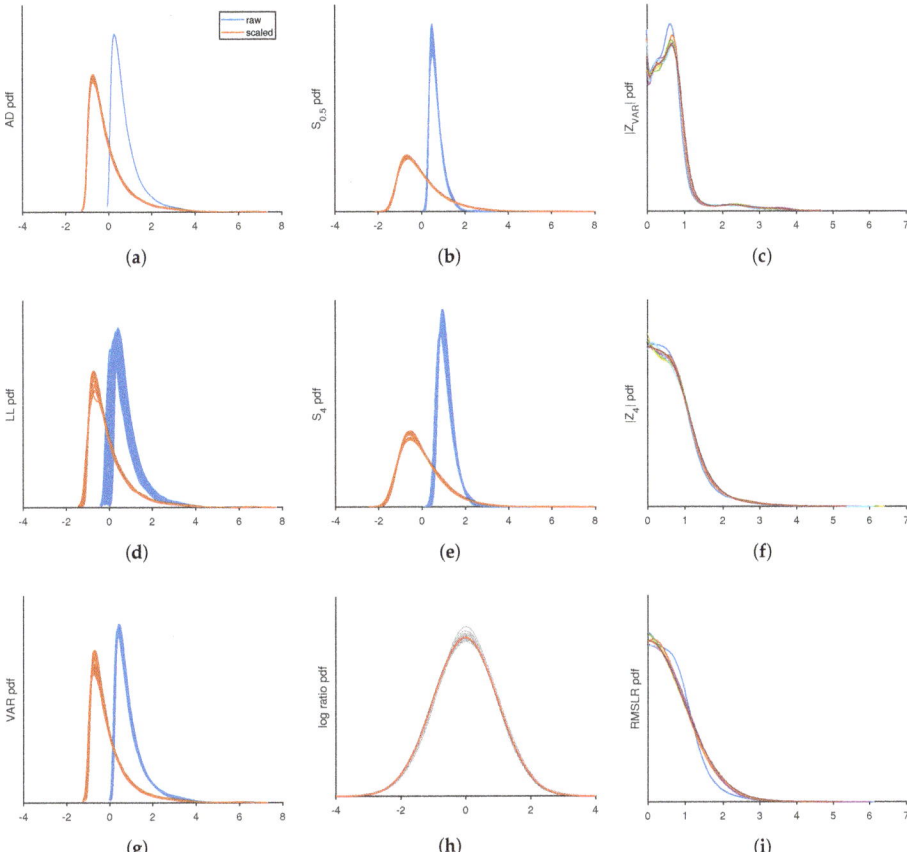

Figure 3. Illustration of sample size invariance in the probability density function for various scoring functions. The sample sizes selected in panels (**a,b,d,e,g,h**) to show data collapse include $N = 9, 11, 12, 14, 17, 20, 24, 33, 49, 95, 110, 124, 142, 166, 199, 249, 332, 497, 990, 1500, 2015, 3298, 5505, 8838, 14,467, 23,684, 38,771, 63,471, 103,905, 272,389, 750,000, 1,000,000, 2,000,000$. The sample sizes selected in panels (**c,f,i**) include $N = 10, 50, 200, 1000, 5000, 20,000, 100,000$. In panel **h**, the results from each system size along with all partitions made within each system size (a total of 111 cases) is plotted as light gray lines. The red line shows the result of a normal distribution, indicating that the scaling is well described by a normal distribution. All other details are described in the text.

Because $S_{LR}^{(i,j)}$ is localized to a pair of blocks, to cover the entire SQR-plot a new scoring function is constructed by taking the root mean square of all distinct pairs of $S_{LR}^{(i,j)}$. For a calculation time proportional to sample size, the size of a block is set proportional to $\sqrt{N}$, which necessarily makes the number of blocks, N_b, proportional to $\sqrt{N}$. The pdf for $RMSLR$ is nearly sample size invariant as shown in Figure 3i. From Table 1, it appears the finite size corrections for $RMSLR$ are complicated. However, as will be discussed below, scale invariance should be preserved for sub-samples of the data, called partitions. It turns out that only $RMSLR$ requires special attention to make partitions scale, where N_p is the number of data points being sub-sampled. Finally, the absolute value of the measures Z_{VAR} and Z_4 are respectively shown in Figure 3c,f. Note that taking an absolute value of a measure that is scale invariant will remain scale invariant.

2.2. Redundant and Complimentary Information

Since the pdf of different scoring functions may be similar or the same, the next question addressed is how do different measures compare when applied to the same SURD? For sample size N, SURD is generated using numerical simulation and each measure is evaluated per realization of $\{r_k\}_N$. For 100,000 random trials per N, a 1 to 1 comparison is made between $Z_a(N)$ versus $Z_b(N)$ with $a \neq b$. Note that by definition, $Z_t(N)$ has a mean of zero and a standard deviation of 1. For reasons that will become clear below, absolute values are taken on the scoring functions. Despite the pdf for $|Z_{VAR}|$, $|Z_{LL}|$ and $|Z_{AD}|$ being practically identical for all sample sizes, scatter plots indicate that the scores are not identical on a 1 to 1 basis. Figure 4a,b plot $|Z_{VAR}|$ and $|Z_{LL}|$ against $|Z_{AD}|$, respectively. Although there is always a tight linear correlation, there is more scatter in the comparison at smaller sample sizes. As $N \to \infty$ the different scores converge to the same value, although the approach to the asymptotic limit for each measure differs. These differences have important implications for application to density estimation as discussed below.

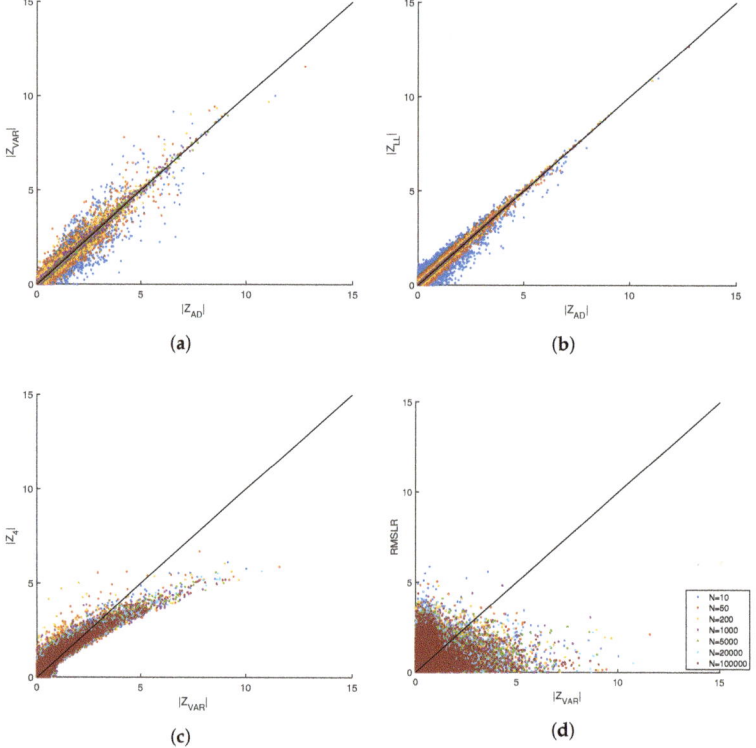

Figure 4. Examples of pairwise comparisons of different measures through scatter plots. (**a**,**b**) show that the $|Z_{AD}|$ measure is statistically the same as the $|Z_{LL}|$ and $|Z_{VAR}|$ measures. (**c**) Shows mild differences between $|Z_4|$ and $|Z_{VAR}|$. (**d**) Shows that the information content between $RMSLR$ and $|Z_{VAR}|$ is very different.

The scatter plot of $|Z_4|$ versus $|Z_{VAR}|$ in Figure 4c shows that these two measures characterize SURD in a fundamentally different way due to the strong deviation of $|Z_4|$ relative to $|Z_{VAR}|$ with modest statistical scatter. The greatest non-linear deviation between the two scores occurs at large values of $|Z_{VAR}|$, corresponding to outliers in SURD. The scatter plot of $RMSLR$ versus $|Z_{VAR}|$ in Figure 4d shows strong random scatter with no discernible deterministic dependence. Hence,

RMSLR and $|Z_{VAR}|$ measure different SURD characteristics. Yet, despite their conspicuous differences, the pdf for $|Z_4|$ and RMSLR are qualitatively similar as shown in Figure 3f,i, respectively.

As demonstrated by scatter plots, various scoring functions characterize SURD in different or similar ways relative to one another. Note that combining measures with complimentary properties can potentially lead to a more sensitive measure. Through reductive analysis, a composite score (CS) is proposed as:

$$CS = |Z_{VAR} + 0.666| + [\max(2.5, |Z_4|, RMSLR) - 2.5] \qquad (2)$$

In constructing CS, the most probable score for Z_{VAR}, near 0.666, is used as a baseline. Then contributions are added from outliers from either $|Z_4|$ or RMSLR, whichever is larger. The last term does not modify the score when no outlier is detected, otherwise the contribution to CS continuously increases starting at zero at just above the threshold for outlier detection.

2.3. Partition Size Invariance

A critical part of the algorithm in the PDFestimator [11] is that the input data sample is partitioned into hierarchical sub-samples by powers of 2 when $N > 1025$. Consequently, the employed scoring function should be sample size invariant for all partitions. Invariance of partition size, N_p, is satisfied by all scoring functions described in this work, as exemplified in Figure 5 for three of the most distinct measures. Furthermore, for any realization of SURD of size N, all partitions within have essentially the same score independent of the type of scoring function.

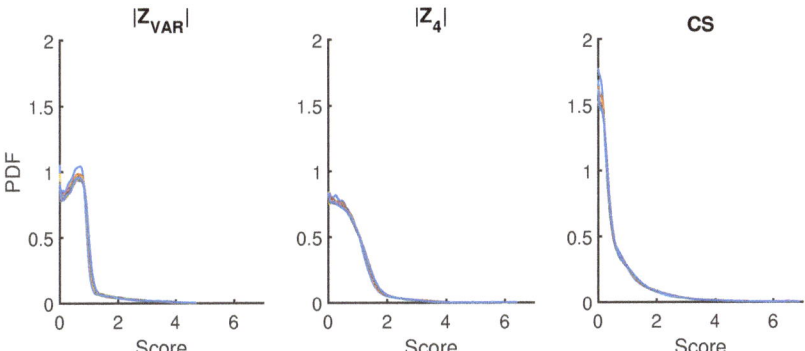

Figure 5. Z_{VAR}, $|Z_4|$ and CS illustrate the three most distinct measures considered. Data collapse based on the probability density for different measures is demonstrated for $N = 10, 50, 200, 1000, 5000, 20,000, 100,000$ in addition to $N_p = 1025, 2049, 4097, 8193, 16,385, 32,769, 65,537$. A different color is used for each sample size.

A necessary requirement for all the scoring functions is that sub-sampling must be uniformly distributed over the data. It is worth noting that S_{AD} (and its corresponding Z_{AD}) is particularly sensitive to the way the uniform sub-sampling is performed within a partition. Due to the form of the S_{AD} equation, it is critical that the selected points are symmetric about the center index in the sort ordering. The number of samples used in a partition is always odd of the form $N_p = 1 + 2^n$. Thus, the median point is included and for each index selected to be in the sub-sample below the median, a corresponding mirror image index above the median is selected. For example, if there are 17 indices in the full sample, indices 1, 4, 9, 14, 17 has the required mirror symmetry. All other scoring functions are not sensitive to breaking mirror symmetry.

2.4. Decoy SURD

For the purpose of quantifying how well a scoring function discriminates between true SURD and random data that is not SURD, a controlled decoy-SURD (dSURD) is generated. Let $\{r_k^o\}$ define SURD and let $\{r_k^d\}$ define dSURD. As described in detail in Section 4.4, a decoy cdf, $F_d(r)$, is constructed to facilitate the 1 to 1 mapping given by $\{r_k^d\} = F_d(\{r_k^o\})$. If $F_d(r) = r$, then the output is identical to the input. A decoy-SURD is controlled by adding a perturbation of the form $F_d(r) = r + \Delta(r)$. By choosing various functional forms for the perturbation and, by controlling the amplitude of the perturbation, it is a simple matter to make a broad spectrum of decoys that range from impossible to markedly obvious to detect at any specified sample size.

In Figure 6, the middle row shows the decoy cdf resulting from the perturbations shown along the top row. This is an example of a moderately hard dSURD because by eye the decoy cdf looks close to a perfect straight line. To make it clear that dSURD is indeed different from SURD, the pdf for each case is shown along the bottom row. For a sufficiently large sample size, statistical resolution will be good enough to resolve these small perturbations, but for smaller sample sizes the perturbation will not be detectable. To demonstrate how statistical resolution increases with larger sample sizes, Figure 7 shows SQR-plots for SURD and its corresponding dSURD for samples sizes of 1000, 5000, 20,000 and 100,000. These three cases are examples of localized perturbations.

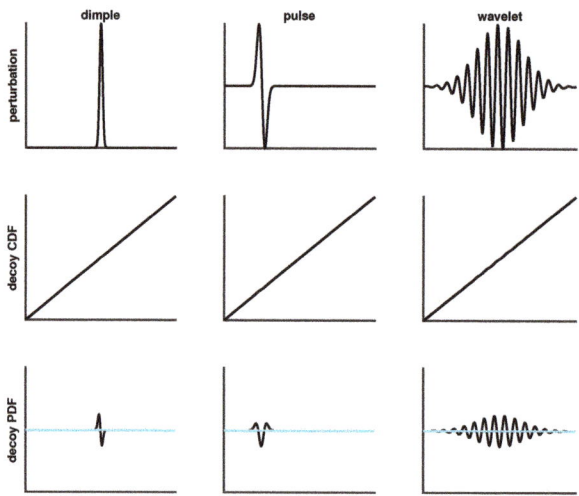

Figure 6. Top row shows three examples of localized perturbations for moderately difficult decoys. Center row shows the corresponding cdf. Bottom row shows the pdf, where the cyan horizontal highlights the probability density function (pdf) for sampled uniform random data (SURD).

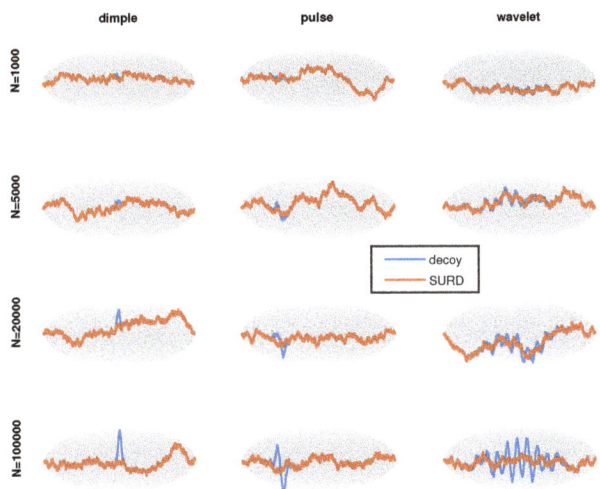

Figure 7. Progression of scaled quantile residual (SQR)-plots for moderately difficult localized decoys as sample size increases.

Three additional perturbations of an extended type are shown in Figure 8 using the same layout. The last column plots the perturbation, cdf and pdf as dashed red lines because "reduced fluctuation" is a special type of perturbation that is also explained in Section 4.4. As the name implies, fluctuations are suppressed, representing a scenario where a pdf estimate over-fits the data. Figure 9 shows the SQR-plots for SURD and its corresponding dSURD of the extended type for samples sizes of 1000, 5000, 20,000 and 100,000. Note that the reduced fluctuation perturbation is equally detectable at any size sample because fluctuations are suppressed by a fixed proportion in relation to true SURD.

By comparing measures applied to dSURD and SURD, it can be expected that the more sensitive scoring function is one that detects a given perturbation at smaller sample sizes compared to other scoring functions. It is also expected that a certain scoring function will be able to detect certain types of perturbations more readily than other types of perturbations. As such, it is likely impossible to find a perfect scoring function that performs best on all decoy types all the time. Nevertheless, for a given diverse set of dSURD examples, the best overall performing scoring functions with the greatest sensitivity or selectivity can be deduced using receiver operator characteristics.

2.5. Receiver Operator Characteristics

Receiver operator characteristics (ROC) are calculated based on simulation data involving 10,000 trials of SURD over a broad range of N samples, and for each SURD, many dSURD mappings are generated for each of the six decoy types shown above. Results are exemplified in Figure 10, showing ROC curves for three different sample sizes and six different decoy types. ROC curves quantify the efficacy of a scoring function in discriminating SURD from dSURD. Figure 10 shows representative results for moderately difficult decoys. As a point of reference, easy, moderate and hard decoys are aimed at requiring about 1000, 10,000 and 100,000 samples to have sufficient statistical resolution to notice dSURD just barely by eye (e.g., see Figures 7 and 9). Only the decoy that reduces fluctuations using a fixed scale factor has the same difficulty for detection independent of sample size.

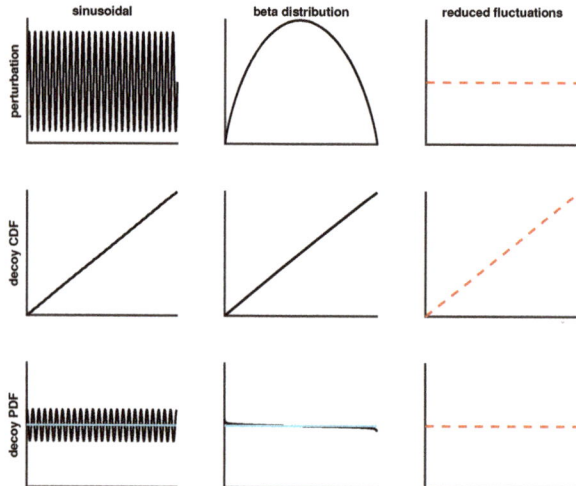

Figure 8. Extended perturbations for moderately difficult decoys. The cyan horizontal line shown on the bottom panels defines the pdf for SURD. The red dashed lines represent suppression of fluctuations.

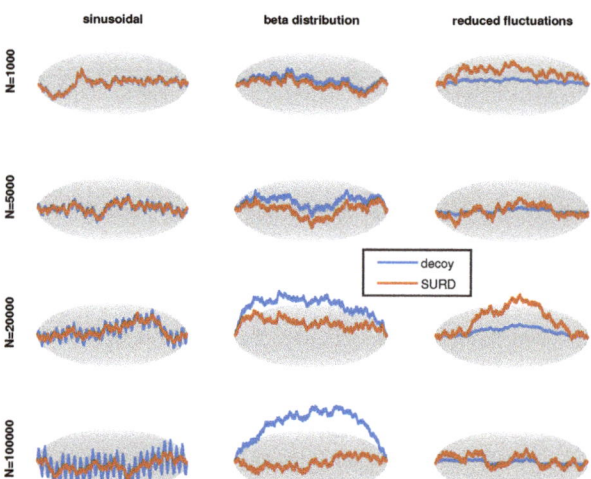

Figure 9. Progression of SQR-plots for moderately difficult extended decoys as sample size increases.

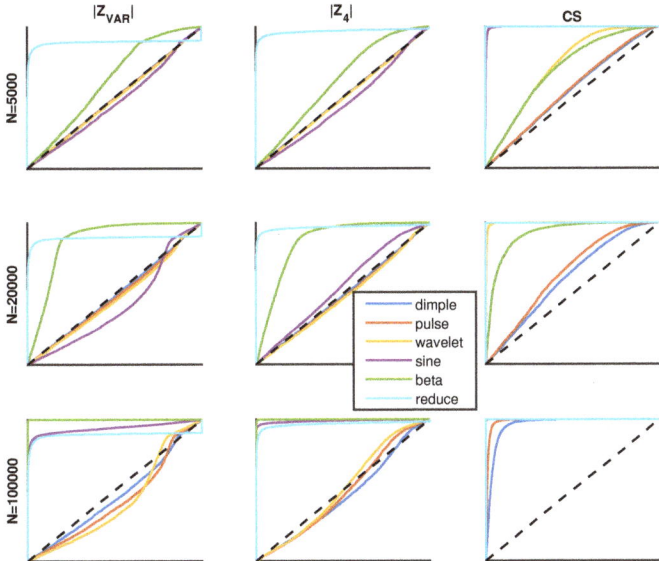

Figure 10. The qualitative features of receiver operator characteristic (ROC) curves are shown for sample sizes of 5000, 20,000 and 100,000 along the top, middle and bottom rows. The left, middle and right columns correspond to the $|Z_{VAR}|$, $|Z_4|$ and CS scoring functions. The (y-axis, x-axis) corresponds to the fraction of true (positives, negatives) having a range from 0 to 1. Each ROC curve compares 6 different decoy types.

It is common practice to quantify ROC curves by their area under the curve (AUC). Table 2 gives all AUC values for all the cases shown in Figure 10. The ROC curves and the results listed in Table 2 clearly show that CS detects decoys better than the other measures. Of course, informed by reductive analysis, this result was purposely intended during the construction of CS given in Equation (2). In summary, it is generally found that Z_{AD}, Z_{LL}, Z_{VAR}, $|Z_{AD}|$, $|Z_{LL}|$, $|Z_{VAR}|$, Z_4, $|Z_4|$ and CS scoring functions are all good measures to distinguish SURD from easy to detect dSURD. However, it is always possible to create decoy SURD that will go undetected by any measure (e.g., Figure 10).

Table 2. Area under the ROC curves shown in Figure 10.

	N = 5000			N = 20,000			N = 100,000		
decoy	$|Z_{VAR}|$	$|Z_4|$	CS	$|Z_{VAR}|$	$|Z_4|$	CS	$|Z_{VAR}|$	$|Z_4|$	CS
dimple	0.50	0.50	0.55	0.49	0.49	0.61	0.46	0.46	0.96
pulse	0.49	0.49	0.55	0.48	0.48	0.65	0.43	0.49	0.99
wavelet	0.49	0.49	0.72	0.47	0.48	0.99	0.42	0.51	1.00
sine	0.47	0.46	1.00	0.42	0.55	1.00	0.94	0.99	1.00
beta	0.62	0.62	0.70	0.87	0.87	0.92	1.00	1.00	1.00
reduced	0.89	0.97	1.00	0.89	0.97	1.00	0.89	0.97	1.00

In general, Z_{AD}, Z_{LL} and Z_{VAR} share similar ROC curves and $|Z_{VAR}|$ and $|Z_4|$ have similar ROC curves. The most sensitive scoring function is CS. The reason Z_t and $|Z_t|$ are considered as two separate cases is now easily explained. First note that Z_t has a mean of zero and a standard deviation of 1. For a decoy type of "reduced fluctuations" that mimics an over-fitting scenario, the ROC curve becomes inverted for any type of measure, Z_t. However, the inversion problem is eliminated when considering $|Z_t|$ because both over-fitting and under-fitting is detected when $|Z_t|$ is large. Finally, only the combined score, CS, readily detects very localized perturbations due to its $RMSLR$ component.

2.6. PDF Estimation Performance

Figure 11 summarizes the comparative statistics for failure rates. The bar plots in Figure 11a report averages across distributions and random samples, for cumulative ranges of sample sizes. As expected, the failure rate increases with sample size. For all scoring methods, average failure rates are typically on the order of 10% for sample sizes less than one million. Failure rate averages are the least for $|Z_4|$ and $|Z_{LL}|$, a trend that holds across sample size. The associated box plots in Figure 11b more clearly demonstrate the computational advantage of $|Z_4|$ and $|Z_{LL}|$ over the other scoring methods. All scoring methods have between 50 and 60 outliers, but $|Z_4|$ and $|Z_{LL}|$ have virtually no failures outside of these extreme values.

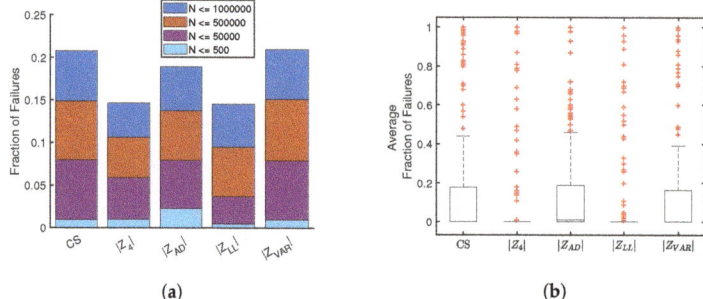

Figure 11. Figure (**a**) cumulative averages of failure rates across four ranges Figure (**b**) distribution of failure rates for each scoring method. Box plots show inner-quartiles and whiskers represent range of data excluding outliers, which are shown as red crosses.

For computational time and Kullback-Leibler (KL) divergence [21], or simply KL, care must be taken to ensure a fair comparison, accounting for failure rates. Thus, a subset of the data is considered for these measurements. Of the 275 test sets (25 distributions at 11 sample sizes), 230 of these contain at least 10 successes out of the 100 trials, across all five scoring methods. The remaining 45 tests contain failure rates greater than 90% for at least one scoring method and are eliminated from further comparison, ensuring an equitable comparison across successful distributions and sample sizes. The results are shown in Figure 12.

Computational time comparisons prove to be the most challenging to pin down, due to wide variations between distributions, sample sizes and random trials. However, Figure 12a demonstrates a clear advantage in the average computational time for $|Z_4|$, across all sample sizes. Once again, the number of outliers, which are compressed for clarity in Figure 12b, is roughly the same across the five scoring methods. However, $|Z_{AD}|$ has a higher range of typical runtimes, as well as higher averages in the smallest sample sizes. The KL-divergence comparisons shown in Figure 12c,d are less variable between scoring methods. A lower divergence between the estimate and the known reference distribution suggests a better estimate is being made. Figure 12c shows a decreasing KL-divergence with increasing sample size for all scoring methods, which demonstrates expected convergence, albeit with diminishing returns for larger sample sizes. Notably, $|Z_{AD}|$ produces slightly lower KL-divergence on average, compared to the other methods.

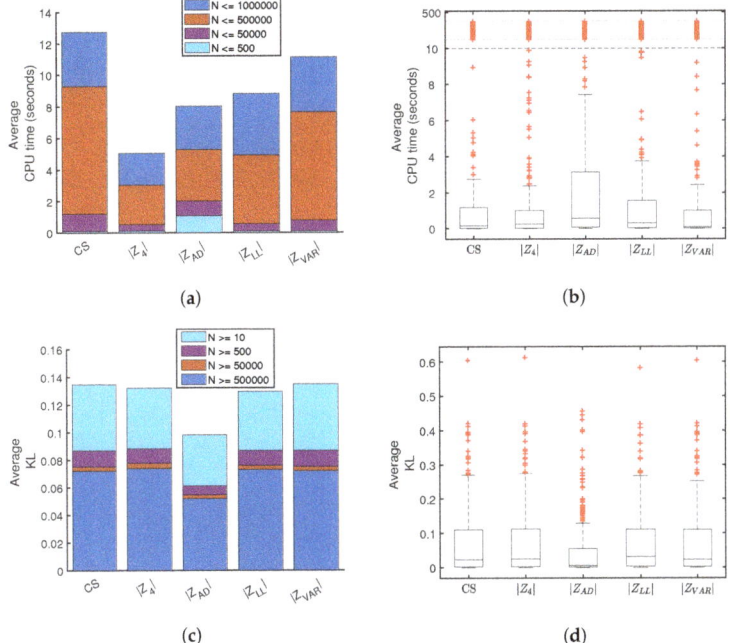

Figure 12. Comparative statistics between five scoring methods averaged over successful solutions. Cumulative averages for (**a**) performance time across four sample size ranges and (**c**) Kullback-Leibler divergence [21]. Panels (**b**,**d**) show box plots for the respective data shown in panels (**a**,**c**). Box plots show inner-quartiles and whiskers represent range of data excluding outliers, which are shown as red crosses.

3. Discussion

Each of the five scoring methods have been evaluated when utilized within the PDFestimator and applied to the same distribution test set in terms of scalability, sensitivity, failure rate and KL-divergence. Each of the proposed measures have strengths and weaknesses in different areas. The $|Z_{AD}|$ measure produces the most accurate scaling and the lowest KL-divergence. The CS measure shows the greatest sensitivity for detecting small deviations from SURD. The $|Z_{LL}|$ method, although not a clear winner in any particular area, is notably well-performing in all tests. These results suggest a possible trade-off between a lower KL-divergence versus longer computational time with the $|Z_{AD}|$ scoring method. However, the slight benefit of a lower KL-divergence is arguably not worth the computational cost, particularly when also considering the higher failure rate. In contrast, the significantly low failure rate and fast performance times are strong arguments in favor of $|Z_4|$ as the preferred scoring method. However, this result is only true when the score of a sensitive measure is minimized, while the threshold to terminate is based on a less sensitive measure (see Section 4.7 in methods for details).

Qualitative analysis is used to elucidate why $|Z_4|$ minimization is the best overall performer. The pdf and SQR for hundreds of different estimates were compared visually and robust trends were observed between the $|Z_{VAR}|$ and $|Z_4|$ methods. Figure 13a is a representative example, showing the density estimates for the Burr distribution at 100,000 samples. Although both estimates were terminated at the same quality level, the smooth curve found for $|Z_4|$ would be subjectively judged superior. However, there is nothing inherently or measurably incorrect about the small wiggles in the $|Z_{VAR}|$ estimate. Note that no smoothness conditions are enforced in the PDFestimator.

The SQR-plot, shown in Figure 13b, is especially insightful in evaluating the differences in this example. The Burr distribution is deceptively difficult to estimate accurately due to a heavy tail on the right. Both $|Z_{VAR}|$ and $|Z_4|$ fall mostly within the expected range, except for the sharp peak to the right corresponding to the long tail. Although the peak is more pronounced for $|Z_{VAR}|$, the more relevant point in this example is the shape of the entire SQR-plot. SQR for $|Z_{VAR}|$ contains scaled residuals close to zero, behavior virtually never observed in true SURD. Hence, this corresponds to over-fitting. This contrast in the SQR-plot between $|Z_{VAR}|$ and $|Z_4|$ is generally true with the following explanation.

The $|Z_4|$ scoring method uses the same threshold scoring as $|Z_{VAR}|$, but simultaneously seeks to minimize the variance from average, thus highly penalizing outliers to the expected z-score. The $|Z_{VAR}|$ method, by contrast, tends to over-fit some areas of the distribution of high density, attempting to compensate for areas of relatively low density where it deviates significantly. This often results in longer run times, many unnecessary Lagrange multipliers, less smooth estimates and unrealistic SQR-plots, as the NMEM algorithm attempts to improve inappropriately. For example, in the test shown in Figure 13, the number of Lagrange multipliers required for the $|Z_{VAR}|$ estimate was 141, whereas $|Z_4|$ required only 19. Therefore, it is easy to see why $|Z_{VAR}|$ took much longer to complete. This phenomenon is a general trend but it is exacerbated in cases where there are large sample sizes on distributions that have a combination of sharp peaks and heavy tails.

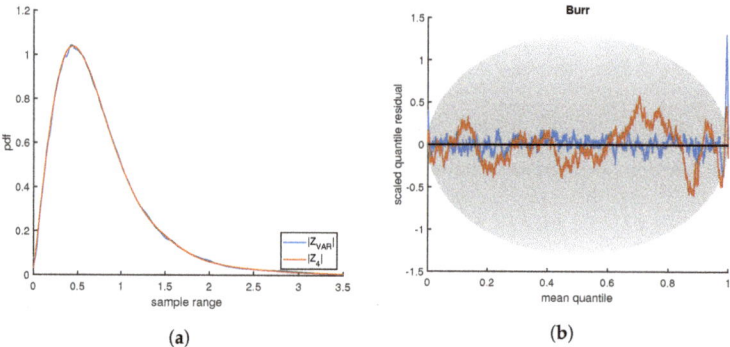

Figure 13. (a) Two density estimates are compared based on two different scoring functions. (b) The corresponding SQR-plots for each density estimate are shown. By eye, both density estimates look exceptionally good, but the SQR-plot has a strong peak representing error in the extreme tail of the distribution. The degree of error depends on the scoring function, but both scoring functions give qualitatively the same results.

A surprising null result of this work is that the *CS* measure, custom designed to have the greatest overall sensitivity and selectivity, failed to be the best overall performer in practice when invoked in the PDFestimator. Although more investigation is required, all comparative results taken together suggest that the *CS* scoring function is the most sensitive but is over-designed for the capability of the random search optimization method currently employed in the PDFestimator. In the progression of improvements on pdf estimation, the results from the initial PDFestimator suggested that a more sensitive scoring function would improve performance. With that aim, more sensitive scoring functions have been determined and performance of the PDFestimator substantially improved. However, it appears the opposite is now true, requiring a shift in attention to optimize the optimizer, with access to a battery of available scoring functions. In preparation, another work (ZM, JF, DJ) optimizes the overall scheme by dividing the data into smaller blocks, which gives much greater speed and higher accuracy, while taking advantage of parallelization.

4. Methods

MATLAB 2019a (MathWorks, Natick, MA, USA) and the density estimation program "PDFestimator" were used to generate all the data presented in this work. The PDFestimator is a C++ program that JF and DJ developed as previously reported [11], which has the original Java program in supporting material. Upgrades on the PDFestimator are continuously being made on the BioMolecular Physics Group (BMPG) GitHub website,Available online: https://github.com/BioMolecularPhysicsGroup-UNCC/PDF-Estimator, where the source code is freely available, including a MATLAB interface to the C++ program. An older C++ version is also available in R, https://cran.r-project.org/web/packages/PDFEstimator/index.html. The version on the public GitHub website is the most recent stable version that has been well tested.

4.1. Generating SURD and Scoring Function Evaluation

MATLAB was employed in numerical simulations to generate SURD. For a sample size N, the sort ordered sequence of numbers $\{r_k\}_N$ was used to evaluate each scoring function being considered. The same realization of SURD was assigned multiple scores to facilitate subsequent cross correlations.

4.2. Method for Partitioning Data

As previously explained in detail [11], sample sizes of $N > 1025$ were partitioned in the PDFestimator to achieve rapid calculations. The lowest and highest random number in the set $\{r_k\}_N$ define the boundaries of each partition. The random number closest to the median was also included. Partitions have an odd number of random numbers due to the recursive process of adding one additional random number between the previously selected random numbers in the current partition. Partition sizes follow the pattern of $3, 5, 9, 17, 33, ...1 + 2^n$. A desired property of scoring functions is that they should maintain size invariance for all partitions. Scores for each measure were tracked for all partitions of size 1026 and greater, including the full data set, which is the last partition. For example, with N=100,000 the scores for partitions of size Np = 1025, 2049, 4097, 8193, 16,385, 32,769, 65,537, 100,000 were calculated. Scores from different partitions were cross correlated in scatter plots.

4.3. Finite Size Corrections

For each partition of size N_p, including the last partition of size N, the scores were transformed to obtain data collapse. For all practical purposes finite size corrections were successfully achieved by shifting the average of a score to zero and normalizing the data by the standard deviation of the raw score. That is to say, the score, $S_t(N_p)$ for N_p samples in the p-th partition, was a random variable. This score was transformed to a Z-value through the procedure $Z_t(N_p) = [S_t(N_p) - \mu_t(N_p, N)] / \sigma_t(N_p, N)$. Operationally, tens of thousands of random sequences of SURD were generated for each scoring function type to empirically estimate $\mu_t(N_p, N)$ and $\sigma_t(N_p, N)$. Note that $\mu_t(N_p, N)$ and $\sigma_t(N_p, N)$ were obtained using basic fitting tools in the MATLAB graphics interface, and these are reported in Table 1.

4.4. Decoy Generation

For each decoy the sort ordered sequence of numbers $\{r_k^o\}_N$ defining SURD was transformed into decoy-SURD, denoted as dSURD. This was accomplished by creating a model decoy cdf, $F_d(r)$. A new set of sort ordered random numbers was created by the 1 to 1 mapping $\{r_k^d\}_N = F_d(\{r_k^o\}_N)$, yielding a dSURD realization per SURD realization. Different decoys were generated based on different types of perturbations, which must meet certain criteria. Let $\Delta(r)$ represent a perturbation to SURD, such that

$$F(r) = r + \Delta(r) \tag{3}$$

For the perturbation to be valid, the pdf given by $f_d(r) = \frac{dF_d(r)}{dr}$ must satisfy $f_d(r) \geq 0$, which implies $1 + \Delta'(r) \geq 0$. The boundary conditions $\Delta(0) = \Delta(1) = 0$ must also be imposed. With these conditions satisfied, decoys of a wide variety could be generated. Four types of decoys were created using this approach, listed in the first 4 rows of Table 3. In this approach, the amplitude of the perturbation is a parameter. A decoy that is marginally difficult to detect at sample size of N_d has $\max(|\Delta|) = 1/\sqrt{N_d}$. It will be challenging to discriminate between SURD and dSURD for $N < N_d$, and markedly distinguishable when $N/N_d \gg 1$.

Two additional types of decoys were also generated. First, $F_d(r)$ is set to a beta distribution cdf, denoted as $F_\beta(r|\alpha, \beta)$. Therefore, the perturbation is given as $\Delta(r) = F_\beta(r|\alpha, \beta) - r$. The α and β parameters were adjusted to tune detection difficulty, by systematically searching for pairs of α and β on a high resolution square grid to find when $\max(\Delta)$ was at a level that was consistent with the targeted sample size, N_d. Second, a decoy can be defined by uniformly reducing fluctuations according to $r_k^d = r_k^o + p(r_k^o - \mu_k)$ where $\mu_k = k/(N+1)$. When $p = 0$ the decoy was the same as SURD, but as $p \to 1$ the decoy retained no fluctuations. In this sense, this decoy type mimics extreme over-fitting, where p controls how much of the fluctuations are reduced.

Table 3. Decoy type summary.

Decoy Name	Perturbation Equation	Parameters
dimple	$\Delta(r) = A \exp\left[\frac{-(r-r_o)^2}{2\sigma^2}\right]$	A, r_o, σ
pulse	$\Delta(r) = A \left(\frac{r_o - r}{\sigma^2}\right) \exp\left[\frac{-(r-r_o)^2}{2\sigma^2}\right]$	A, r_o, σ
wavelet	$\Delta(r) = A \sin(m\pi r) \exp\left[\frac{-(r-r_o)^2}{2\sigma^2}\right]$	A, r_o, σ, m
sine	$\Delta(r) = A \sin(m\pi r)$	A, m
beta distribution	$\Delta(r) = F_\beta(r|\alpha, \beta) - r$	α, β
reduced fluctuations	$\Delta_k = p(r_k^o - \mu_k)$	p

4.5. ROC Curves

All ROC curves were generated according to the definition that the fraction of true positives (FTP) were plotted on the y-axis versus the fraction of false positives (FFP) plotted on the x-axis [22]. Note that alternative definitions for ROC are possible. To calculate FTP and FFP, a threshold score must be specified. If a score is below this threshold, the sort ordered sequence of numbers is predicted to be SURD. Conversely, if a score exceeds the threshold, the prediction is not SURD. As such, there are four possible outcomes. First, true SURD can be predicted as SURD or not, respectively, corresponding to a true positive (TP) or a false negative (FN). Second, dSURD can be predicted as SURD or not, respectively corresponding to a false positive (FP) or true negative (TN). All possible outcomes are tallied, such that FTP = TP/(TP + FN) and FFP = FP/(FP + TN). For a given threshold value, this calculation determines one point on the ROC curve. By considering a continuous range of possible thresholds, the entire ROC curve is constructed.

Procedurally, the data used to calculate the fractions of true and false positives that come from numerical simulations in MATLAB comprised 10,000 random SURD and dSURD pairs for sample sizes, $N = 10, 50, 200, 1000, 5000, 20{,}000$ and $100{,}000$. About 60 different types of decoys were considered with diverse sets of parameters.

4.6. Distribution Test Set

To benchmark the effect of a scoring function on the performance of the PDFestimator, a diverse collection of distributions was selected and these are listed in Table 4. A MATLAB script was created to utilize built in functions dealing with statistical distributions to generate random samples of specified size. The random samples were subsequently processed by the PDFestimator to estimate the pdf, but

4.7. PDF Estimation Method

Each alternative scoring function, $\{|Z_{AD}|, |Z_{LL}|, |Z_{VAR}|, |Z_4|, CS\}$ was implemented in the PDFestimator and were evaluated separately. Factors confounding comparisons in performance include sample size, distribution type, selection of key factors to evaluate and consistency across multiple trials. To provide a quantitative synopsis of the strengths and weaknesses of the proposed scoring methods, large numbers of trials were conducted on the distribution test set listed in Table 4. The distribution test set increases atypical failures amongst the estimates because it is necessary to consider extreme scenarios to identify breaking points in each of the scoring methods. Nevertheless, easier distributions, such as Gaussian, uniform and exponential, were included. To wit, good performance of an estimator when applied to challenging cases should not suffer when applied to easier distributions.

As an inverse problem, density estimation applied to multiple random samples of the same size for any given distribution will generally produce variation amongst the estimates. For small samples, the pdf estimate must resist over-fitting, whereas large sample sizes create computational challenges that must trade between speed and accuracy. To monitor these issues, a large range of sample sizes were tested, each with 100 trials of an independently generated input sample data set. Specifically, 100 random samples were generated for each of the 25 distributions, for each of the following 11 sample sizes with $N = 10, 50, 100, 500, 1000, 5000, 10,000, 50,000, 100,000, 500,000, 1,000,000$. This produced a total of 27,500 test cases, each of which were estimated using five scoring methods. Statistics were collected and averaged over each of the 100 random sample sets.

Three key quantities were calculated for a quantitative comparison of the scoring methods—failure rate, computational time and Kullback-Leibler (KL) divergence [21]. It was found that the KL-divergence distance was not sensitive to the different scoring functions. Alternative information measures [23,24] could be considered in future work. Failure rate is expressed as a fraction of failures out of 100 random samples. The KL-divergence measures the difference between the estimate against the known reference distribution. Computational times and KL-divergences were averaged only for successful solutions and thus were not impacted by failures. A failure is automatically determined by the PDFestimator when a score does not reach a minimum threshold.

During an initial testing phase, it was found that the measures Z_t and $|Z_t|$ for $t =$ AD, LL, and VAR all worked successfully, which is not surprising considering the original measure, Z_{LL}, works markedly well. However, for the more sensitive measures, Z_4, $|Z_4|$ and CS, the PDFestimator failed consistently because the score rarely reached its target threshold, at least within a reasonable time. Therefore, a hybrid method was developed that minimizes a sensitive measure as usual, but the $|Z_{VAR}|$ measure was invoked to determine when to terminate. In tests of $|Z_t|$ for $t =$ AD, LL or VAR, these measures were optimized and were simultaneously used as a stopping condition with a threshold of 0.66 corresponding to the 40% level in the cdf, which was the same level used previously [11]. All these measures have the same pdf and cdf, and thus the same threshold value. This threshold was used for $|Z_{VAR}|$ as a stopping condition when different scoring functions are minimized.

Table 4. List of distribution types and corresponding parameters used to generate random data samples. Parameter and variable names correspond to the labeling scheme of MATLAB. For mixture distributions, subscripts indicate the distribution used to create the mixture with ordinal numbering, and under the *Scale Parameter* column, for mixture distributions p_i is the mixing weight.

Distribution Name	Shape Parameter	Scale Parameter	Location Parameter
Beta	$a = 0.5$	$b = 1.5$	
Beta	$a = 2$	$b = 0.5$	
Beta	$a = 0.5$	$b = 0.5$	
Bimodal Normal	$\sigma_1 = 0.8$	$p_1 = 0.65$	$\mu_1 = 2$
	$\sigma_2 = 0.3$	$p_2 = 0.35$	$\mu_2 = 6$
Birnbaum-Saunders	$\gamma = 0.5$	$\beta = 1.5$	
Birnbaum-Saunders and Stable	$\gamma_1 = 0.5$	$\beta_1 = 1.5$	$\delta_2 = 7$
	$\alpha_2 = 0.5$	$\gamma_2 = 1$	
	$\beta_2 = 0.5$		
Burr	$c = 2$	$\alpha = 1$	
	$k = 2$		
Exponential		$\mu = 1$	
Extreme-Value		$\sigma = 2$	$\mu = 1$
Gamma	$k = 1$	$\sigma = 2$	$\mu = 2$
Generalized-Extreme-Value	$a = 2$	$b = 2$	$b = 2$
Generalized-Pareto	$k = 2$	$\sigma = 1$	$\theta = 0$
Half Normal		$\sigma = 1$	$\mu = 0$
Inverse Gaussian	$\lambda = 1$	$\mu = 5$	
Normal		$\sigma = 1$	$\mu = 1$
Normal Contaminated	$\sigma_1 = 2$	$p_1 = 0.5$	$\mu_1 = 5$
	$\sigma_2 = 0.25$	$p_2 = 0.5$	$\mu_2 = 5$
Stable	$\alpha = 0.5$	$\gamma = 1$	$\delta = 4$
	$\beta = 0.05$		
Stable	$\alpha = 0.2$	$\gamma = 1$	$\delta = 4$
	$\beta = 0.05$		
Stable	$\alpha_1 = 0.5$	$\gamma_1 = 1$	$\delta_1 = 2$
	$\beta_1 = 0.05$	$\gamma_2 = 1$	$\delta_2 = 5$
	$\alpha_2 = 0.5$	$p_1 = 0.25$	
	$\beta_2 = 0.05$	$p_2 = 0.75$	
Stable	$\alpha_1 = 0.5$	$\gamma_1 = 1$	$\delta_1 = 2$
	$\beta_1 = 0.05$	$\gamma_2 = 1$	$\delta_2 = 5$
	$\alpha_2 = 0.5$	$\gamma_3 = 1$	$\delta_3 = 8$
	$\beta_2 = 0.05$	$p_1 = 0.25$	
	$\beta_3 = 0.05$	$p_2 = 0.5$	
		$p_3 = 0.25$	
Trimodal Normal	$\sigma_1 = 0.5$	$p_1 = 0.\overline{3}$	$\mu_1 = 4$
	$\sigma_2 = 0.25$	$p_2 = 0.\overline{3}$	$\mu_2 = 5$
	$\sigma_3 = 0.5$	$p_3 = 0.\overline{3}$	$\mu_3 = 6$
t-Location Scale	$\nu = 1$	$\sigma = 0.5$	$\mu = 4$
Uniform	$l = 4$		
	$u = 8$		
Uniform-Mix	$l_1 = 1$	$p_1 = 0.1\overline{6}$	
	$l_2 = 3.5$	$p_2 = 0.6\overline{6}$	
	$l_3 = 7$	$p_3 = 0.3\overline{6}$	
	$u_1 = 2$		
	$u_2 = 5.5$		
	$u_3 = 9$		
Uniform Periodic	$l_1 = 1$	$p_1 = 0.1\overline{6}$	
	$l_2 = 2.5$	$p_2 = 0.1\overline{6}$	
	$l_3 = 4$	$p_3 = 0.1\overline{6}$	
	$l_4 = 5.5$	$p_4 = 0.1\overline{6}$	
	$l_5 = 7$	$p_5 = 0.1\overline{6}$	
	$l_6 = 8.5$	$p_6 = 0.1\overline{6}$	
	$u_1 = 2$		
	$u_2 = 3.5$		
	$u_3 = 5$		
	$u_4 = 6.5$		
	$u_5 = 8$		
	$u_6 = 9.5$		
Weibull	$b = 2$	$a = 1$	

5. Conclusions

Several conclusions can be drawn from the large body of results presented. (1) The scaled quantile residual (SQR) is instrumental in assessing the quality of a pdf by means of visual inspection. The advantage of an SQR-plot over a traditional QQ-plot is that the displayed information is not only universal (distribution free), but importantly, sample size invariant; (2) It is possible to construct myriad scoring functions that are universal and sample size invariant based on quantitatively characterizing SQR. In particular, various measures can be developed based on mathematical properties of single order statistics (SOS) and/or double order statistics (DOS); (3) Finite size corrections can generally be applied to scoring functions so that their asymptotic properties can be utilized for finite size samples, as low as $N = 9$; (4) Surprisingly, the scoring functions based on the Anderson-Daring test, quasi log-likelihood of SOS and the variance of SOS z-score —when applied to sampled uniform random data (SURD) share identical pdf for their scores for all practical purposes. Moreover, the scores are invariant across sample size and for different size partitions that sub-sample the input data; (5) The concept of decoy-SURD is introduced and a few methods are given for creating decoy-SURD (dSURD). The purpose of dSURD is to quantify the sensitivity and selectivity of a proposed scoring function using Receiver Operator Characteristics (ROC) or other means, such as machine learning. The usefulness of dSURD to quantify uncertainty in density estimation parallels the use of decoys in the field of protein structure prediction. That is, better scoring functions can be developed by focusing on how they discriminate between true SURD and dSURD; (6) Implementing a more sensitive scoring function in a method that estimates a pdf from random sampled data does not necessarily imply the process of estimation will be improved. There are many confounding factors that determine the ultimate performance characteristics of an algorithm for density estimation, since speed and accuracy need to be balanced for a practical software tool; (7) Minimizing either the Z_4 or $|Z_4|$ scores greatly improved the performance of the PDFestimator, a C++ program for univariate density estimation, compared to the initially used scoring function Z_{LL}.

In closing, a few research directions that can stem from this work are highlighted. Interestingly, the mean log ratio of nearest neighbor differences in sort ordered SURD, when taken from two disjoint subsets, is normally distributed (at least to a very good approximation). Unaware of an existing proof of this result, the empirical result suggests that a proof should be sought given that the literature contains many works that derive the pdf for ratios of random numbers that are distributed in a specific way. The results presented here can be applied to the problem of constructing a more sensitive distribution free "test for goodness of fit." Essentially, this was the main objective that was addressed but here the emphasis was on how to better quantify uncertainty for the process of estimating a pdf for a random sample of data. Going forward, the universal sample size invariant measures developed here can be employed to test the similarity of two random samples of data.

Author Contributions: D.J. formalized the project objectives, proposed most measures and all decoy types. D.J. wrote the MATLAB code to scale all measures for SURD, generate decoy-SURD and discriminate SURD from decoy-SURD using ROC curves. Z.M. selected the distribution test set, wrote the MATLAB code to generate the random samples, and performed preliminary tests on the PDFestimator. A.G. evaluated ROC curves and scatter plots for the proposed scoring functions applied to SURD and decoy-SURD across sample sizes. J.F. motivated the initial work by reductive data analysis using methods of data collapse and scaling. J.F. modified the PDFestimator as needed, performed all simulations involving the PDFestimator, and performed all data analysis regarding comparative density estimation performance. D.J., J.F. and Z.M. wrote the paper.

Funding: This research received no external funding.

Acknowledgments: We thank Michael Grabchak for several discussions that helped direct this work.

Conflicts of Interest: The authors declare no conflict of interest.

References

1. Jacobs, D.J. Best Probability Density Function for Random Sampled Data. *Entropy* **2009**, *11*, 1001–1024.
2. Xiang, N.; Cai, S.; Yang, S.; Zhong, Z.; Zheng, F.; He, J.; Wu, Y. Statistical Analysis of Gait Maturation in Children Using non-parametric Probability Density Function Modeling. *Entropy* **2013**, *15*, 753–766. [CrossRef]
3. Bee, M. A Maximum Entropy Approach to Loss Distribution Analysis. *Entropy* **2013**, *15*, 1100–1117. [CrossRef]
4. Popkov, Y.; Popkov, A. New Methods of Entropy-Robust Estimation for Randomized Models under Limited Data. *Entropy* **2014**, *16*, 675–698. [CrossRef]
5. Wei, T.; Song, S. Confidence Interval Estimation for Precipitation Quantiles Based on Principle of Maximum Entropy. *Entropy* **2019**, *21*, 315. [CrossRef]
6. Crehuet, R.; Buigues, P.J.; Salvatella, X.; Lindorff-Larsen, K. Bayesian-Maximum-Entropy Reweighting of IDP Ensembles Based on NMR Chemical Shifts. *Entropy* **2019**, *21*, 898. [CrossRef]
7. Yu, L.; Su, Z. Application of Kernel Density Estimation in Lamb Wave-Based Damage Detection. *Math. Probl. Eng.* **2012**, *2012*. [CrossRef]
8. Baxter, M.J.; Beardah, C.C.; Westwood, S. Sample Size and Related Issues in the Analysis of Lead Isotope Data. *J. Archaeol. Sci.* **2000**, *27*, 973–980. [CrossRef]
9. DiNardo, J.; Fortin, N.M.; Lemieux, T. Labor market institutions and the distribution of wages, 1973–1992: A semiparametric approach. *Econometrica* **1996**, *64*, 1001. [CrossRef]
10. Cranmer, K. Kernel estimation in high-energy physics. *Comput. Phys. Commun.* **2001**, *136*, 198–207. [CrossRef]
11. Farmer, J.; Jacobs, D. High throughput non-parametric probability density estimation. *PLoS ONE* **2018**, *13*, e0196937. [CrossRef] [PubMed]
12. Devroye, L. *Non-Uniform Random Variate Generation*; Springer-Verlag: Berlin, Germany, 1986.
13. Nason, G.; Arnold, B.C.; Balakrishnan, N.; Nagaraja, H.N. A First Course in Order Statistics. *Statistician* **1994**, *43*, 329. [CrossRef]
14. Feng, X.; Liang, Y.; Shi, X.; Xu, D.; Wang, X.; Guan, R. Overfitting Reduction of Text Classification Based on AdaBELM. *Entropy* **2017**, *19*, 330. [CrossRef]
15. Anderson, T.W.; Darling, D.A. A Test of Goodness of Fit. *J. Am. Stat. Assoc.* **1954**, *49*, 765–769.
16. Engmann, S.; Cousineau, D. Comparing distributions: The two-sample Anderson–Darling test as an alternative to the Kolmogorov–Smirnov test. *J. Appl. Quant. Methods* **2011**, *6*, 1–17.
17. Murali, R.; Chen, Y.; Vemuri, B.C.; Wang, F. Cumulative residual entropy: A new measure of information. *IEEE Trans. Inf. Theory* **2004**, *50*, 1220–1228. [CrossRef]
18. Crescenzo, A.D.; Longobardi, M. Some properties and applications of cumulative Kullback–Leibler information. *Appl. Stochastic Models Bus. Ind.* **2015**, *31*, 875–891. [CrossRef]
19. Laguna, H.G.; Salazar, S.J.C.; Sagar, R.P. Entropic Kullback-Leibler type distance measures for quantum distributions. *Int. J. Quantum Chem.* **2019**, *119*, 875–891. [CrossRef]
20. Lewis, P.A.W. Distribution of the Anderson-Darling Statistic. *Ann. Math. Statist.* **1961**, *32*, 1118–1124. [CrossRef]
21. Kullback, S.; Leibler, R.A. On Information and Sufficiency. *Ann. Math. Stat.* **1951**, *22*, 79–86. [CrossRef]
22. Streiner, D.L.; Cairney, J. What's under the ROC? An Introduction to Receiver Operating Characteristics Curves. *Can. J. Psychiatry* **2007**, *52*, 121–128.
23. Fisher, R.A. Theory of Statistical Estimation. *Math. Proc. Camb. Philos. Soc.* **1925**, *22*, 700–725. [CrossRef]
24. Lin, J. Divergence measures based on the Shannon entropy. *IEEE Trans. Inf. Theory* **1991**, *37*, 145–151. [CrossRef]

© 2019 by the authors. Licensee MDPI, Basel, Switzerland. This article is an open access article distributed under the terms and conditions of the Creative Commons Attribution (CC BY) license (http://creativecommons.org/licenses/by/4.0/).

Article

Prior Sensitivity Analysis in a Semi-Parametric Integer-Valued Time Series Model

Helton Graziadei [1,*], Antonio Lijoi [2], Hedibert F. Lopes [3] and Paulo C. Marques F. [3] and Igor Prünster [2]

1. Instituto de Matemática e Estatística, Universidade de São Paulo, São Paulo 05508-090, Brazil
2. Department of Decision Sciences and BIDSA, Bocconi University, via Röntgen 1, 20136 Milano, Italy; antonio.lijoi@unibocconi.it (A.L.); igor.pruenster@unibocconi.it (I.P.)
3. Insper Institute of Education and Research, Rua Quatá 300, São Paulo 04546-042, Brazil; HedibertFL@insper.edu.br (H.F.L.); PauloCMF1@insper.edu.br (P.C.M.F.)
* Correspondence: hltgraziadei@gmail.com

Received: 26 November 2019; Accepted: 3 January 2020; Published: 6 January 2020

Abstract: We examine issues of prior sensitivity in a semi-parametric hierarchical extension of the INAR(p) model with innovation rates clustered according to a Pitman–Yor process placed at the top of the model hierarchy. Our main finding is a graphical criterion that guides the specification of the hyperparameters of the Pitman–Yor process base measure. We show how the discount and concentration parameters interact with the chosen base measure to yield a gain in terms of the robustness of the inferential results. The forecasting performance of the model is exemplified in the analysis of a time series of worldwide earthquake events, for which the new model outperforms the original INAR(p) model.

Keywords: time series of counts; Bayesian hierarchical modeling; Bayesian nonparametrics; Pitman–Yor process; prior sensitivity; clustering; Bayesian forecasting

---

1. Introduction

Integer-valued time series are relevant to many fields of knowledge, ranging from finance and econometrics to ecology and meteorology. An extensive number of models for this kind of data has been proposed since the introduction of the INAR(1) model in the pioneering works of McKenzie [1] and Al-Osh and Alzaid [2] (see also the book by Weiss [3]). A higher-order INAR(p) model was considered in the work of Du and Li [4].

In this paper, we generalize the Bayesian version of the INAR(p) model studied by Neal and Kypraios [5]. In our model, the innovation rates are allowed to vary through time, with the distribution of the innovation rates being modeled hierarchically by means of a Pitman–Yor process [6]. In this way, we account for potential heterogeneity in the innovation rates as the process evolves through time, and this feature is automatically incorporated in the Bayesian forecasting capabilities of the model.

The semi-parametric form of the model demands a robustness analysis of our inferential conclusions as we vary the hyperparameters of the Pitman–Yor process. We investigate this prior sensitivity issue carefully and find ways to control the hyperparameters in order to achieve robust results.

This paper is organized as follows. In Section 2, we construct a generalized INAR(p) model with variable innovation rates. The likelihood function of the generalized model is derived and a data augmentation scheme is developed, which gives a specification of the model in terms of conditional distributions. This data augmented representation of the model enables the derivation in Section 4 of full conditional distributions in simple analytical form, which are essential for the stochastic simulations in Section 5. Section 3 recollects the main properties of the Pitman–Yor process which are used to define the PY-INAR(p) model in Section 4, including its clustering properties. In building the PY-INAR(p), we propose a form for the prior distribution of the thinning parameters vector which improves on the choice made for the Bayesian INAR(p) model studied in [5]. In Section 5, we investigate the robustness of the inference with respect to changes in the Pitman–Yor process hyperparameters. Using the full conditional distributions of the innovation rates derived in Section 4, we inspect the behavior of the model as we concentrate or spread the mass of the Pitman–Yor base measure. This leads us to a graphical criterion that identifies an elbow in the posterior expectation of the number of clusters as we vary the hyperparameters of the base measure. Once we have control over the base measure, we study its interaction with the concentration and discount hyperparameters, showing how to make choices that yield robust results. In the course of this development, we use geometrical tools to inspect the clustering of the innovation rates produced by the model. Section 6 puts the graphical criterion to work for simulated data. In Section 7, using a time series of worldwide earthquake events, we finish the paper comparing the forecasting performance of the PY-INAR(p) model against the original INAR(p) model, with favorable results.

2. A Generalization of the INAR(p) Model

We begin by generalizing the original INAR(p) model of Du and Li [4] as follows.

Let $\{Y_t\}_{t\geq 1}$ be an integer-valued time series, and, for some integer $p \geq 1$, let the *innovations* $\{Z_t\}_{t\geq p+1}$, given positive parameters $\{\lambda_t\}_{t\geq p+1}$, be a sequence of conditionally independent Poisson(λ_t) random variables. For a given vector of parameters $\alpha = (\alpha_1, \ldots, \alpha_p) \in [0,1]^p$, let $\mathscr{F}_i = \{B_{ij}(t) : j \geq 0, t \geq 2\}$ be a family of conditionally independent and identically distributed Bernoulli(α_i) random variables. For $i \neq k$, suppose that $\mathscr{F}_i$ and $\mathscr{F}_k$ are conditionally independent, given α. Furthermore, assume that the innovations $\{Z_t\}_{t \geq p+1}$ and the families $\mathscr{F}_1, \ldots, \mathscr{F}_p$ are conditionally independent, given α and λ. The generalized INAR(p) model is defined by the functional relation

$$Y_t = \alpha_1 \circ Y_{t-1} + \cdots + \alpha_p \circ Y_{t-p} + Z_t,$$

for $t \geq p+1$, in which $\circ$ denotes the binomial thinning operator, defined by $\alpha_i \circ Y_{t-i} = \sum_{j=1}^{Y_{t-i}} B_{ij}(t)$, if $Y_{t-i} > 0$, and $\alpha_i \circ Y_{t-i} = 0$, if $Y_{t-i} = 0$. In the homogeneous case, when all the λ_t's are assumed to be equal, we recover the original INAR(p) model.

When $p = 1$, this model can be interpreted as specifying a birth-and-death process, in which, at epoch t, the number of cases Y_t is equal to the new cases Z_t plus the cases that survived from the previous epoch; the role of the binomial thinning operator being to remove a random number of the Y_{t-1} cases present at the previous epoch $t-1$ (see [7] for an interpretation of the order p case as a birth-and-death process with immigration).

Let $y = (y_1, \ldots, y_T)$ denote the values of an observed time series. For simplicity, we assume that $Y_1 = y_1, \ldots, Y_p = y_p$ with probability one. The joint distribution of $Y_1, \ldots, Y_T$, given parameters α and $\lambda = (\lambda_{p+1}, \ldots, \lambda_T)$, can be factored as

$$\Pr\{Y_1 = y_1, \ldots, Y_T = y_T \mid \alpha, \lambda\} = \prod_{t=p+1}^{T} \Pr\{Y_t = y_t \mid Y_{t-1} = y_{t-1}, \ldots, Y_{t-p} = y_{t-p}, \alpha, \lambda_t\}.$$

Since, with probability one, $\alpha_i \circ Y_{t-i} \leq Y_{t-i}$ and $Z_t \geq 0$, the likelihood function of the generalized INAR(p) model is given by

$$L_y(\alpha, \lambda) = \prod_{t=p+1}^{T} \sum_{m_{1,t}=0}^{\min\{y_t, y_{t-1}\}} \cdots \sum_{m_{p,t}=0}^{\min\{y_t - \sum_{j=1}^{p-1} m_{j,t}, y_{t-p}\}} \left(\prod_{i=1}^{p} \binom{y_{t-i}}{m_{i,t}} \alpha_i^{m_{i,t}} (1-\alpha_i)^{y_{t-i} - m_{i,t}} \right) \times$$

$$\left(\frac{e^{-\lambda_t} \lambda_t^{y_t - \sum_{j=1}^{p} m_{j,t}}}{(y_t - \sum_{j=1}^{p} m_{j,t})!} \right).$$

For some epoch t and $i = 1, \ldots, p$, suppose that we could observe the values of the latent maturations $M_{i,t}$. Postulate that $M_{i,t} \mid Y_{t-i} = y_{t-i}, \alpha_i \sim \text{Binomial}(y_{t-i}, \alpha_i)$, so that the conditional probability function of $M_{i,t}$ is given by

$$p(m_{i,t} \mid y_{t-i}, \alpha_i) = \Pr\{M_{i,t} = m_{i,t} \mid Y_{t-i} = y_{t-i}, \alpha_i\}$$
$$= \binom{y_{t-i}}{m_{i,t}} \alpha_i^{m_{i,t}} (1-\alpha_i)^{y_{t-i} - m_{i,t}} \mathbb{I}_{\{0, \ldots, y_{t-i}\}}(m_{i,t}).$$

Furthermore, suppose that

$$p(y_t \mid m_{1,t}, \ldots, m_{p,t}, \lambda_t) = \Pr\{Y_t = y_t \mid M_{1,t} = m_{1,t}, \ldots, M_{p,t} = m_{p,t}, \lambda_t\}$$
$$= \frac{e^{-\lambda_t} \lambda_t^{y_t - \sum_{j=1}^{p} m_{j,t}}}{(y_t - \sum_{j=1}^{p} m_{j,t})!} \mathbb{I}_{\{\sum_{j=1}^{p} m_{j,t}, \sum_{j=1}^{p} m_{j,t}+1, \ldots\}}(y_t).$$

Using the law of total probability and the product rule, we have that

$$p(y_t \mid y_{t-1}, \ldots, y_{t-p}, \alpha, \lambda_t) = \sum_{m_{1,t}=0}^{y_{t-1}} \cdots \sum_{m_{p,t}=0}^{y_{t-p}} p(y_t, m_{1,t}, \ldots, m_{p,t} \mid y_{t-1}, \ldots, y_{t-p}, \alpha, \lambda_t)$$

$$= \sum_{m_{1,t}=0}^{y_{t-1}} \cdots \sum_{m_{p,t}=0}^{y_{t-p}} p(y_t \mid m_{1,t}, \ldots, m_{p,t}, \lambda_t) \times \prod_{i=1}^{p} p(m_{i,t} \mid y_{t-i}, \alpha_i).$$

Since

$$\mathbb{I}_{\{\sum_{j=1}^{p} m_{j,t}, \sum_{j=1}^{p} m_{j,t}+1, \ldots\}}(y_t) = \mathbb{I}_{\{0, \ldots, y_t\}}\left(\sum_{j=1}^{p} m_{j,t}\right)$$
$$= \mathbb{I}_{\{0, \ldots, y_t\}}(m_{1,t}) \times \cdots \times \mathbb{I}_{\{0, \ldots, y_t - \sum_{j=1}^{p-1} m_{j,t}\}}(m_{p,t})$$

and

$$\mathbb{I}_{\{\sum_{j=1}^{p} m_{j,t}, \sum_{j=1}^{p} m_{j,t}+1, \ldots\}}(y_t) \times \mathbb{I}_{\{0, \ldots, y_{t-i}\}}(m_{i,t}) = \mathbb{I}_{\{0, 1, \ldots, \min\{y_t - \sum_{j \neq i} m_{j,t}, y_{t-i}\}\}}(m_{i,t}),$$

we recover the original likelihood of the generalized INAR(p), showing that the introduction of the latent maturations $M_{i,t}$ with the specified distributions is a valid data augmentation scheme (see [8,9] for a general discussion of data augmentation techniques).

In the next section, we review the needed definitions and properties of the Pitman–Yor process.

3. Pitman–Yor Process

Let the random probability measure $\mathbb{G} \sim \text{DP}(\tau, G_0)$ be a Dirichlet process [10–12] with concentration parameter τ and base measure G_0. If the random variables $X_1, \ldots, X_n$, given $\mathbb{G} = G$, are conditionally independent and identically distributed as G, then it follows that

$$\Pr\{X_{n+1} \in B \mid X_1 = x_1, \ldots, X_n = x_n\} = \frac{\tau}{\tau + n} G_0(B) + \frac{1}{\tau + n} \sum_{i=1}^{n} \mathbb{I}_B(x_i),$$

for every Borel set B. If we imagine the sequential generation of the X_i's, for $i = 1, \ldots, n$, the former expression shows that a value is generated anew from G_0 with probability proportional to τ, or we repeat one the previously generated values with probability proportional to its multiplicity. Therefore, almost surely, realizations of a Dirichlet process are discrete probability measures, perhaps with denumerable infinite support, depending on the nature of G_0. Also, this data-generating process, known as the Pólya–Blackwell–MacQueen urn, implies that the X_i's are "softly clustered", in the sense that in one realization of the process the elements of a subset of the X_i's may have exactly the same value.

The Pitman–Yor process [6] is a generalization of the Dirichlet process which results in a model with added flexibility. Essentially, the Pitman–Yor process modifies the expression of the probability associated with the Pólya-Blackwell-MacQueen urn introducing a new parameter so that the posterior predictive probability becomes

$$\Pr\{X_{n+1} \in B \mid X_1 = x_1, \ldots, X_n = x_n\} = \frac{\tau + k\sigma}{\tau + n} G_0(B) + \frac{1}{\tau + n} \sum_{i=1}^{n} \left(1 - \frac{\sigma}{n_i}\right) I_B(x_i),$$

in which $0 \leq \sigma < 1$ is the discount parameter, $\tau > -\sigma$, k is the number of distinct elements in $\{X_1, \ldots, X_n\}$, and n_i is the number of elements in $\{X_1, \ldots, X_n\}$ which are equal to X_i, for $i = 1, \ldots, n$. It is well known that $E[\mathbb{G}(B)] = G_0(B)$ and

$$\mathrm{Var}[\mathbb{G}(B)] = \left(\frac{1-\sigma}{\tau+1}\right) G_0(B)(1 - G_0(B)),$$

for every Borel set B. Hence, $\mathbb{G}$ is centered on the base probability measure G_0, while τ and σ control the concentration of $\mathbb{G}$ around G_0. We use the notation $\mathbb{G} \sim \mathrm{PY}(\tau, \sigma, G_0)$. When $\sigma = 0$, we recover the Dirichlet process as a special case. The PY process is also defined for $\sigma < 0$ and $\tau = |\sigma|m$, for some positive integer m. For our purposes, it is enough to consider the case of non-negative σ.

Pitman [6] derived the distribution of the number of clusters K (the number of distinct X_i's), conditionally on both the concentration parameter τ and the discount parameter σ, as

$$\Pr\{K = k \mid \tau, \sigma\} = \frac{\prod_{i=1}^{k-1}(\tau + i\sigma)}{\sigma^k \times (\tau + 1)_{n-1}} \times \mathscr{C}(n, k; \sigma),$$

in which $(x)_n = \Gamma(x+n)/\Gamma(x)$ is the rising factorial and $\mathscr{C}(n, k; \sigma)$ is the generalized factorial coefficient [13].

In the next section, we use a Pitman–Yor process to model the distribution of the innovation rates in the generalized INAR(p) model.

4. PY-INAR(p) Model

The PY-INAR(p) model is as a hierarchical extension of the generalized INAR(p) model defined in Section 2. Given a random measure $\mathbb{G} \sim \mathrm{PY}(\tau, \sigma, G_0)$, in which G_0 is a Gamma(a_0, b_0) distribution, let the innovation rates $\lambda_{p+1}, \ldots, \lambda_T$ be conditionally independent and identically distributed with distribution $\Pr\{\lambda_t \in B \mid \mathbb{G} = G\} = G(B)$.

To complete the PY-INAR(p) model, we need to specify the form of the prior distribution for the vector of thinning parameters $\alpha = (\alpha_1, \ldots, \alpha_p)$. By comparison with standard results from the theory of the AR(p) model [14], Du and Li [4] found that in the INAR(p) model the constraint $\sum_{i=1}^{p} \alpha_i < 1$ must be fulfilled to guarantee the non-explosiveness of the process. In their Bayesian analysis of the INAR(p) model, Neal and Kypraios [5] considered independent beta distributions for the α_i's. Unfortunately, this choice is problematic. For example, in the particular case when the α_i's have independent uniform distributions, it is possible to show that $\Pr\{\sum_{i=1}^{p} \alpha_i < 1\} = 1/p!$, implying that we would be concentrating most of the prior mass on the explosive region even for moderate values of the model order p. We circumvent this problem using a prior distribution for α that places all of its mass on the nonexplosive region and still allows us to derive the full conditional distributions of the α_i's in simple closed form. Specifically, we take the prior distribution of α to be a Dirichlet distribution with hyperparameters $(a_1, \ldots, a_p; a_{p+1})$, and corresponding density

$$\pi(\alpha) = \frac{\Gamma\left(\sum_{i=1}^{p+1} a_i\right)}{\prod_{i=1}^{p+1} \Gamma(a_i)} \prod_{i=1}^{p+1} \alpha_i^{a_i - 1},$$

in which $a_i > 0$, for $i = 1, \ldots, p+1$, and $\alpha_{p+1} = 1 - \sum_{i=1}^{p} \alpha_i$.

Let $m = \{m_{i,t} : i = 1, \ldots, p, t = p+1, \ldots, T\}$ denote the set of all maturations, and let $\mu_\mathbb{G}$ be the distribution of $\mathbb{G}$. Our strategy to derive the full conditionals distributions of the model parameters and latent variables is to consider the marginal distribution

$$p(y, m, \alpha, \lambda) = \int p(y, m, \alpha, \lambda \mid G) \, d\mu_\mathbb{G}(G)$$

$$= \left\{ \prod_{t=p+1}^{T} p(y_t \mid m_{1,t}, \ldots, m_{p,t}, \lambda_t) \prod_{i=1}^{p} p(m_{i,t} \mid y_{t-i}, \alpha_i) \right\}$$

$$\times \pi(\alpha) \times \int \prod_{t=p+1}^{T} p(\lambda_t \mid G) \, d\mu_\mathbb{G}(G).$$

From this expression, using the results in Section 3, the derivation of the full conditional distributions is straightforward. In the following expressions, the symbol $\propto$ denotes proportionality up to a suitable normalization factor, and the label "all others" designate the observed counts y and all the other latent variables and model parameters, with the exception of the one under consideration.

Let $\lambda_{\setminus t}$ denote the set $\{\lambda_{p+1}, \ldots, \lambda_T\}$ with the element λ_t removed. Then, for $t = p+1, \ldots, T$, we have

$$\lambda_t \mid \text{all others} \sim w_t \times \text{Gamma}(y_t - m_t + a_0, b_0 + 1) + \sum_{r \neq t} \left(1 - \frac{\sigma}{n_r}\right) \lambda_r^{y_t - m_t} e^{-\lambda_r} \delta_{\{\lambda_r\}},$$

in which the weight

$$w_t = \frac{(\tau + k_{\setminus t} \sigma) \times b_0^{a_0} \times \Gamma(y_t - m_t + a_0)}{\Gamma(a_0) \times (b_0 + 1)^{y_t - m_t + a_0}},$$

n_r is the number of elements in $\lambda_{\setminus t}$ which are equal to λ_r, and $k_{\setminus t}$ is the number of distinct elements in $\lambda_{\setminus t}$. In this mixture, we suppressed the normalization constant that makes all weights add up to one.

Making the choice $a_{p+1} = 1$, we have

$$\alpha_i \mid \text{all others} \sim \text{TBeta}\left(a_i + \sum_{t=p+1}^{T} m_{i,t}, 1 + \sum_{t=p+1}^{T} (y_{t-i} - m_{i,t}), 1 - \sum_{j \neq i} \alpha_j\right),$$

for $i = 1, \ldots, p$, in which TBeta denotes the right truncated Beta distribution with support $(0, 1 - \sum_{j \neq i}^{p} \alpha_j)$.

For the latent maturations, we find

$$p(m_{i,t} \mid \text{all others}) \propto \frac{1}{(m_{i,t})! (y_t - \sum_{j=1}^{p} m_{j,t})! (y_{t-i} - m_{i,t})!} \left(\frac{\alpha_i}{\lambda_t (1 - \alpha_i)}\right)^{m_{i,t}}$$

$$\times \mathbb{I}_{\{0, 1, \ldots, \min\{y_t - \sum_{j \neq i} m_{j,t}, y_{t-i}\}\}}(m_{i,t}).$$

To explore the posterior distribution of the model, we build a Gibbs sampler [15] using these full conditional distributions. Escobar and West [16] showed, in a similar context, that we can improve mixing by resampling simultaneously the values of all λ_t's inside the same cluster at the end of each iteration of the Gibbs sampler. Letting $(\lambda_1^*, \ldots, \lambda_k^*)$ be the k unique values among $(\lambda_{p+1}, \ldots, \lambda_T)$, define the number of occupants of cluster j by $\nu_j = \sum_{t=p+1}^{T} \mathbb{I}_{\{\lambda_j^*\}}(\lambda_t)$, for $j = 1, \ldots, k$. It follows that

$$\lambda_j^* \mid \text{all others} \sim \text{Gamma}\left(a_0 + \sum_{t=p+1}^{T}\left(y_t - \sum_{i=1}^{p} m_{i,t}\right) \cdot \mathbb{I}_{\{\lambda_j^*\}}(\lambda_t), b_0 + \nu_j\right).$$

for $j = 1, \ldots, k$. At the end of each iteration of the Gibbs sampler, we update the values of all λ_t's inside each cluster by the corresponding λ_j^* using this distribution.

5. Prior Sensitivity

As it is often the case for Bayesian models with nonparametric components, a choice of the prior parameters for the PY-INAR(p) model which yields robustness of the posterior distribution is nontrivial [17].

The first aspect to be considered is the fact that the base measure G_0 plays a crucial role in the determination of the posterior distribution of the number of clusters K. This can be seen directly by inspecting the form of the full conditional distributions derived in Section 4. Recalling that G_0 is a gamma distribution with mean a_0/b_0 and variance a_0/b_0^2, from the full conditional distribution of λ_t one may note that the probability of generating, on each iteration of the Gibbs sampler, a value for λ_t anew from G_0 is proportional to

$$\frac{(\tau + k_{\setminus t}\sigma) \times b_0^{a_0} \times \Gamma(y_t - m_t + a_0)}{\Gamma(a_0)(b_0 + 1)^{y_t - m_t + a_0}}.$$

Therefore, supposing that all the other terms are fixed, if we concentrate the mass of G_0 around zero by making $b_0 \to \infty$, this probability decreases to zero. This is not problematic, because it is hardly the case that we want to make such a drastic choice for G_0. The behavior in the other direction is more revealing, since taking $b_0 \downarrow 0$, in order to spread the mass of G_0, also makes the limit of this probability to be zero. Due to this behavior, we need to establish a criterion to choose the hyperparameters of the base measure which avoids these extreme cases.

In our analysis, it is convenient to have a single hyperparameter regulating how the mass of G_0 is spread over its support. For a given $\lambda_{\max} > 0$, we find numerically the values of a_0 and b_0 which minimize the Kullback-Leibler divergence between G_0 and a uniform distribution on the interval $[0, \lambda_{\max}]$. This Kullback-Leibler divergence can be computed explicitly as

$$-\log \lambda_{\max} - a_0 \log b_0 + \log \Gamma(a_0) - (a_0 - 1)(\log \lambda_{\max} - 1) + \frac{b_0 \lambda_{\max}}{2}.$$

In this new parameterization, our goal is to make a sensible choice for $\lambda_{\max}$. It is worth emphasizing that by this procedure we are not truncating the support of G_0, but only using the uniform distribution on the interval $[0, \lambda_{\max}]$ as a reference for our choice of the base measure hyperparameters a_0 and b_0.

Our proposal to choose $\lambda_{\max}$ goes as follows. We fix some value $0 \leq \sigma < 1$ for the discount parameter and choose an integer k_0 as the prior expectation of the number of clusters K, which, using the results at the end of Section 3, can be computed explicitly as

$$E[K] = \begin{cases} \tau \times (\psi(\tau + T - p) - \psi(\tau)) & \text{if } \sigma = 0; \\ ((\tau + \sigma)_{T-p}/(\sigma \times (\tau + 1)_{T-p-1})) - \tau/\sigma & \text{if } \sigma > 0, \end{cases}$$

in which $\psi(x)$ is the digamma function (see [6] for a derivation of this result). Next, we find the value of the concentration parameter τ by solving $E[K] = k_0$ numerically. After this, for each λ_{max} in a grid of values, we run the Gibbs sampler and compute the posterior expectation of the number of clusters $E[K \mid y]$. Finally, in the corresponding graph, we look for the value of λ_{max} located at the "elbow" of the curve, that is, the value of λ_{max} at which the values of $E[K \mid y]$ level off.

6. Simulated Data

As an explicit example of the graphical criterion in action, we used the functional form of a first-order model with thinning parameter $\alpha = 0.15$ to simulate a time series of length $T = 1000$, for which the distribution of the innovations is a symmetric mixture of three Poisson distributions with parameters 1, 8, and 15. Figure 1 shows the formations of the elbows for two values of the discount parameter: $\sigma = 0.5$ and $\sigma = 0.75$.

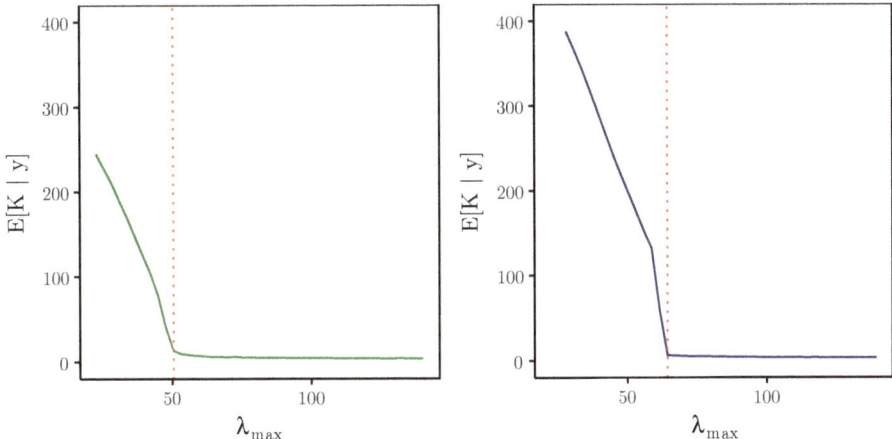

Figure 1. Formation of the elbows for $\sigma = 0.5$ (left) and $\sigma = 0.75$ (right). The red dotted lines indicate the chosen values of λ_{max}.

For the simulated time series, Figures 2–5 display the behavior of the posterior distributions obtained using the elbow method for $(k_0, \sigma) \in \{4, 10, 16, 30\} \times \{0, 0.25, 0.5, 0.75\}$. These figures make the relation between the choice of the value of the discount parameter σ and the achieved robustness of the posterior distribution quite explicit: as we increase the value of the discount parameter σ, the posterior becomes insensitive to the choice of k_0. In particular, for $\sigma = 0.75$, the posterior mode is always near 3, which is the number of components used in the distribution of the innovations of the simulated time series.

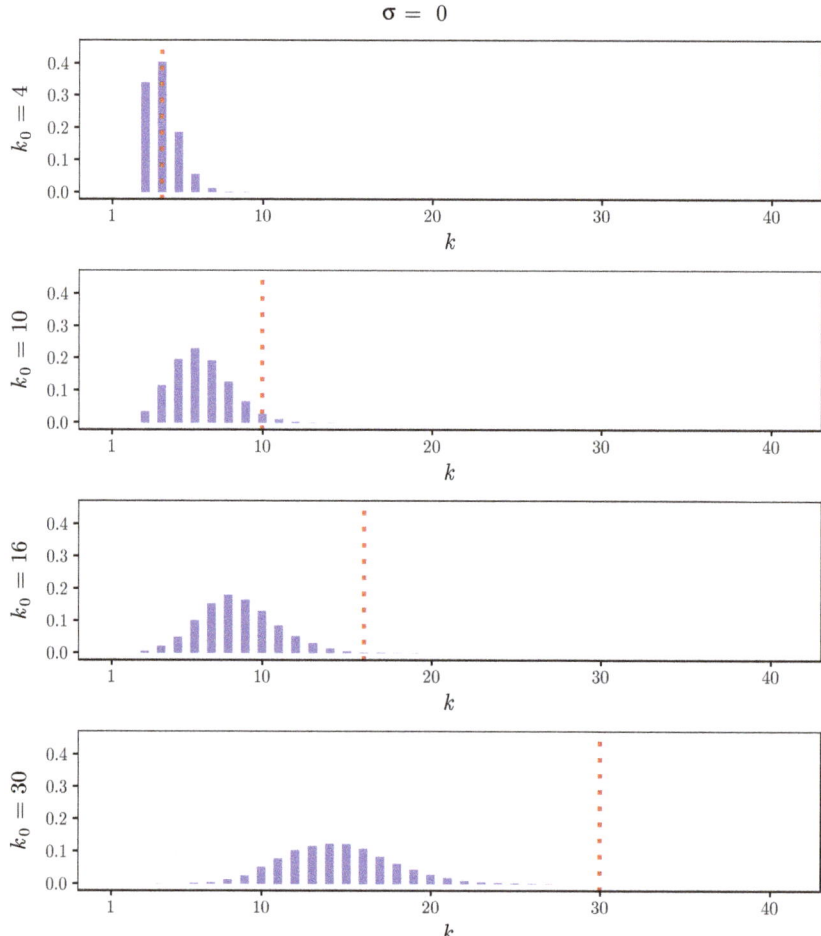

Figure 2. Posterior distributions of the number of clusters K for the simulated time series with $\sigma = 0$ and $k_0 = 4, 10, 16, 30$. The red dotted lines indicate the value of k_0.

Once we understand the influence of the prior parameters on the robustness of the posterior distribution, an interesting question is how to get a point estimate for the distribution of clusters, in the sense that each λ_t, for $t = p + 1, \ldots, T$, would be assigned to one of the available clusters.

From the Gibbs sampler, we can easily get a Monte Carlo approximation for the probabilities $d_{rt} = \Pr\{\lambda_r \neq \lambda_t \mid y\}$, for $r, t = p + 1, \ldots, T$. These probabilities define a dissimilarity matrix $D = (d_{rt})$ among the innovation rates. Although D is not a distance matrix, we can use it as a starting point to represent the innovation rates in a two-dimensional Euclidean space using the technique of metric multidimensional scaling (see [18] for a general discussion). From this two-dimensional representation, we use hierarchical clustering techniques to build a dendrogram, which is appropriately cut in order to define three clusters, allowing us to assign a single cluster label to each innovation rate.

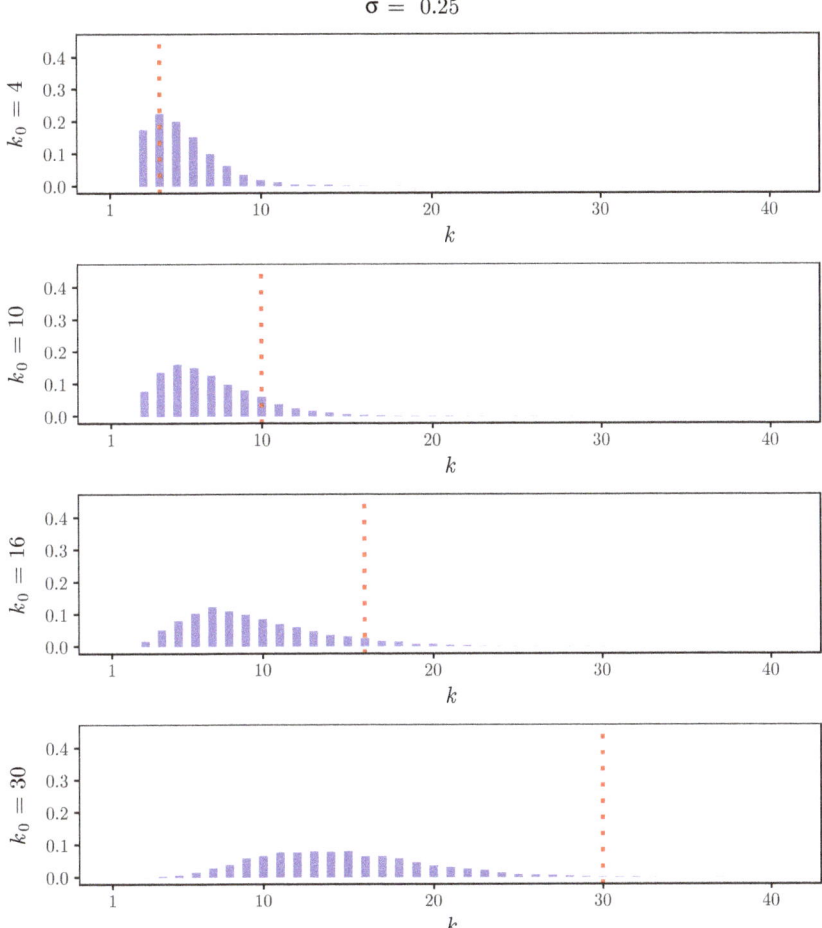

Figure 3. Posterior distributions of the number of clusters K for the simulated time series with $\sigma = 0.25$ and $k_0 = 4, 10, 16, 30$. The red dotted lines indicate the value of k_0.

Table 1 displays the confusion matrix of this assignment, showing that 83% of the innovations were grouped correctly in the clusters which correspond to the mixture components used to simulate the time series.

Table 1. Confusion matrix for the cluster assignments.

	True		
Predicted	1	2	3
1	297	32	0
2	11	217	42
3	0	84	316

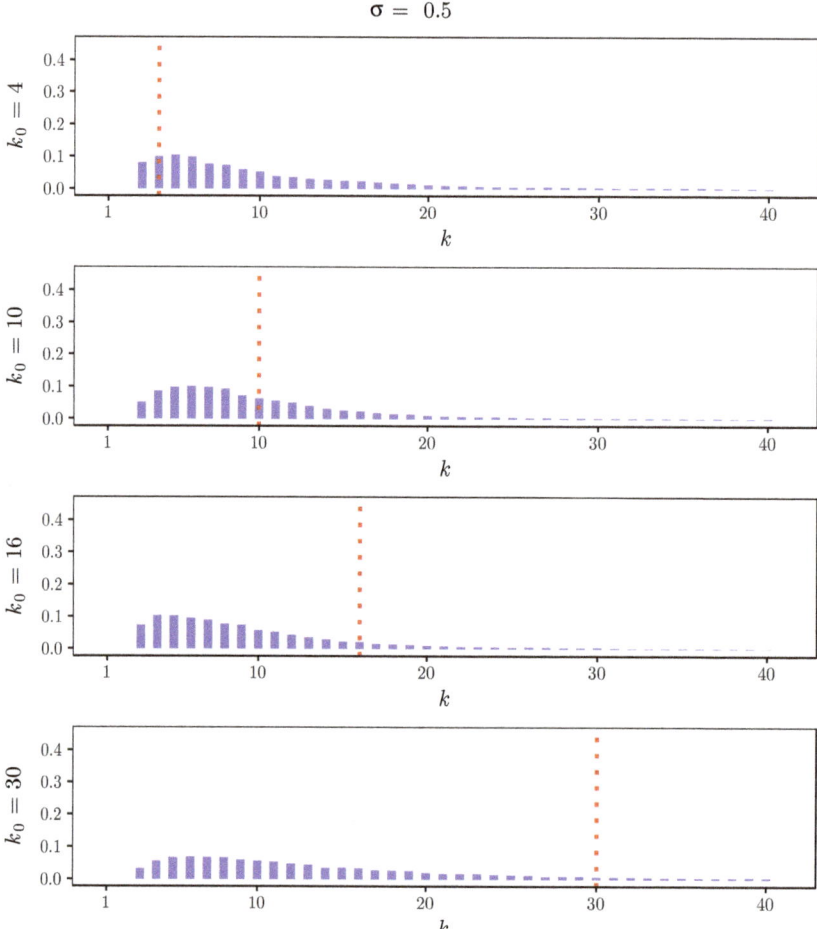

Figure 4. Posterior distributions of the number of clusters K for the simulated time series with $\sigma = 0.5$ and $k_0 = 4, 10, 16, 30$. The red dotted lines indicate the value of k_0.

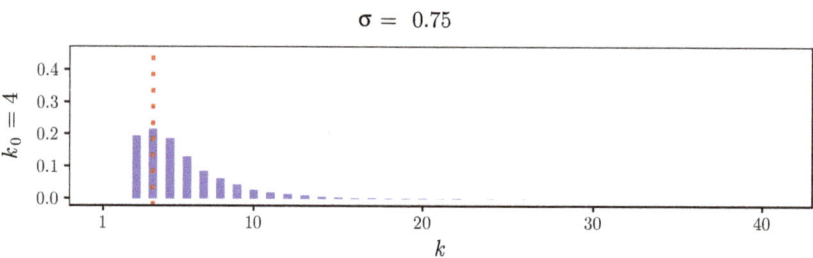

Figure 5. *Cont.*

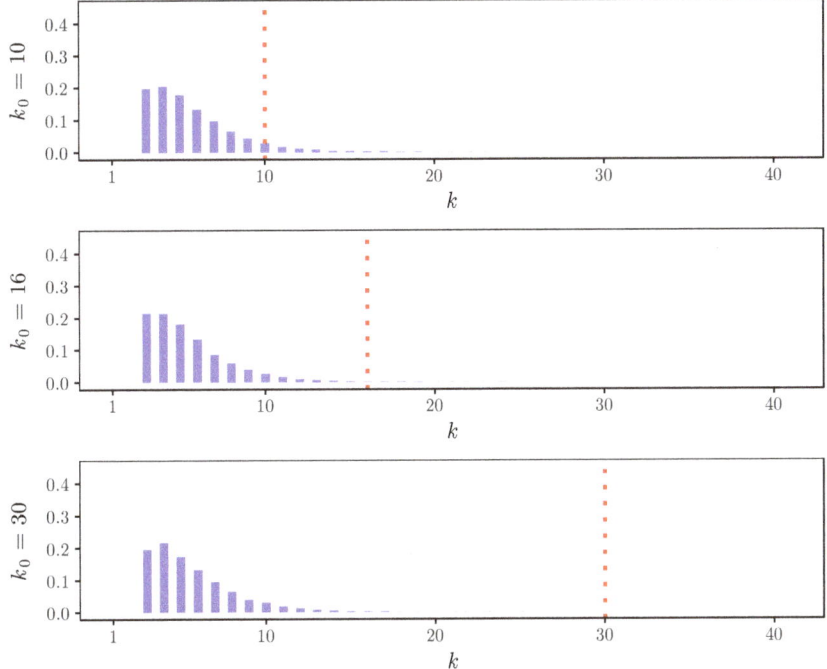

Figure 5. Posterior distributions of the number of clusters K for the simulated time series with $\sigma = 0.75$ and $k_0 = 4, 10, 16, 30$. The red dotted lines indicate the value of k_0.

7. Earthquake Data

In this section, we analyze a time series of yearly worldwide earthquakes events of substantial magnitude (equal or greater than 7 points on the Richter scale) from 1900 to 2018 (http://www.usgs.gov/natural-hazards/earthquake-hazards/earthquakes).

The forecasting performances of the INAR(p) and the PY-INAR(p) models are compared using a cross-validation procedure in which the models are trained with data ranging from the beginning of the time series up to a certain time, and predictions are made for epochs outside this training range.

Using this cross-validation procedure, we trained the INAR(p) and the PY-INAR(p) models with orders $p = 1, 2$, and 3, and made one-step-ahead predictions. Table 2 shows the out-of-sample mean absolute errors (MAE) for the INAR(p) and the PY-INAR(p) models. In this table, the MAE's are computed predicting the counts for the last 36 months. For the three model orders, the PY-INAR(p) model yields a smaller MAE than the original INAR(p) model.

Table 2. Out-of-sample MAE's for the INAR(p) and the PY-INAR(p) models, with orders $p = 1, 2$, and 3. The last column shows the relative variations of the MAE's for the PY-INAR(p) models with respect to the corresponding MAE's for the INAR(p) models.

	INAR	PY-INAR	$\Delta_{\text{PY-INAR}}$
$p=1$	3.861	3.583	-0.072
$p=2$	3.583	3.417	-0.046
$p=3$	3.972	3.305	-0.202

Author Contributions: Theoretical development: H.G., A.L., H.F.L., P.C.M.F, I.P. Software development: H.G., P.C.M.F. All authors have read and agreed to the published version of the manuscript.

Funding: Helton Graziadei and Hedibert F. Lopes thank FAPESP (Fundação de Amparo à Pesquisa do Estado de São Paulo) for financial support through grants numbers 2017/10096-6 and 2017/22914-5. Antonio Lijoi and Igor Prünster are partially supported by MIUR, PRIN Project 2015SNS29B.

Conflicts of Interest: The authors declare no conflict of interest.

References

1. McKenzie, E. Some simple models for discrete variate time series. *J. Am. Water Resour. Assoc.* **1985**, *21*, 645–650. [CrossRef]
2. Al-Osh, M.; Alzaid, A. First-order integer-valued autoregressive (INAR(1)) process: Distributional and regression properties. *Stat. Neerl.* **1988**, *42*, 53–61. [CrossRef]
3. Weiß, C. *An Introduction to Discrete-Valued Time Series*; John Wiley & Sons: Hoboken, NJ, USA, 2018.
4. Du, J.G.; Li, Y. The integer-valued autoregressive (INAR(p)) model. *J. Time Ser. Anal.* **1991**, *12*, 129–142. [CrossRef]
5. Neal, P.; Kypraios, T. Exact Bayesian inference via data augmentation. *Stat. Comput.* **2015**, *25*, 333–347. [CrossRef]
6. Pitman, J. Combinatorial stochastic processes. Technical Report 621; Springer: Berlin/Heidelberg, Germany, 2002. [CrossRef]
7. Dion, J.; Gauthier, G.; Latour, A. Branching processes with immigration and integer-valued time series. *Serdica Math. J.* **1995**, *21*, 123–136.
8. Van Dyk, D.; Meng, X.L. The art of data augmentation. *J. Comput. Graph. Stat.* **2001**, *10*, 1–50. [CrossRef]
9. Tanner, M.; Wong, W. The calculation of posterior distributions by data augmentation. *J. Am. Stat. Assoc.* **1987**, *82*, 528–540. [CrossRef]
10. Ferguson, T. A Bayesian analysis of some nonparametric problems. *Ann. Stat.* **1973**, *1*, 209–230. [CrossRef]
11. Schervish, M.J. *Theory of Statistics*; Springer: Berlin/Heidelberg, Germany, 1995; pp. 52–60.
12. Hjort, N.; Holmes, C.; Müller, P.; Walker, S. *Bayesian Nonparametrics*; Cambridge University Press: Cambridge, UK, 2010; Voume 28.
13. Lijoi, A.; Mena, R.H.; Prünster, I. Bayesian nonparametric estimation of the probability of discovering new species. *Biometrika* **2007**, *94*, 769–786. [CrossRef]
14. Hamilton, J. *Time Series Analysis*; Princeton University Press: Princeton, NJ, USA, 1994; Volume 2, pp. 43–71.
15. Gamerman, D.; Lopes, H. *Markov Chain Monte Carlo: Stochastic Simulation for Bayesian Inference*; Chapman & Hall/CRC: Boca Raton, FL, USA, 2006.
16. Escobar, M.; West, M. Computing nonparametric hierarchical models. In *Practical Nonparametric and Semiparametric Bayesian Statistics*; Dey, D., Müller, P., Sinha, D., Eds.; Springer: Berlin/Heidelberg, Germany, 1998; Chapter 1, pp. 1–22. [CrossRef]
17. Canale, A.; Prünster, I. Robustifying Bayesian nonparametric mixtures for count data. *Biometrics* **2017**, *73*, 174–184. [CrossRef] [PubMed]
18. Friedman, J.; Hastie, T.; Tibshirani, R. *The Elements of Statistical Learning*; Springer: Berlin/Heidelberg, Germany, 2009; pp. 570–572.

© 2020 by the authors. Licensee MDPI, Basel, Switzerland. This article is an open access article distributed under the terms and conditions of the Creative Commons Attribution (CC BY) license (http://creativecommons.org/licenses/by/4.0/).

Article

The Decomposition and Forecasting of Mutual Investment Funds Using Singular Spectrum Analysis

Paulo Canas Rodrigues [1,2,*], Jonatha Pimentel [1] and Patrick Messala [1] and Mohammad Kazemi [3]

1. Department of Statistics, Federal University of Bahia, 40170-110 Salvador, Brazil; jsppimentel9@gmail.com (J.P.); patrickmessala@gmail.com (P.M.)
2. CAST, Faculty of Information Technology and Communication Sciences, Tampere University, FI-33014 Tampere, Finland
3. Department of Statistics, Faculty of Mathematical Sciences, Shahrood University of Technology, P.O. Box 3619995161 Shahroud, Iran; m.kazemie64@gmail.com
* Correspondence: paulocanas@gmail.com or paulo.canas@ufba.br

Received: 7 November 2019; Accepted: 7 January 2020; Published: 9 January 2020

Abstract: Singular spectrum analysis (SSA) is a non-parametric method that breaks down a time series into a set of components that can be interpreted and grouped as trend, periodicity, and noise, emphasizing the separability of the underlying components and separate periodicities that occur at different time scales. The original time series can be recovered by summing all components. However, only the components associated to the signal should be considered for the reconstruction of the noise-free time series and to conduct forecasts. When the time series data has the presence of outliers, SSA and other classic parametric and non-parametric methods might result in misleading conclusions and robust methodologies should be used. In this paper we consider the use of two robust SSA algorithms for model fit and one for model forecasting. The classic SSA model, the robust SSA alternatives, and the autoregressive integrated moving average (ARIMA) model are compared in terms of computational time and accuracy for model fit and model forecast, using a simulation example and time series data from the quotas and returns of six mutual investment funds. When outliers are present in the data, the simulation study shows that the robust SSA algorithms outperform the classical ARIMA and SSA models.

Keywords: singular spectrum analysis; robust singular spectrum analysis; time series forecasting; mutual investment funds

1. Introduction

Mutual investment funds provide management services to institutional and individual investors, besides great liquidity for financial investments made in them and low transactional costs [1,2]. These funds can be of fixed or variable income and allow to diversify the assets while reducing unsystematic risk. Fixed income mutual investment funds are of low risk, whereas variable-income mutual investment funds vary in terms of risk but also in terms of returns. In this study, we were interested in analyzing the quotas and returns of six of the largest Brazilian based mutual investment funds—three purely based on stocks: (i) Alaska Black, (ii) APEX Long Biased, and (iii) Brasil Capital; and three balanced funds (usually combining a stock component, a bond component, and sometimes a money market component in a single portfolio): (iv) ADAM Strategy, (v) Gavea Macro, and (vi) SPX Nimitz.

A natural framework for analyzing mutual investment funds, due to its underlying structure, is a time series method.

Singular spectrum analysis (SSA) is a powerful non-parametric technique for time series analysis and forecasting, which incorporates elements of classical time series analysis, multivariate statistics, and matrix algebra. Its main aim is to decompose the original time series into a set of components that can be interpreted as trend components, seasonal components, and noise components [3–6]. SSA has proven both wide usefulness and applicability across many applications [7–17], being that its scope of application ranges from parameter estimation to time series filtering, synchronization analysis, and forecasting [18].

The SSA methodology for model fit can be summarized in four steps: (i) embedding, which maps the original univariate time series into a trajectory matrix; (ii) singular value decomposition (SVD), which helps decomposing the trajectory matrix into the sum of rank-one matrices; (iii) eigentriple grouping, which helps deciding which of the components are associated to the signal and which are associated to the noise; and (iv) diagonal averaging, which maps the rank-one matrices, associated to the signal, back to time series that can be interpreted as trend, seasonal, or other meaningful components.

SSA results and interpretation, similarly to many other classical time series methods, can be sensitive to data contamination with outliers [19,20]. In those cases, even a small percentage of outliers can make a big difference on the results for model fit and model forecast. Very few attempts have been made in order to access the effect of the presence of outliers in the data while conducting a SSA. One study [21,22] presented some preliminary results on the effect of outliers in singular spectrum analysis, and [23] made a first attempt to robustify the SSA by considering an SVD based on a robust L_1 norm [24] instead of the L_2 norm used in the classical algorithm, which they used for model fit.

In this paper we go one step further than [23] and propose a new robust algorithm for SSA that considers the SVD based on the Huber function [25]. Moreover, we propose two robust SSA forecasting algorithms, one based on the the L_1 norm and another based on the Huber function. Comparisons are made between the classical SSA algorithm, the robust SSA algorithm based on the L_1 norm (RLSSA), the robust SSA algorithm based on the Huber function (RHSSA), and the classical autoregressive integrated moving average (ARIMA) model, in terms of computational time and accuracy for model fit and model forecast. These comparisons for decomposing and forecasting time series were done by considering a simulation example and the six mutual investment funds mentioned above.

The rest of this paper is organized as follows. Section 2 provides the materials and methods containing the data description, a brief introduction to the ARIMA and SSA methodologies, and the details of the proposed robust SSA algorithm that uses the SVD based on the Huber function. Section 3 presents the results and discussion, wherein the ARIMA, SSA, and robust SSA algorithms are compared in terms of model fit and model forecast, using the six mutual investment funds and the simulation example. The paper closes in Section 4, wherein some conclusions are drawn.

2. Materials and Methods

2.1. Data

In this paper we consider a dataset that includes daily observations of six mutual investment funds, three based purely on stocks and three balanced funds:

Stock funds

- Alaska Black: 3 January 2017–30 August 2019 (N = 666 observations).
- APEX Long Biased: 15 April 2013–30 August 2019 (N = 1604 observations).
- Brasil Capital: 27 August 2012–30 August 2019 (N = 1760 observations).

Balanced funds

- ADAM Strategy: 29 April 2016–30 August 2019 (N = 838 observations).
- Gavea Macro: 30 June 2008–30 August 2019 (N = 2809 observations).
- SPX Nimitz: 01 December 2010–30 August 2019 (N = 2199 observations).

The datasets were collected from https://infofundos.com.br/carteira.

2.2. ARIMA Model

The autoregressive integrated moving average (ARIMA) models are among the most widely used techniques for time series analysis and forecasting. Such a model depends on three parameters: p is the number of lagged observations in the model, i.e., the autoregressive (AR) order; d is the number of times that the original observations are differenced, i.e., the integrated (I) degree; and q is the size of the moving average window, i.e., the order of the moving average (MA) [26]. This parametric model can then be written as $ARIMA(p,d,q)$, with p, d, and q non-negative integers. Given a time series $Y_N = y_1, \ldots, y_N$, the $ARIMA(p,d,q)$ model can be written as:

$$(1 - \phi_1 B^1 - \cdots - \phi_p B^p)(1-B)^d y_t = c + (1 + \theta_1 B^1 + \cdots + \theta_q B^q)\varepsilon_t, \tag{1}$$

where $\phi_1, \ldots, \phi_p$ are the parameters or coefficients of the p autoregressive terms; B is the time lag operator, or backward shift, which is a linear operator denoted by B^k such that $L^k y_t = y_{t-k}, t \in \mathbb{Z}$; y_t is the observation at the time point t; $c = \mu(1 - \phi_1 - \cdots - \phi_p)$; μ is the mean of $(1-B)^d y_t$; $\beta_1, \ldots, \beta_q$ are the parameters or coefficients of the q moving average terms; and ε_t is an error term, usually white noise with variance σ^2.

Alternatively, the model can be written as:

$$(1 - \phi_1 B - \cdots - \phi_p B^p)(1-B)^d (y_t - \mu t^d/d!) = (1 + \theta_1 B + \cdots + \theta_q B^q)\varepsilon_t, \tag{2}$$

which is the parametrization used in the "arima" function of the software R [27].

2.3. Singular Spectrum Analysis

Singular spectrum analysis is a non-parametric technique for model fit and model forecasting that decomposes a time series into a number of components that are summed and interpreted as trend, periodicity, and noise. Similarly to many other time series techniques, SSA can be used for solving a wide range of problems, some of the most relevant being its ability to smooth the original time series, and to separate the signal (i.e., trend and oscillatory components with different amplitudes) from the noise components. Therefore, SSA can be used to analyze and reconstruct smoother noise-free time series that can then be used for model forecasting.

SSA is divided into two interconnected stages: decomposition and reconstruction of the time series. These stages are divided into two sets each, forming a total of four steps: embedding, singular value decomposition (SVD), grouping, and diagonal averaging. The complete algorithm for model fit is described in the following sub-section. Further details can be found in, e.g., [5,6,28].

2.3.1. Decomposition

In the first stage, the (univariate) time series is converted into a high-dimensional matrix called a trajectory matrix, which is then decomposed into the sum of rank-one matrices based on the SVD.

(1) Embedding:

Consider a non-zero time series $Y_N = \{y_1, \ldots, y_n\}$ with size $N > 2$. Let $L(1 < L < N)$ be an integer value called window length and K an integer such that the trajectory matrix includes all values; i.e., $K = N - L + 1$. The embedding step is achieved by mapping the original time series into a sequence of K vectors with length L:

$$Y_i = (y_i, \cdots, y_{i+L-1})^T, 1 \leq i \leq K. \tag{3}$$

Then, the trajectory matrix $\mathbf{X}$, that includes the vectors Y_i, $i = 1, \ldots, K$, in its columns can be written as:

$$\mathbf{X} = [Y_1, \cdots, Y_K] = (y_{ij})_{i,j=1}^{LK} = \begin{bmatrix} y_1 & y_2 & \cdots & y_K \\ y_2 & y_3 & \cdots & y_{K+1} \\ \vdots & \vdots & \ddots & \vdots \\ y_L & y_{L+1} & \cdots & y_N \end{bmatrix}. \tag{4}$$

(2) Singular value decomposition:

Let $\mathbf{S} = \mathbf{XX}^T$, $\mathbf{U}_1, \ldots, \mathbf{U}_L$ be the eigenvectors of $\mathbf{S}$, and $\lambda_1 \geq \cdots \geq \lambda_L$, its corresponding eigenvalues. If d is the number of non-null eigenvalues of $\mathbf{S}$, and considering $\mathbf{V}_i = \mathbf{X}^T \mathbf{U}_i \sqrt{\lambda_i}$, we can decompose the trajectory matrix $\mathbf{X}$ as:

$$\mathbf{X} = \sum_{i=1}^{d} \mathbf{X}_i = \sum_{i=1}^{d} \sqrt{\lambda_i} \mathbf{U}_i \mathbf{V}_i^T. \tag{5}$$

The decomposition stage can be accomplished either by the eigendecomposition of $\mathbf{X}^T \mathbf{X}$ or by the SVD of $\mathbf{X}$ ($\mathbf{X} = \mathbf{UDV}^T$, $\mathbf{D} = diag(\sqrt{\lambda_1}, \ldots, \sqrt{\lambda_d})$). A comparison between both decompositions can be found in [29].

2.3.2. Reconstruction

In the second stage, after a separating signal from noise components, a diagonal averaging procedure is conducted in the matrices associated to the signal resulting into the sum of time series components that can then be interpreted as trend or oscillatory components:

(1) Eigentriple grouping:

This step consists of identifying the first r eigentriples associated with the signal and discarding the $d - r$ eigentriples associated with the noise. Formally, let $I = 1, \ldots, r$ and $I^c = r + 1, \ldots, d$. The goal of this step is to choose I such that the trajectory matrix can be written as:

$$\mathbf{X}_I = \sum_{i \in I} \sqrt{\lambda_i} \mathbf{U}_i \mathbf{V}_i^T + \epsilon, \tag{6}$$

where ϵ is the noise term.

The number of eigentriples to conduct the reconstruction is often decided based on w-correlations. We shall say that two series $Y^{(1)}$ and $Y^{(2)}$ are approximately separable if all correlations between the rows and the columns of the corresponding trajectory matrices obtained from series $Y^{(1)}$ and $Y^{(2)}$ are close to zero. In [5] they considered other characteristics of the quality of separability; namely, the weighted correlation or **w**-correlation, which is a natural measure of deviation of two series $Y_T^{(1)}$ and $Y_T^{(2)}$ from **w**-orthogonality:

$$\rho_{12}^{(w)} = \frac{(Y_T^{(1)}, Y_T^{(2)})_w}{\|Y_T^{(1)}\|_w \|Y_T^{(1)}\|_w}, \tag{7}$$

where $\|Y_T^{(i)}\|_w = \sqrt{\left(Y_T^{(i)}, Y_T^{(i)}\right)_w}$, $i = 1, 2$, and $(Y_T^{(1)}, Y_T^{(2)})_w = \sum_{t=1}^{T} w_t y_t^{(1)} y_t^{(2)}$ with $w_t = \min\{t, L, T - t + 1\}$. If the absolute value of the **w**-correlation is small, the two series are almost **w**-orthogonal. If the absolute value of the **w**-correlation is large, the series are far from being **w**-orthogonal and are, therefore, badly separable. Further explanation and intuition about this measure can be found in [5,28]. Other proposals for this choice were proposed by, e.g., [30,31].

(2) Diagonal averaging:

In this step, using anti-diagonal averaging on the matrices included in $\mathbf{X}_I$, the noise-free time series is reconstructed. First, the approximate trajectory matrix $\mathbf{X}_I$ is transformed into a Hankel matrix. Let $A_s = \{(l,k) : l+k = s, 1 \leq l \leq L, 1 \leq k \leq K\}$ and $\#(A_s)$ be the number of elements in A_s. The element $\tilde{x}_{ij}$ of the new Hankel matrix $\tilde{\mathbf{X}}$ is given by:

$$\tilde{x}_{ij} = \sum_{(l,k) \in A_s} \frac{x_{lk}}{\#(A_s)}. \tag{8}$$

Next, the Hankel matrix $\tilde{\mathbf{X}}_I$ is transformed into a new series of dimension N, and the original time series $\mathbf{Y}_N$ can be approximated by:

$$\tilde{y}_i = \begin{cases} \tilde{x}_{i1} & \text{for } i = 1, \ldots, L, \\ \tilde{x}_{Lj} & \text{for } i = L+1, \ldots, N, \end{cases} \tag{9}$$

where $j = i - L + 1$.

The reconstructed noise-fee time series can then be used for out-of-sample forecasting.

2.4. Robust SSA

Despite knowing that SSA has shown to be superior to traditional model-based methods in many applications, the singular value decomposition (second step of the SSA algorithm) is highly sensitive to data contamination with outliers. Very few studies were made in order to access effects of outliers in SSA and to generalize this methodology [21,22]. A first attempt to robustify the SSA by considering an SVD based on a robust L_1 norm [24] instead of the L_2 norm used in the classical algorithm, was proposed by [23]. That robust generalization was compared with the classical SSA algorithm for model fit by these authors. In this subsection we review that robust SSA algorithm proposed by [23] and propose a new robust algorithm for SSA that considers the SVD based on the Huber function [25] and also propose an algorithm for robust SSA model forecasting. While the robust algorithms based on the L_1 norm are very popular, they have difficulties in handling heavy tail outliers. The robust algorithms based on the Huber function combine the sum of squares loss and the least absolute deviation loss, that is, a quadratic on small errors, but grows linearly for large errors. As a result, the Huber loss function is not only more robust against outliers but also more adaptive for different types of data [32]. Further details and comparisons between the L_1 and Huber loss functions, among others, can be found in [33]. The R source code is available upon request from the first author of this paper.

2.4.1. Robust SSA Based on the L_1 Norm

The robust SSA algorithm proposed by [23] replaces the classical SVD based on the least squares L_2 norm, by the robust SVD algorithm based on the L_1 norm [24]. This robust SVD is performed iteratively, starting with an initial estimate of the first left singular vector U_1 and leading to an outlier-resistant approach that also allows for missing data. The robust SVD based on the L_1 norm is implemented under the function "robustSVD()" from the R package "pcaMethods".

2.4.2. Robust SSA based on the Huber Function

Here we propose a new alternative to robustify the SSA algorithm, where the least squares SVD in the step two is replaced by the robust SVD based on the Huber function [25]. The Huber loss function [34] can be defined as:

$$L_\delta(a) = \begin{cases} \frac{1}{2}a^2 & \text{if } |a| \leq \delta \\ \delta\left(|a| - \frac{1}{2}\delta\right) & \text{if } |a| > \delta \end{cases}, \tag{10}$$

where δ is a parameter that controls the robustness level, and a smaller value of δ usually leads to more robust estimation.

The robust SVD based on the Huber function is a special case of robust regularized SVD and can be obtained with the function "RobRSVD" of the "RobRSVD" R package, in the following way: RobRSVD (data, rough = TRUE, uspar = 0, vspar = 0). In this R implementation, the authors consider $\delta = 1.345$, the value commonly used in robust regression that produces 95% efficiency for normal errors [35]. However, numerical studies suggested that the RobRSVD function is not very sensitive to the choice of δ [25]. More details about this robust SVD can be found in [25].

2.5. Robust SSA Forecasting Algorithm

The standard recurrent SSA forecasting algorithm assumes that a given observation can be written as a linear combination of the $L-1$ previous observations [5,6,30]. The coefficients of those linear combinations in the classical SSA forecasting algorithm are obtained based on the left singular vectors, U, of the trajectory matrix $\mathbf{X}$. This is valid for SSA because of the orthogonality of the vectors in U and of the full rank decomposition of $\mathbf{X}$, which is not the case for the robust SVD algorithms because of their construction and specific properties. To overcome this limitation for the robust SSA algorithms and to be able to obtain out-of-sample forecasts using a robust SSA algorithm, a three stages approach can be conducted:

(i) Use the robust SSA algorithm to obtain a robust approximation for the signal in the trajectory matrix; i.e., conduct the two stages of the robust SSA algorithms, decomposition (using the robust SVD algorithm) and reconstruction, to obtain the noise free (i.e., the signal) trajectory matrix $\widetilde{\mathbf{X}}$;
(ii) Apply the standard SVD to the matrix $\widetilde{\mathbf{X}}$ obtained in (i) and obtain U_j^∇, the vector of the first $L-1$ components of U_j and π_j, the last component of the vector U_j, $j = 1, \cdots, r$. Then, we can write the coefficient vector $\hat{a}$ as

$$\hat{a} = (\hat{a}_{L-1}, \cdots, \hat{a}_1)' = \frac{1}{1-\gamma^2} \sum_{j=1}^{r} \pi_j U_j^\nabla, \qquad (11)$$

where $\gamma^2 = \sum_{j=1}^{r} \pi_j^2$.
(iii) The h-steps-ahead out-of-sample recurrent robust SSA forecasts $\hat{y}_{N+1}, \ldots, \hat{y}_{N+h}$, can be obtained as

$$\hat{y}_t = \begin{cases} \tilde{y}_t, & \text{for } t = 1, \cdots, N \\ \sum_{j=1}^{L-1} \hat{a}_j \hat{y}_{t-j}, & \text{for } t = N+1, \cdots, N+h \end{cases} \qquad (12)$$

where $\tilde{y}_1, \ldots, \tilde{y}_N$, are the fitted values for the reconstructed time series, as obtained from the robust SSA algorithm in (i).

2.6. Accuracy Measures

There are several methods and measures for assessing model accuracy based on the behavior of model errors. Here, there are two types of errors:

- Sample errors, called tuning errors;
- Out-of-sample errors, called forecast errors.

Typically, the root mean squared error (RMSE) is used as a criterion for accessing the precision of a model. The RMSE to investigate the quality of the model fit can be written as:

$$RMSE = \sqrt{\frac{1}{N} \sum_{t=1}^{N} (y_t - \tilde{y}_t)^2}, \qquad (13)$$

where y_t are the observed values and $\widetilde{y}_t$ the fitted values by the considered model/algorithm (i.e., ARIMA, SSA, robust SSA).

To investigate the forecasting accuracy, let us assume that the last g observations are used as a reference (i.e., as test set). Let $N_0 = N - h - g$. The RMSE to investigate the quality of the forecasting model can be written as:

$$RMSE = \sqrt{\frac{1}{g} \sum_{t=N_0+h+1}^{N} (y_t - \widetilde{y}_t)^2}, \qquad (14)$$

where y_t are the last g observed values and $\widetilde{y}_t$ the respective h-steps-ahead forecast values.

3. Results and Discussion

In this section, comparisons are made between the classical ARIMA model, the classical SSA algorithm, and the robust SSA algorithms, in terms of computational time and accuracy for model fit and model forecast. These comparisons for decomposing and forecasting time series are done by considering a simulation example and the time series of six mutual investment funds.

Table 1 shows the descriptive statistics for the six mutual investment funds, including the minimum, maximum, and mean returns, being clear that Alaska Black is the fund that shows the largest variation and with the highest mean daily return. On the other end there are Gavea Macro and SPX Nimitz, which show the smallest variations among the considered funds, and low mean returns.

In addition to the descriptive measures, Figure 1 shows the behavior of the six investment funds over time. From these plots, it is possible to observe that all funds have an overall growing tendency, with similar patterns for Gavea Macro and SPX Nimitz.

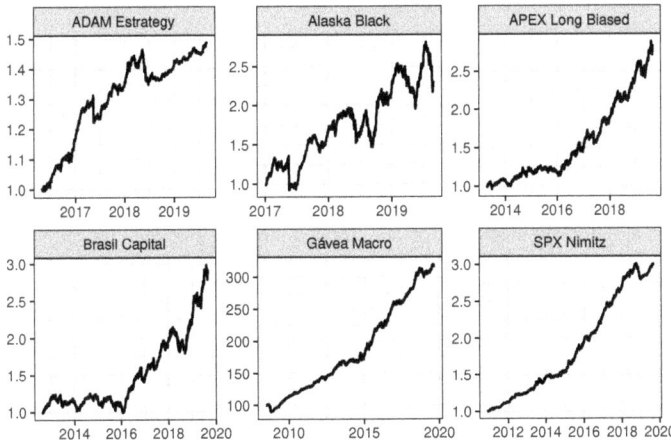

Figure 1. Time series for the returns of the six mutual investment funds, ADAM Strategy, Alaska Black, APEX Long Biased, Brasil Capital, Gávea Macro and SPX Nimitz, from left to right and from top to bottom. The vertical axes show the quota values; i.e., the total net assets of a fund divided by the total number of quotas existing.

Table 1. Descriptive measures for returns of the six mutual investment funds.

Investment Fund	Minimum	Mean	Maximum	Standard deviation
ADAM Strategy	−6.26%	0.05%	1.63%	0.0045%
Alaska Black	−29.62%	0.16%	9.80%	0.0240%
APEX Long Biased	−8.60%	0.07%	3.72%	0.0085%
Brasil Capital	−7.55%	0.07%	3.42%	0.0094%
Gavea Macro	−2.22%	0.04%	2.36%	0.0033%
SPX Nimitz	−1.92%	0.05%	1.42%	0.0030%

3.1. Model Fit

The models/algorithms under comparison for model fit are: (i) ARIMA, (ii) SSA, (iii) robust SSA based on the L_1 norm (RLSSA), and (iv) robust SSA based on the Huber function (RHSSA).

The parameters of the ARIMA model for each of the six mutual investment funds were estimated with the function "auto.arima" from the R package "forecast" [36].

For the SSA and robust SSA algorithms, there are two choices to be made by the researcher: (i) the window length L; and (ii) the number of eigentriples used for reconstruction r. Three values of L were chosen for each time series, as defined in Table 2—$L_1 = N/20$, $L_2 = N/2$, and L_p—being the L_p obtained from the periodogram, based on the largest cycle for each time series [37] (i.e., about one trimester for ADAM Strategy, one semester for Alaska Black, one year for APEX Long Biased, one quadrimeter for Brasil Capital, one quadrimeter for Gavea Macro, and one quadrimester for SPX Nimitz), and N being the time series length. The choice of the number of eigentriples used for reconstruction r, for each of the considered window lengths and each of the time series, was done by taking into consideration the the w-correlations among components [5]. Figure 2 shows the w-correlation matrices for each of the six mutual investment funds, considering an window length $L = N/20$, and Figure A1 of the appendix shows the w-correlation matrices for each of the six mutual investment funds, considering an window length $L = N/2$. The w-correlation matrices can be obtained with the function "wcor" of the R package "Rssa" [38] and the number of eigentriples r should be chosen in order to maximize the separability between signal and noise components; i.e., maximize the w-correlation among signal components, maximize the w-correlation among noise components, and minimize the w-correlation between signal and noise components. A summary of the number of eigentriples used for the reconstruction of each time series for each of the window length considered can be seen in Table 2.

Since one of the objectives in SSA is to decompose the original time series into interpretable components such as trend and seasonality, plus the noise component that is then discarded, Figure 3 shows the original time series for the Alaska Black mutual investment fund, its trend component (sum of individual trend components), its seasonal component (sum of individual seasonal components), and its residuals (sum of the remaining components associated to noise), considering an window length $L = N/20 = 33$ and $r = 12$ eigentriples for reconstruction. Similar SSA decompositions for ADAM Strategy, APEX Long Biased, Brasil Capital, ADAM Strategy, Gavea Macro, and SPX Nimitz—considering the values of window length L_1 and r_1 eigentriples used for reconstruction, as defined in Table 2—can be found in Figures A2–A6 of the appendix, respectively.

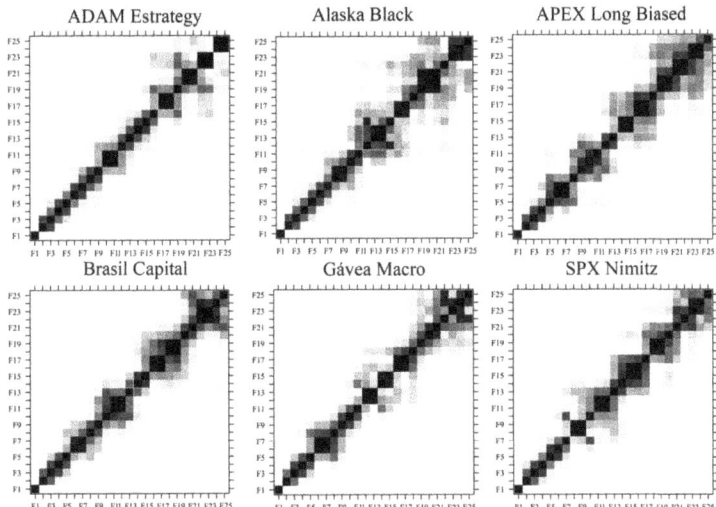

Figure 2. W-correlation matrices for each of the six mutual investment funds, ADAM Strategy, Alaska Black, APEX Long Biased, Brasil Capital, Gávea Macro and SPX Nimitz, from left to right and from top to bottom, considering an window length $L = N/20$.

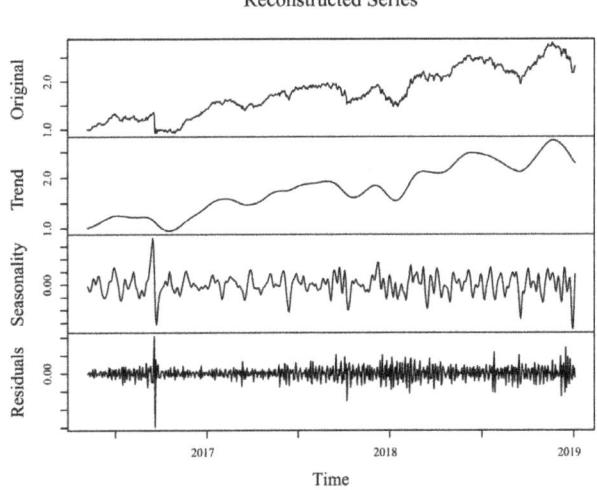

Figure 3. Decomposition of the original time series for the Alaska Black mutual investment fund (top panel), with a trend component (sum of individual trend components, second panel), a seasonal component (sum of individual seasonal components, third panel), and a residual (sum of the remaining components associated to noise, bottom panel), considering an window length $L = N/20 = 33$ and $r = 12$ eigentriples for reconstruction.

Table 2. Window length L_1, L_2, and L_p, and number of eigentriples r considered for model fit and model forecast for each of the mutual investment funds.

Investment Fund	n	L_1	r_1	L_2	r_2	L_p	r_p
ADAM Strategy	838	41	17	419	18	60	13
Alaska Black	666	33	12	333	11	125	8
APEX Long Biased	1604	80	14	802	11	250	11
Brasil Capital	1760	88	12	880	12	80	13
Gavea Macro	2809	140	12	1404	12	80	12
SPX Nimitz	2199	109	8	1099	8	80	11

In order to evaluate and compare the ability for model fit using the four models, ARIMA, SSA, robust SSA based on the L_1 norm (RLSSA), and robust SSA based on the Huber function (RHSSA), the root mean square error (RMSE) was calculated for each time series. Table 3 shows the RMSE for model fit by each of the four models applied to each of the six mutual investment funds, considering a window length $L = N/2$ (Table 2). Table 4 shows the RMSE for model fit by each of the four models applied to each of the six mutual investment funds, considering a window length $L = N/20$ (Table 2). Table 5 shows the RMSE for model fit by each of the four models applied to each of the six mutual investment funds, considering a window length obtained based on the largest cycle for each time series (Table 2). From the analyzes of these tables, we can conclude that the ARIMA model shows an overall better performance when the window length in the SSA related algorithms is set to be half of the time series (Table 3). However, when the window length is set to be $L_1 = N/20$ or L_p (i.e., equal to the length of the largest cycle), the classical SSA provides the best results, while the ARIMA model and the robust SSA algorithms alternate for the second best performances. For all choices of window length, the two robust SSA algorithms behaved similarly.

Table 3. Root mean square error for each of the six mutual investment funds, considering each of the four models, ARIMA, SSA, robust SSA based on the L_1 norm (RLSSA), and robust SSA based on the Huber function (RHSSA), for the window length $L_2 = N/2$ and considering r_2 engentriples for reconstruction as defined in Table 2.

Investment Fund	ARIMA	SSA	RLSSA	RHSSA
ADAM Strategy	0.0057	0.0075	0.0088	0.0076
Alaska Black	0.0402	0.0450	0.0508	0.0476
APEX Long Biased	0.0160	0.0294	0.0318	0.0320
Brasil Capital	0.0170	0.0338	0.0429	0.0346
Gavea Macro	0.6756	1.9758	2.1486	2.0016
SPX Nimitz	0.0063	0.0197	0.0239	0.0207

Table 4. Root mean square error for each of the six mutual investment funds, considering each of the four models, ARIMA, SSA, robust SSA based on the L_1 norm (RLSSA), and robust SSA based on the Huber function (RHSSA), for the window length $L_1 = N/20$ and considering r_1 engentriples for reconstruction as defined in Table 2.

Investment Fund	ARIMA	SSA	RLSSA	RHSSA
ADAM Strategy	0.0057	0.0024	0.0034	0.0034
Alaska Black	0.0402	0.0190	0.0244	0.0234
APEX Long Biased	0.0160	0.0107	0.0124	0.0116
Brasil Capital	0.0170	0.0124	0.0143	0.0133
Gavea Macro	0.6756	0.6508	0.7716	0.7432
SPX Nimitz	0.0063	0.0066	0.0078	0.0077

Table 5. Root mean square error for each of the six mutual investment funds, considering each of the four models, ARIMA, SSA, robust SSA based on the L_1 norm (RLSSA), and robust SSA based on the Huber function (RHSSA), for the window length L_p (i.e., the length of the largest cycle) and considering r_p engentriples for reconstruction as defined in Table 2.

Investment Fund	ARIMA	SSA	RLSSA	RHSSA
ADAM Strategy	0.0057	0.0038	0.0046	0.0045
Alaska Black	0.0402	0.0415	0.0482	0.0459
APEX Long Biased	0.0160	0.0185	0.0196	0.0190
Brasil Capital	0.0170	0.0123	0.0139	0.0132
Gavea Macro	0.6756	0.5049	0.5997	0.5986
SPX Nimitz	0.0063	0.0049	0.0058	0.0057

Tables 6–8 show the computational times for each combination of model/algorithm and mutual investment fund, as presented in Tables 3–5, respectively. From the analyzes of these tables, we can conclude that the best performance was obtained by the ARIMA and SSA algorithms. The computational time, for the classic and robust SSA algorithms, increases with the increase of the length L. Moreover, for larger trajectory matrices (i.e., considering $L = N/2$) the robust SSA algorithm based on the Huber function has a lower computational time than the robust SSA algorithm based on the L_1 norm (Table 6). However, when the trajectory matrices are more rectangular (i.e., considering $L = N/20$, Table 7, or $L = L_p$, Table 8), the robust SSA algorithm based on the L_1 norm has a much lower computational time (comparable to the ARIMA and SSA computational times) than the robust SSA algorithm based on the Huber function).

Table 6. Computational time, in minutes, for each of the six mutual investment funds, considering each of the four models, ARIMA, SSA, robust SSA based on the L_1 norm (RLSSA), and robust SSA based on the Huber function (RHSSA), for the window length $L_2 = N/2$ and considering r_2 engentriples for reconstruction as defined in Table 2.

Investment Fund	ARIMA	SSA	RLSSA	RHSSA
ADAM Strategy	0.0010	0.0052	15.563	14.232
Alaska Black	0.0018	0.0042	7.5859	6.8834
APEX Long Biased	0.0175	0.0320	195.27	61.031
Brasil Capital	0.0226	0.0366	287.80	83.821
Gavea Macro	0.0057	0.1584	1605.2	632.84
SPX Nimitz	0.0022	0.0618	616.75	120.83

Table 7. Computational time, in minutes, for each of the six mutual investment funds, considering each of the four models, ARIMA, SSA, robust SSA based on the L_1 norm (RLSSA), and robust SSA based on the Huber function (RHSSA), for the window length $L_1 = N/20$ and considering r_1 engentriples for reconstruction as defined in Table 2.

Investment Fund	ARIMA	SSA	RLSSA	RHSSA
ADAM Strategy	0.0010	0.0025	0.1257	68.384
Alaska Black	0.0018	0.0031	0.0669	16.794
APEX Long Biased	0.0175	0.0039	1.2952	530.43
Brasil Capital	0.0226	0.0048	1.9145	629.79
Gavea Macro	0.0057	0.0088	10.823	1441.1
SPX Nimitz	0.0022	0.0050	3.7450	375.29

Table 8. Computational time, in minutes, for each of the six mutual investment funds, considering each of the four models, ARIMA, SSA, robust SSA based on the L_1 norm (RLSSA), and robust SSA based on the Huber function (RHSSA), for the window length L_p (i.e., the length of the longest cycle) and considering r_p engentriples for reconstruction as defined in Table 2.

Investment Fund	ARIMA	SSA	RLSSA	RHSSA
ADAM Strategy	0.0010	0.0024	0.3371	65.149
Alaska Black	0.0018	0.0026	1.6994	3.3270
APEX Long Biased	0.0175	0.0078	26.826	115.14
Brasil Capital	0.0226	0.0099	2.0020	804.16
Gavea Macro	0.0057	0.0126	3.4485	1718.4
SPX Nimitz	0.0022	0.0078	3.4937	905.16

Figure 4 shows the original time series and the model fit by the SSA model with $L = N/20$ and by the ARIMA model. We can confirm that both fits are almost overlapped and very near to the original time series, which was expected from the small RMSE showed in Table 4.

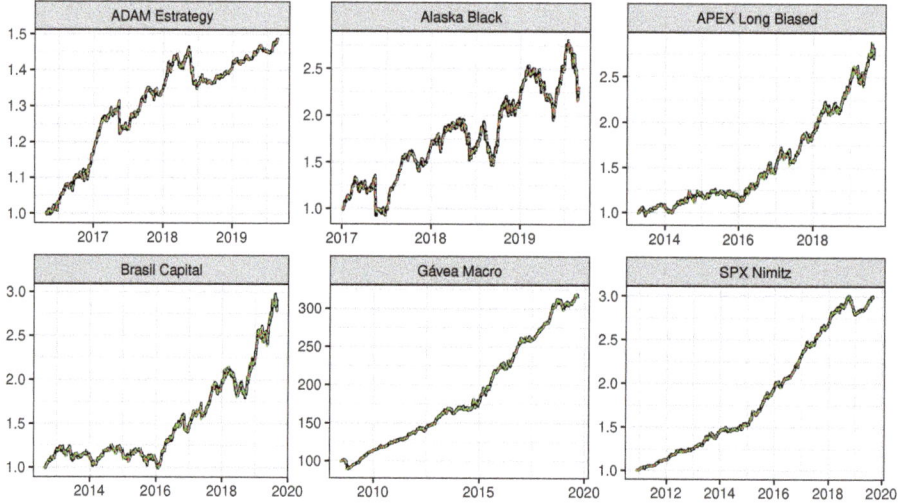

Figure 4. Original time series (black line); smoothed time series after applying the SSA considering $L = N/20$, with the number of eigentriples r as they are defined in Table 2 (red line); and model fit by the ARIMA model (green line), for each of the six mutual investment funds, ADAM Strategy, Alaska Black, APEX Long Biased, Brasil Capital, Gávea Macro and SPX Nimitz, from left to right and from top to bottom. The vertical axes show the quota values.

3.2. Model Forecasting

In this section we compare the forecasting abilities of ARIMA, SSA with $L = N/2$, SSA with $L = N/20$, SSA with $L = L_p$ based on the largest cycle for each time series, and robust SSA based on the L_1 norm with $L = N/20$ and L_p. The decision for not considering the robust SSA algorithm based on the Huber function was because of its similarity in terms of RMSE with the robust SSA based on the L_1 norm (Tables 3–5) and the much higher computational time (Tables 6–8). A similar argument was considered for not presenting the results for the robust SSA algorithm based on the L_1 norm with $L = N/2$.

Table 9 shows the RRMSE for model forecasting for each of the six mutual investment funds, considering each of the four models, ARIMA, SSA with $L = N/2$, SSA with $L = N/20$, SSA with $L = L_p$, and robust SSA based on the L_1 norm (RLSSA) with $L = N/20$ and L_p, considering the window length and engentriples used for reconstruction as defined in Table 2. These values were obtained based on the forecasting of the $g = 12$ observations from each time series, obtained for one, five, and ten steps ahead out-of-sample forecast; i.e., one day ahead, one week ahead, and two weeks ahead.

The overall best performance was obtained with the classic SSA algorithm that considers a lower value for the window length, either $L = N/20$ or $L = L_p$, followed closely by ARIMA and the robust SSA algorithm based on the L_1 norm. The ARIMA model obtained the best performance in three cases for one-step-ahead forecasting, and the robust SSA algorithm based on the L_1 norm with $L = N/20$ yielded the best performance in a couple of time series for five-steps-ahead forecasting. As expected, the RMSE shows an overall increase when increasing the number of steps ahead to be forecast. A possible justification for the similarity between the SSA and robust SSA algorithm can be explained by the possible lack of outliers in the data. Table 10 shows the computational time for model forecasting for each of the six mutual investment funds, considering each of the five models shown in Table 9. As expected, after analyzing the computational times for model fit (Tables 6–8), the best performance in terms of computational time for model forecasting was obtained by the the ARIMA and SSA (with lower values for the window length) models and the worse by the robust SSA algorithm based on the L_1 norm.

Table 9. Root mean square error for model forecasting for each of the six mutual investment funds, considering the models ARIMA, SSA with $L = N/2$, SSA with $L = N/20$, SSA with L_p, robust SSA based on the L_1 norm (RLSSA) with $L = N/20$, and RSSA with L_p, and their respective engentriples, as defined in Table 2.

Investment Fund	ARIMA	SSA $\frac{N}{2}$	SSA $\frac{N}{20}$	SSA L_p	RLSSA $\frac{N}{20}$	RLSSA L_p
one-step-ahead						
ADAM Strategy	0.0027	0.0036	0.0029	0.0047	0.0048	0.0048
Alaska Black	0.0712	0.2118	0.0638	0.1357	0.1138	0.178
APEX Long Biased	0.0426	0.1778	0.0544	0.0646	0.0663	0.0576
Brasil Capital	0.0436	0.0496	0.0590	0.0573	0.0545	0.0512
Gavea Macro	1.1670	2.3104	1.5536	1.2571	1.1532	1.6582
SPX Nimitz	0.0081	0.0278	0.0061	0.0061	0.0061	0.0074
five-step-ahead						
ADAM Strategy	0.0056	0.0047	0.0058	0.0038	0.0089	0.0057
Alaska Black	0.2031	0.2990	0.1800	0.1848	0.2120	0.2365
APEX Long Biased	0.1184	0.1965	0.0578	0.0724	0.0830	0.0577
Brasil Capital	0.1277	0.0481	0.0704	0.0669	0.0693	0.0615
Gavea Macro	2.4007	2.8585	2.0509	1.8165	1.2367	2.3534
SPX Nimitz	0.0275	0.0292	0.0075	0.0077	0.0076	0.0108
ten-step-ahead						
ADAM Strategy	0.0057	0.0087	0.0055	0.0086	0.0111	0.0091
Alaska Black	0.2958	0.3795	0.2201	0.0263	0.3311	0.3329
APEX Long Biased	0.2012	0.2162	0.0929	0.0706	0.1020	0.0555
Brasil Capital	0.1998	0.0460	0.1100	0.1101	0.0844	0.0700
Gavea Macro	3.2948	3.6784	2.6578	2.7515	2.8015	2.5541
SPX Nimitz	0.0467	0.0314	0.0166	0.0120	0.0103	0.0170

Table 10. Computational time, in minutes, for the model for each of the six mutual investment funds, considering the models ARIMA, SSA with $L = N/2$, SSA with $L = N/20$, SSA with L_p, robust SSA based on the L_1 norm (RLSSA) with $L = N/20$, and RSSA with L_p, and their respective engentriples, as defined in Table 2.

Investment Fund	ARIMA	SSA $\frac{N}{2}$	SSA $\frac{N}{20}$	SSA L_p	RLSSA $\frac{N}{20}$	RLSSA L_p
			one-step-ahead			
ADAM Strategy	0.0123	0.1231	0.0277	0.0253	39.768	58.804
Alaska Black	0.0222	0.0549	0.0183	0.0267	30.516	45.948
APEX Long Biased	0.2106	0.4888	0.0613	0.1752	176.18	692.18
Brasil Capital	0.2712	0.8409	0.0644	0.0648	212.60	295.20
Gavea Macro	0.0681	2.7338	0.1687	0.0976	698.34	857.58
SPX Nimitz	0.0265	1.2750	0.0774	0.0740	420.23	584.59
			five-step-ahead			
ADAM Strategy	0.0129	0.0879	0.0222	0.0256	44.019	56.524
Alaska Black	0.0181	0.0531	0.0150	0.0246	32.351	58.674
APEX Long Biased	0.2203	0.4909	0.0682	0.1840	250.85	675.41
Brasil Capital	0.2620	0.6400	0.0764	0.0675	314.59	290.72
Gavea Macro	0.0702	2.7839	0.1460	0.1034	988.02	858.96
SPX Nimitz	0.0348	1.3029	0.0805	0.0755	537.94	572.93
			ten-step-ahead			
ADAM Strategy	0.0089	0.0924	0.0344	0.0261	45.729	46.518
Alaska Black	0.0156	0.0469	0.0184	0.0263	28.140	54.289
APEX Long Biased	0.1775	0.5057	0.0678	0.1906	198.27	638.13
Brasil Capital	0.2103	0.6628	0.0726	0.0679	244.06	307.30
Gavea Macro	0.0532	2.6942	0.1724	0.1060	761.49	520.66
SPX Nimitz	0.0243	1.2388	0.0634	0.0786	407.61	316.60

3.3. Simulation Example

To verify the hypothesis raised in the previous subsection that the similarity between the results from SSA and the robust SSA algorithm can be due to the lack of outliers in the time series, in this subsection we present a simulation example where the methods are compared while analyzing a time series contaminated with outlying observations. The synthetic data were obtained by generating random values from the following function, and then we transformed them into a time series (right-hand plot in Figure 5):

$$f(t) = \exp\{0.02t + 0.5\sin(2\pi t/5)\} + \epsilon, \quad t = 1,...,100,$$

where ϵ is the noise generated from the $N(0, 0.1)$. A total of 100 simulated time series were considered.

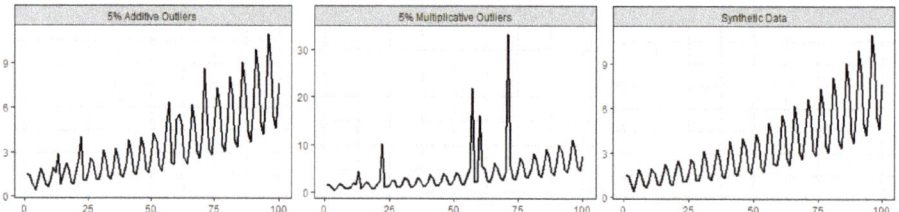

Figure 5. Synthetic data without contamination (**right**), data with 5% additive outliers (**left**), and data with 5% multiplicative outliers (**center**). The vertical axes show the simulated value and the horizontal axes show the index of the simulated observation.

The data contamination, for illustration purposes, was made by considering additive outliers and magnitude increase outliers in the following way:

- *Additive outliers:* 2%, 5%, and 10% of the time points y_i are randomly chosen to be replaced by $2 + y_i$; i.e., the values of y_i are increased by a constant value of 2, resulting in a mild contamination scenario (e.g., (left-hand plot in Figure 5));
- *Magnitude increase:* 2%, 5%, and 10% of the time points y_i are randomly chosen to be replaced by $5 \times y_i$; i.e., the time point magnitude of y_i is increased by a factor of 5, resulting in an a quite extreme contamination scenario (e.g., central plot in Figure 5).

Table 11 shows the mean of the root mean square errors for model fit, computed for each of the four models, ARIMA, SSA, robust SSA based on the L_1 norm, and robust SSA based on the Huber function, for the simulated data, based on 100 runs, using $L = 24$ and $r = 5$, and considering both contamination scenarios with 2, 5, and 10% outliers. As expected, when there is no data contamination, the classic SSA model is the most appropriated. For the mild contamination scenario with additive outliers, the robust SSA algorithms outperform both ARIMA and SSA models, the better performance being more evident when the percentage of the outliers increases. For the more extreme contamination scenario with multiplicative outliers, a similar patters was obtained, the RLSSA being the best robust algorithm, in this simulation example.

Appendix B includes a second simulation scenario where robust SSA algorithm based on the Huber function (RHSSA) outperforms the classic ARIMA and SSA models and the robust SSA algorithm based on the L_1 norm (RLSSA).

Table 12 shows mean of the root mean square errors for model forecasting ($M = 1, 5, 10$ steps-ahead), computed for each of ARIMA, SSA, and robust SSA based on the L_1 norm, for the simulated data, based on 100 runs, using $L = 24$ and $r = 5$. The results for the robust SSA based on the Huber function were not included because of their computational cost and out-performance when compared with the robust SSA based on the L_1 norm. Again, as expected, the SSA model yielded the best performance for no data contamination. For scenarios with data contamination, the best performance was obtained by the robust SSA forecasting algorithm, with a very large decrease in RMSE in many scenarios.

Table 11. Mean of the root mean square errors for model fit, computed for each of the four models, ARIMA, SSA, robust SSA based on the L_1 norm, and robust SSA based on the Huber function, for the simulated data, based on 100 runs, using $L = 24$ and $r = 5$.

% of Data Contamination	Shift	ARIMA	SSA	RLSSA	RHSSA
0%	-	0.715	0.083	0.109	0.127
2%	$y_i + 2$	0.612	0.149	0.119	0.133
5%	$y_i + 2$	0.640	0.236	0.134	0.148
10%	$y_i + 2$	0.675	0.364	0.179	0.232
2%	$y_i \times 5$	1.206	1.235	0.126	0.389
5%	$y_i \times 5$	1.828	2.289	0.167	0.929
10%	$y_i \times 5$	2.384	3.404	0.425	1.463

Table 12. Mean of the root mean square errors for model forecasting ($M = 1, 5$, and 10 steps-ahead), computed for each of the four models, ARIMA, SSA, robust SSA based on the L_1 norm, and robust SSA based on the Huber function, for the simulated data, based on 100 runs, using $L = 24$ and $r = 5$.

M	% of Cont.	Shift	Method ARIMA	SSA	RLSSA
M = 1	0%	-	1.685	0.125	0.245
	5%	$y_i + 2$	0.843	0.475	0.330
	10%	$y_i + 2$	0.793	0.596	0.426
	5%	$y_i \times 5$	3.960	8.461	0.358
	10%	$y_i \times 5$	4.359	9.692	0.652
M = 5	0%	-	1.631	0.122	0.222
	5%	$y_i + 2$	0.984	0.475	0.307
	10%	$y_i + 2$	0.768	0.586	0.413
	5%	$y_i \times 5$	3.789	538.447	0.323
	10%	$y_i \times 5$	3.853	17.670	0.720
M = 10	0%	-	1.381	0.127	0.244
	5%	$y_i + 2$	1.320	0.601	0.358
	10%	$y_i + 2$	1.148	0.698	0.474
	5%	$y_i \times 5$	3.486	22.695 *	4.015
	10%	$y_i \times 5$	3.694	622.783	2.320

* 10% trimed mean. The mean value is 1.566×10^6.

4. Conclusions

In this paper we considered the problem of model fit and model forecasting in time series. In particular, we analyzed six mutual investment funds. Following up on [23], who proposed a robust SSA algorithm by replacing the standard least squares SVD by a robust SVD algorithm based on the L_1 norm [24] for model fit, we proposed another robust SSA algorithm where the robust SVD based on the Huber function is considered [25]. Moreover, we propose a forecasting strategy for the robust SSA algorithms, based on the linear recurrent SSA forecasting algorithm.

Comparisons were made between the classical SSA algorithm, the robust SSA algorithms, and the classical ARIMA model, both in terms of computational time and accuracy for model fit and model forecast. Those comparisons were made by using daily observations of six mutual investment funds, and a synthetic data set where the time series were contaminated with outlying observations.

For model fit of the six mutual investment funds, the best results were obtained for the SSA model when the window length L was set to be equal to the length of the time series divided by 20, or when the window length is defined as the length of the largest cycle in the time series. The ARIMA model and the robust SSA algorithms alternated for the second best performance. For model forecasting of the six mutual investment funds, the best overall performance was obtained for the classic SSA model considering a lower value for the window length, $L = N/20$ or L_p, followed closely by the ARIMA model and the robust SSA algorithm based on the L_1 norm.

Based on the similarity between the results from the classic SSA model and the robust SSA algorithms, both for model fit and model forecasting, one may assume that the time series data from the six mutual investment funds had no or little data contamination. To access that hypothesis and to better illustrate the usefulness of the robust SSA algorithms, using a scenario with known and controlled outliers, a simulation study and its results were presented in this article. For both mild and and more extreme contamination scenarios, the robust SSA algorithms clearly outperformed the classical AMMI and SSA models, both for model fit and for model forecasting. Another important advantage of the robust SSA algorithms, because of their use of the robust SVD, is that they allow for missing values.

In terms of computational time, the SSA model gives the best performance, the robust algorithms being the most time consuming. A possible future development to reduce the computational time in the robust SSA algorithms is to consider a similar strategy as in [39], where a randomized SVD algorithm was used to speed up the SSA algorithm.

The usefulness of the proposed approach, regarding the forecasting case, can be assessed based on forecasting competitions (e.g., [40]) or large scale forecasting studies (see, e.g., [41]).

The methodology and results presented in this paper are of great generality and can be applied to other time series applications.

Author Contributions: Conceptualization, P.C.R.; Formal analysis, P.C.R., J.P. and P.M.; Methodology, P.C.R. and M.K.; Software, P.C.R., J.P., P.M. and M.K.; Supervision, P.C.R.; Visualization, J.P. and P.M.; Writing—original draft, P.C.R., J.P., P.M. and M.K.; Writing—review and editing, P.C.R., J.P. and M.K. All authors have read and agreed to the published version of the manuscript.

Funding: This research received no external funding.

Acknowledgments: The authors thank the associate editor and three anonymous reviewers for providing helpful suggestions which contributed to the improvement of the paper.

Conflicts of Interest: The authors declare no conflict of interest.

Abbreviations

The following abbreviations are used in this manuscript:

ARIMA	autoregressive integrated moving average
SSA	singular spectrum analysis
SVD	singular value decomposition
RHSSA	robust SSA algorithm based on the Huber function
RLSSA	robust SSA algorithm based on the L_1 norm
RMSE	root mean squared error

Appendix A

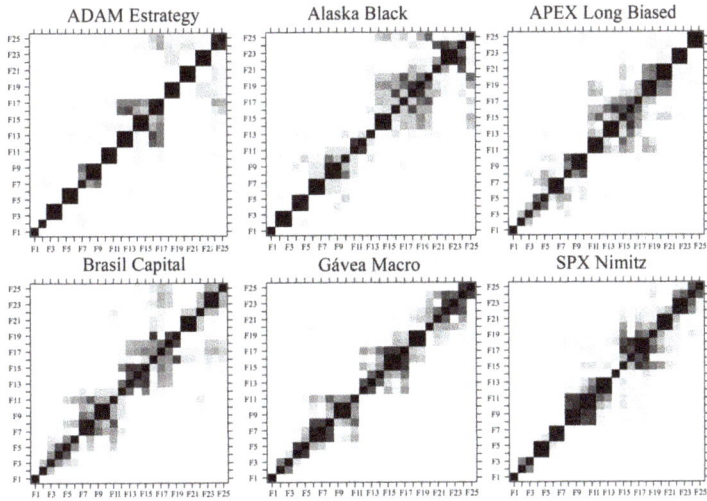

Figure A1. W-correlation matrices for each of the six mutual investment funds, ADAM Strategy, Alaska Black, APEX Long Biased, Brasil Capital, Gávea Macro and SPX Nimitz, from left to right and from top to bottom, considering an window length $L = N/2$.

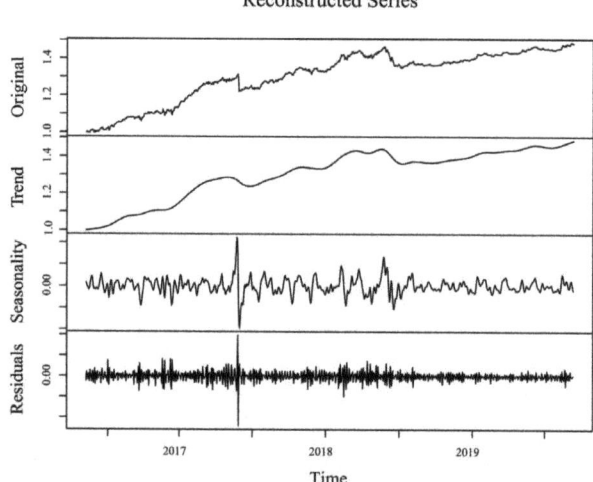

Figure A2. Decomposition of the original time series for the ADAM Strategy mutual investment fund (top panel), with a trend component (sum of individual trend components, second panel), a seasonal component (sum of individual seasonal components, third panel), and a residual (sum of the remaining components associated to noise, bottom panel), considering an window length $L = N/20 = 41$ and $r = 17$ eigentriples used for reconstruction.

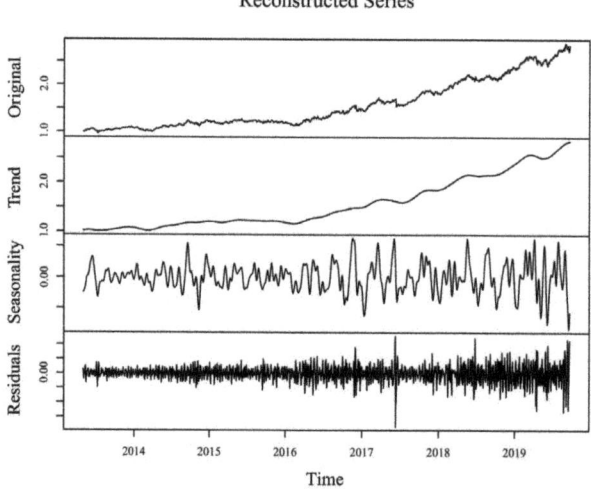

Figure A3. Decomposition of the original time series for the APEX Long Biased mutual investment fund (top panel), with a trend component (sum of individual trend components, second panel), a seasonal component (sum of individual seasonal components, third panel), and a residual (sum of the remaining components associated to noise, bottom panel), considering an window length $L = N/20 = 80$ and $r = 14$ eigentriples used for reconstruction.

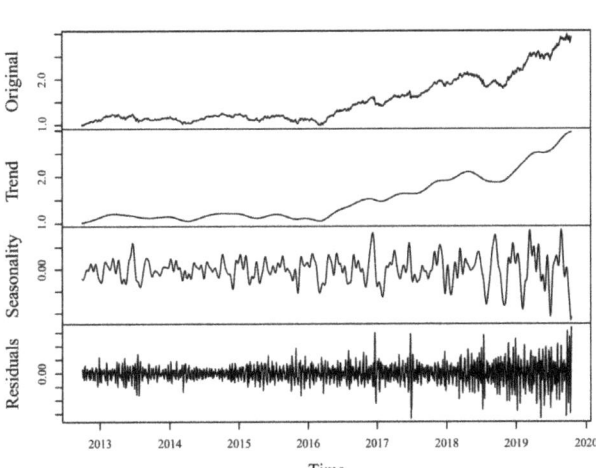

Figure A4. Decomposition of the original time series for the Brasil Capital mutual investment fund (top panel), with a trend component (sum of individual trend components, second panel), a seasonal component (sum of individual seasonal components, third panel), and a residual (sum of the remaining components associated to noise, bottom panel), considering an window length $L = N/20 = 88$ and $r = 12$ eigentriples used for reconstruction.

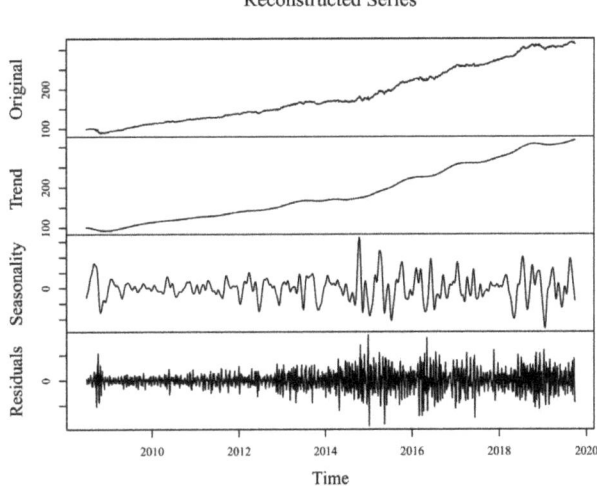

Figure A5. Decomposition of the original time series for the Gavea Macro mutual investment fund (top panel), with a trend component (sum of individual trend components, second panel), a seasonal component (sum of individual seasonal components, third panel), and a residual (sum of the remaining components associated to noise, bottom panel), considering an window length $L = N/20 = 140$ and $r = 12$ eigentriples used for reconstruction.

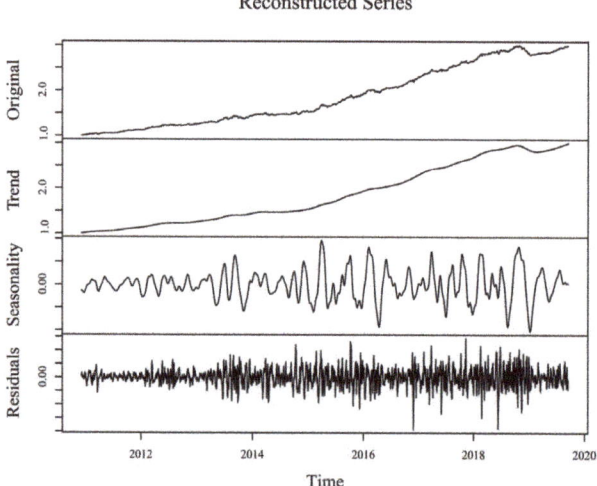

Figure A6. Decomposition of the original time series for the SPX Nimitz mutual investment fund (top panel), with a trend component (sum of individual trend components, second panel), a seasonal component (sum of individual seasonal components, third panel), and a residual (sum of the remaining components associated to noise, bottom panel), considering an window length $L = N/20 = 109$ and $r = 8$ eigentriples used for reconstruction.

Appendix B

A second synthetic dataset was obtained by generating random values from the following function and then transforming them into a time series:

$$f(t) = \cos(2\pi wt + \phi) + \epsilon, \quad t = 1, ..., 100,$$

with $w = 3/8$, $\phi = \pi/8$ and ϵ the noise generated from the $N(0, 0.1)$ (right-hand side of Figure A7). A total of 100 simulated time series were considered.

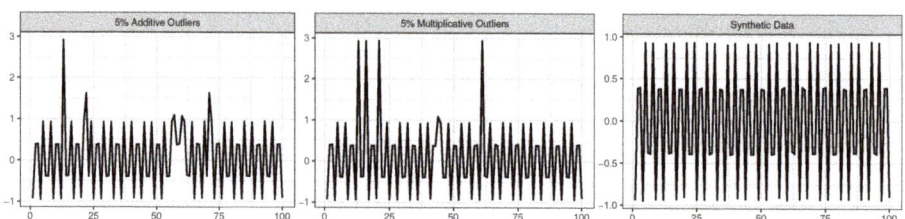

Figure A7. Synthetic data without contamination (**right**), data with 5% additive outliers (**left**), and data with 5% multiplicative outliers (**center**). The vertical axes show the simulated value and the horizontal axes show the index of the simulated observation.

The data contamination was done in the same manner as described before. An example of 5% additive outliers scenario can be found on the left-hand plot of Figure A7, and an example of 5% multiplicative outliers scenario can be found on the central plot of Figure A7. The results for the root mean square errors for model fit, computed for each of the four models, ARIMA, SSA, robust SSA based on the L_1 norm, and robust SSA based on the Huber function, can be found in Table A1.

Table A1. Mean of the root mean square errors for model fit, computed for each of the four models, ARIMA, SSA, robust SSA based on the L_1 norm, and robust SSA based on the Huber function, for the simulated data, based on 100 runs, using $L = 24$ and $r = 2$.

% of Data Contamination	Shift	ARIMA	SSA	RLSSA	RHSSA
0%	-	0.1045	0.0097	0.0099	0.0104
2%	$y_i + 2$	0.277	0.071	0.058	0.019
5%	$y_i + 2$	0.351	0.113	0.096	0.032
10%	$y_i + 2$	0.465	0.161	0.197	0.055
2%	$y_i \times 5$	0.279	0.108	0.026	0.018
5%	$y_i \times 5$	0.386	0.193	0.052	0.040
10%	$y_i \times 5$	0.484	0.338	0.075	0.098

References

1. Varga, G.; Wengert, M. A industria de fundos de investimentos no Brasil. *Rev. Econ. Adm.* **2011**, *10*, 66–109. [CrossRef]
2. Maestri, C.O.N.M.; Malaquias, R.F. Exposition to factors of the investment funds market in Brazil. *Rev. Contab. Financ.* **2017**, *28*, 61–76. [CrossRef]
3. Broomhead, D.S.; King, G.P. Extracting qualitative dynamics from experimental data. *Phys. D Nonlinear Phenom.* **1986**, *20*, 217–236. [CrossRef]
4. Fraedrich, K. Estimating the Dimensions of Weather and Climate Attractors. *J. Atmos. Sci.* **1986**, *43*, 419–432. [CrossRef]
5. Golyandina, N.; Nekrutkin, V.; Zhigljavsky, A. *Analysis of Time Series Structure: SSA and Related Techniques*; Chapman & Hall/CRC: New York, NY, USA, 2001.
6. Golyandina, N.; Zhigljavsky, A. *Singular Spectrum Analysis for Time Series*; Springer Science and Business Media: Berlin/Heidelberger, Germany, 2013.
7. Hassani, H. Singular spectrum analysis: Methodology and comparison. *J. Data Sci.* **2007**, *5*, 239–257.
8. Hassani, H.; Zhigljavsky, A. Singular spectrum analysis: methodology and application to economics data. *J. Syst. Sci. Complex.* **2009**, *22*, 372–394. [CrossRef]
9. Mahmoudvand, R.; Alehosseini, F.; Rodrigues, P.C. Forecasting mortality rate by singular spectrum analysis. *RevStat-Stat. J.* **2015**, *13*, 193–206.
10. Mahmoudvand, R.; Rodrigues, P.C. Missing value imputation in time series using singular spectrum analysis. *Int. J. Energy Stat.* **2016**, *4*, 1650005. [CrossRef]
11. Groth, A.; Ghil, M. Synchronization of world economic activity. *Chaos: An Interdisciplinary. J. Nonlinear Sci.* **2017**, *27*, 127002.
12. Mahmoudvand, R.; Konstantinides, D.; Rodrigues, P.C. Forecasting mortality rate by multivariate singular spectrum analysis. *Appl. Stoch. Models Bus. Ind.* **2017**, *33*, 717–732. [CrossRef]
13. Zabalza, J.; Qing, C.; Yuen, P.; Sun, G.; Zhao, H.; Ren, J. Fast implementation of two-dimensional singular spectrum analysis for effective data classification in hyperspectral imaging. *J. Frankl. Inst.* **2018**, *355*, 1733–1751. [CrossRef]
14. Mahmoudvand, R.; Rodrigues, P.C.; Yarmohammadi, M. Forecasting daily exchange rates: A comparison between SSA and MSSA. *RevStat-Stat. J.* **2019**, *17*, 599–616.
15. Mahmoudvand, R.; Rodrigues, P.C. Predicting the Brexit outcome using singular spectrum analysis. *J. Comput. Stat. Model.* **2019**, *1*, 9–15.
16. Ge, M.; Lv, Y.; Zhang, Y.; Yi, C.; Ma, Y. An effective bearing fault diagnosis technique via local robust principal component analysis and multi-scale permutation entropy. *Entropy* **2019**, *21*, 959. [CrossRef]
17. Sulandari, W.; Subanar; Lee, M.H.; Rodrigues, P.C. Indonesian electricity load forecasting using singular spectrum analysis. *Energy* **2020**, *190*, 116408. [CrossRef]
18. Mahmoudvand, R.; Rodrigues, P.C. Prediction intervals for the vector SSA forecasting algorithm in a median based singular spectrum analysis. *Comput. Math. Methods* **2020**. [CrossRef]
19. Reisen, V.A.; Molinares, F.F. Robust estimation in time series with long and short memory properties. *Ann. Math. Inform.* **2012**, *39*, 207–224.

20. Rodrigues, P.C.; Monteiro, A.; Lourenço, V.M. A Robust additive main effects and multiplicative interaction model for the analysis of genotype-by-environment data. *Bioinformatics* **2016**, *32*, 58–66.
21. Hassani, H.; Mahmoudvand, R.; Omer, H.N.; Silva, E.S. A preliminary investigation into the effect of outlier(s) on singular spectrum analysis. *Fluct. Noise Lett.* **2014**, *13*, 1450029. [CrossRef]
22. Rodrigues, P.C.; Mahmoudvand, R. Correlation analysis in contaminated data by singular spectrum analysis. *Qual. Reliab. Eng. Int.* **2016**, *32*, 2127–2137. [CrossRef]
23. Rodrigues, P.C.; Lourenço, V.M.; Mahmoudvand, R. A robust approach to singular spectrum analysis. *Qual. Reliab. Eng. Int.* **2018**, *34*, 1437–1447. [CrossRef]
24. Hawkins, D.M.; Liu, L.; Young, S. Robust singular value decomposition. *Natl. Inst. Stat. Sci.* **2001**, *122*, 1–12.
25. Zhang, L.; Shen, H.; Huang, J.Z. Robust regularized singular value decomposition with application to mortality data. *Ann. Appl. Stat.* **2013**, *7*, 1540–1561. [CrossRef]
26. Brockwell, P.J.; Davis, R.A. *Introduction to Time Series and Forecasting*; Springer: New York, NY, USA, 1996.
27. Ripley, B.D. Time Series in R 1.5.0. R News, 2/2, 2–7. Available online: https://www.r-project.org/doc/Rnews/Rnews_2002-2.pdf (accessed on 6 January 2020).
28. Rodrigues, P.C.; Mahmoudvand, R. The benefits of multivariate singular spectrum analysis over the univariate version. *J. Frankl. Inst.* **2018**, *355*, 544–564. [CrossRef]
29. Ghil, M.; Allen, M.R.; Dettinger, M.D.; Ide, K.; Kondrashov, D.; Mann, M.E.; Robertson, A.W.; Saunders, A.; Tian, Y.; Varadi, F.; et al. Advanced spectral methods for climate time series. *Rev. Geophys.* **2002**, *40*, 3.1–3.41. [CrossRef]
30. Mahmoudvand, R.; Rodrigues, P.C. A new parsimonious recurrent forecasting model in singular spectrum analysis. *J. Forecast.* **2018**, *37*, 191–200. [CrossRef]
31. Rodrigues, P.C.; Mahmoudvand, R. A new approach for the vector forecast algorithm in singular spectrum analysis. *Commun. Stat. Simul. Comput.* **2020**. [CrossRef]
32. Wen, Q.; Gao, J.; Song, X.; Sun, L.; Tan, J. RobustTrend: A Huber loss with a combined first and second order difference regularization for time series trend filtering. In Proceedings of the Twenty-Eighth International Joint Conference on Artificial Intelligence, Macao, China, 10–16 August 2019; pp. 3856–3862.
33. Bouwmans, T.; Aybat, N.S.; Zahzah, E. *Handbook of Robust Low-Rank and Sparse Matrix Decomposition: Applications in Image and Video Processing*; CRC Press: New York, NY, USA, 2016.
34. Huber, P.J. Robust estimation of a location parameter. *Ann. Math. Stat.* **1964**, *35*, 73–101. [CrossRef]
35. Huber, P.J.; Ronchetti, E.M. *Robust Statistics*; Wiley: Hoboken, NJ, USA, 2009.
36. Hyndman, R.J.; Khandakar, Y. Automatic time series forecasting: The forecast package for R. *J. Stat. Softw.* **2008**, *26*, 1–22.
37. de Carvalho, M.; Rua, A. Real-Time Nowcasting the US Output Gap: Singular Spectrum Analysis at Work. *Int. J. Forecast.* **2017**, *33*, 185–198. [CrossRef]
38. Golyandina, N.; Korobeynikov, A.; Shlemov, A.; Usevich, K. Multivariate and 2D Extensions of Singular Spectrum Analysis with the Rssa Package. *J. Stat. Softw.* **2015**, *67*. Available online: https://www.jstatsoft.org/article/view/v067i02 (accessed on 6 January 2020). [CrossRef]
39. Rodrigues, P.C.; Tuy, P.G.S.E.; Mahmoudvand, R. Randomized singular spectrum analysis for long time series. *J. Stat. Comput. Simul.* **2018**, *88*, 1921–1935. [CrossRef]
40. Hyndman, R.J. A brief history of forecasting competitions. *Int. J. Forecast.* **2020**, *36*, 7–14. [CrossRef]
41. Papacharalampous, G.; Tyralis, H.; Koutsoyiannis, D. Comparison of stochastic and machine learning methods for multi-step ahead forecasting of hydrological processes. *Stoch. Environ. Res. Risk Assess.* **2019**, *33*, 481–514. [CrossRef]

© 2020 by the authors. Licensee MDPI, Basel, Switzerland. This article is an open access article distributed under the terms and conditions of the Creative Commons Attribution (CC BY) license (http://creativecommons.org/licenses/by/4.0/).

Article

Channels' Confirmation and Predictions' Confirmation: From the Medical Test to the Raven Paradox

Chenguang Lu

Intelligence Engineering and Mathematics Institute, Liaoning Technical University, Fuxin 123000, China; survival99@gmail.com

Received: 24 January 2020; Accepted: 25 March 2020; Published: 26 March 2020

Abstract: After long arguments between positivism and falsificationism, the verification of universal hypotheses was replaced with the confirmation of uncertain major premises. Unfortunately, Hemple proposed the Raven Paradox. Then, Carnap used the increment of logical probability as the confirmation measure. So far, many confirmation measures have been proposed. Measure F proposed by Kemeny and Oppenheim among them possesses symmetries and asymmetries proposed by Elles and Fitelson, monotonicity proposed by Greco et al., and normalizing property suggested by many researchers. Based on the semantic information theory, a measure b^* similar to F is derived from the medical test. Like the likelihood ratio, measures b^* and F can only indicate the quality of channels or the testing means instead of the quality of probability predictions. Furthermore, it is still not easy to use b^*, F, or another measure to clarify the Raven Paradox. For this reason, measure c^* similar to the correct rate is derived. Measure c^* supports the Nicod Criterion and undermines the Equivalence Condition, and hence, can be used to eliminate the Raven Paradox. An example indicates that measures F and b^* are helpful for diagnosing the infection of Novel Coronavirus, whereas most popular confirmation measures are not. Another example reveals that all popular confirmation measures cannot be used to explain that a black raven can confirm "Ravens are black" more strongly than a piece of chalk. Measures F, b^*, and c^* indicate that the existence of fewer counterexamples is more important than more positive examples' existence, and hence, are compatible with Popper's falsification thought.

Keywords: relative entropy; cross-entropy; uncertain reasoning; inductive logic; confirmation measure; semantic information; medical test; raven paradox

1. Introduction

A universal judgment is equivalent to a hypothetical judgment or a rule, such as "All ravens are black" is equivalent to "For every x, if x is a raven, then x is black". Both can be used as a major premise for a syllogism. Deductive logic needs major premises; however, some major premises for empirical reasoning must be supported by inductive logic. Logical empiricism affirmed that a universal judgment can be verified finally by sense data. Popper said against logical empiricism that a universal judgment could only be falsified rather than be verified. However, for a universal or hypothetical judgment that is not strict, and is therefore uncertain, such as "Almost all ravens are black", "Ravens are black", or "If a man's Coronavirus test is positive, then he is very possibly infected", we cannot say that one counterexample can falsify it. After long arguments, Popper and most logical empiricists reached the identical conclusion [1,2] that we may use evidence to confirm universal judgments or major premises that are not strict or uncertain.

In 1945, Hemple [3] proposed the confirmation paradox or the Raven Paradox. According to the Equivalence Condition in the classical logic, "If x is a raven, then x is black" (Rule I) is equivalent to "If x is not black, then x is not a raven" (Rule II). A piece of white chalk supports the Rule II, and hence, also supports the Rule I. However, according to the Nicod criterion [4], a black raven supports the Rule I, a non-black raven undermines the Rule I, and a non-raven thing, such as a black cat or a piece of white chalk, is irrelevant to the Rule I. Hence, there exists a paradox between the Equivalence Condition and the Nicod criterion.

To quantize confirmation, both Carnap [1] and Popper [2] proposed their confirmation measures. However, only Carnap's confirmation measures are famous. So far, researchers have proposed many confirmation measures [1,5–13]. The induction problem seemly has become the confirmation problem. To screen reasonable confirmation measures, Elles and Fitelson [14] proposed **symmetries and asymmetries** as desirable properties; Crupi et al. [8] and Greco et al. [15] suggested **normalization** (for measures between −1 and 1) as a desirable property; Greco et al. [16] proposed **monotonicity** as a desirable property. We can find that only measures F (proposed by Kemeny and Oppenheim) and Z among popular confirmation measures possess these desirable properties. Measure Z was proposed by Crupi et al. [8] as the normalization of some other confirmation measures. It is also called the certainty factor proposed by Shortliffe and Buchanan [7].

When the author of this paper researched semantic information theory [17], he found that an uncertain prediction could be treated as the combination of a clear prediction and a tautology; the combining proportion of the clear prediction could be used as the degree of belief; the degree of belief optimized with a sampling distribution could be regarded as a confirmation measure. This measure is denoted by b^*; it is similar to measure F and also possesses the above-mentioned desirable properties.

Good confirmation measures should possess not only mathematically desirable properties but also practicabilities. We can use medical tests to check their practicabilities. We use the degree of belief to represent the degree to which we believe a major premise and use the degree of confirmation to denote the degree of belief that is optimized by a sample or some examples. The former is subjective, whereas the latter is objective. A medical test provides the test-positive (or the test-negative) to predict if a person or a specimen is infected (or uninfected). Both the test-positive and the test-negative have degrees of belief and degrees of confirmation. In medical practices, there exists an important issue: if two tests provide different results, which test should we believe? For example, when both Nucleic Acid Test (NAT) and CT (Computed Tomography) are used to diagnose the infection of Novel Coronavirus Disease (COVID-19), if the result of NAT is negative and the result of CT is positive, which should we believe? According to the sensitivity and the specificity [18] of a test and the prior probability of the infection, we can use any confirmation measure to calculate the degrees of confirmation of the test-positive and the test-negative. Using popular confirmation measures, can we provide reasonable degrees of confirmation to help us choose a better result from NAT-negative and CT-positive? Can these degrees of confirmation reflect the probability of the infection?

This paper will show that only measures that are the functions of the likelihood ratio, such as F and b^*, can help us to diagnose the infection or choose a better result that can be accepted by the medical society. However, measures F and b^* do not reflect the probability of the infection. Furthermore, using F, b^*, or another measure, it is still difficult to eliminate the Raven Paradox.

Recently, the author found that the problem with the Raven Paradox is different from the problem with the medical diagnosis. Measures F and b^* indicate how good the testing means are instead of how good the probability predictions are. To clarify the Raven Paradox, we need a confirmation measure that can indicate how good a probability prediction is. The confirmation measure c^* is hence derived. We call c^* a prediction confirmation measure and call b^* a channel confirmation measure. The distinction between Channels' confirmation and predictions' confirmation is similar to yet different from the distinction between Bayesian confirmation and Likelihoodist confirmation [19]. Measure c^* accords with the Nicod criterion and undermines the Equivalence Condition, and hence can be used to eliminate the Raven Paradox.

The main purposes of this paper are:

- to distinguish channel confirmation measures that are compatible with the likelihood ratio and prediction confirmation measures that can be used to assess probability predictions,
- to use a prediction confirmation measure c^* to eliminate the Raven Paradox, and
- to explain that confirmation and falsification may be compatible.

The confirmation methods in this paper are different from popular methods, since:

- Measures b^* and c^* are derived by the semantic information method [17,20] and the maximum likelihood criterion rather than defined directly.
- Confirmation and statistical learning mutually support so that the confirmation measures can be used not only to assess major premises but also to make probability predictions.

The main contributions of this paper are:

- It clarifies that we cannot use one confirmation measure for two different tasks: (1) to assess (communication) channels, such as medical tests as testing means, and (2) to assess probability predictions, such as to assess "Ravens are black".
- It provides measure c^* that manifests the Nicod criterion and hence provides a new method to clarify the Raven Paradox.

The rest of this paper is organized as follows. Section 2 includes background knowledge. It reviews existing confirmation measures, introduces the related semantic information method, and clarifies some questions about confirmation. Section 3 derives new confirmation measures b^* and c^* with the medical test as an example. It also provides many confirmation formulas for major premises with different antecedents and consequents. Section 4 includes results. It gives some cases to show the characteristics of new confirmation measures, to compare various confirmation measures by applying them to the diagnosis of COVID-19, and to show how an increased example affects the degrees of confirmation with different confirmation measures. Section 5 discusses why we can only eliminate the Raven Paradox by measure c^*. It also discusses some conceptual confusion and explains how new confirmation measures are compatible with Popper's falsification thought. Section 5 ends with conclusions.

2. Background

2.1. Statistical Probability, Logical Probability, Shannon's Channel, and Semantic Channel

First we distinguish logical probability and statistical probability. Logical probability of a hypothesis (or a label) is the probability in which the hypothesis is judged to be true, whereas its statistical probability is the probability in which the hypothesis or the label is selected.

Suppose that ten thousand people go through a door. For everyone denoted by x, entrance guards judge if x is elderly. If two thousand people are judged to be elderly, then the logical probability of the predicate "x is elderly" is 2000/10,000 = 0.2. If the task of entrance guards is to select a label for every person from four labels: "Child", "Youth", "Adult", and "Elderly", there may be one thousand people who are labeled "Elderly". The statistical probability of "Elderly" should be 1000/10,000 = 0.1. Why are not two thousand people are labeled "Elderly"? The reason is that some elderly people are labeled "Adult". A person may make two labels be true, such as a 65 years old person makes both "Adult" and "Elderly" be true. That is why the logical probability of a label is often greater than its statistical probability. An extreme example is that the logical probability of a tautology, such as "x is elderly or not elderly", is 1, whereas its statistical probability is almost 0 in general because a tautology is rarely selected. Statistical probability is normalized (the sum is 1), whereas logical probability is not normalized in general [17]. Therefore, we use two different symbols "P" and "T" to distinguish statistical probability and logical probability.

We now consider the Shannon channel [21] between human ages and labels "Child", "Adult", "Youth", "Middle age", "Elderly", and the like.

Let X be a random variable to denote an age and Y be a random variable to denote a label. X takes a value $x \in \{ages\}$; Y takes a value $y \in \{$"Child", "Adult", "Youth", "Middle age", "Elderly", ...$\}$. Shannon calls the prior probability distribution $P(X)$ (or $P(x)$) the source, and calls $P(Y)$ the destination. There is a Shannon channel $P(Y|X)$ from X to Y. It is a transition probability matrix:

$$P(Y|X) \Leftrightarrow \begin{bmatrix} P(y_1|x_1) & P(y_1|x_2) & \ldots & P(y_1|x_m) \\ P(y_2|x_1) & P(y_2|x_2) & \ldots & P(y_2|x_m) \\ \ldots & \ldots & \ldots & \ldots \\ P(y_n|x_1) & P(y_n|x_2) & \ldots & P(y_n|x_m) \end{bmatrix} \Leftrightarrow \begin{bmatrix} P(y_j|x) \\ P(y_j|x) \\ \ldots \\ P(y_n|x) \end{bmatrix}, \quad (1)$$

where $\Leftrightarrow$ indicates equivalence. This matrix consists of a group of conditional probabilities $P(y_j|x_i)$ ($j = 0, 1, \ldots, n; i = 0, 1, \ldots, m$) or a group of transition probability functions (so called by Shannon [21]), $P(y_j|x)$ ($j = 0, 1, \ldots, n$), where y_j is a constant, and x is a variable.

There is also a semantic channel that consists of a group of truth functions. Let $T(\theta_j|x)$ be the truth function of y_j, where θ_j is a model or a set of model parameters, by which we construct $T(\theta_j|x)$. The θ_j is also explained as a fuzzy sub-set of the domain of x [17]. For example, y_j = "x is young". Its truth function may be

$$T(\theta_j|x) = \exp[-(x-20)^2/25], \quad (2)$$

where 20 and 25 are model parameters. For y_k = "x is elderly", its truth function may be a logistic function:

$$T(\theta_k|x) = 1/[1 + \exp[-0.2(x-65)]], \quad (3)$$

where 0.2 and 65 are model parameters. The two truth functions are shown in Figure 1.

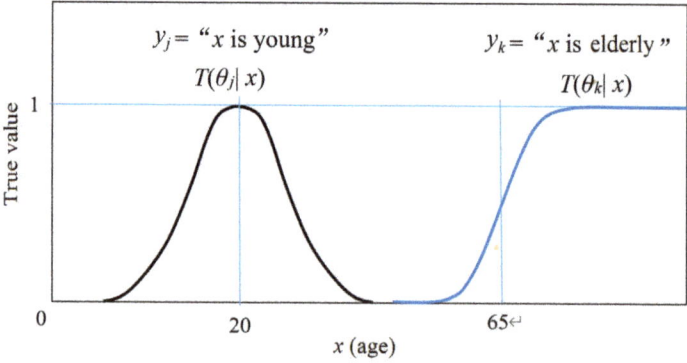

Figure 1. The truth functions of two hypotheses about ages.

According to Tarski's truth theory [22] and Davidson's truth-conditional semantics [23], a truth function can represent the semantic meaning of a hypothesis. Therefore, we call the matrix, which consists of a group of truth functions, a semantic channel:

$$T(\theta|X) \Leftrightarrow \begin{bmatrix} T(\theta_1|x_1) & T(\theta_1|x_2) & \ldots & T(\theta_1|x_m) \\ T(\theta_2|x_1) & T(\theta_2|x_2) & \ldots & T(\theta_2|x_m) \\ \ldots & \ldots & \ldots & \ldots \\ T(\theta_n|x_1) & T(\theta_n|x_2) & \ldots & T(\theta_n|x_m) \end{bmatrix} \Leftrightarrow \begin{bmatrix} T(\theta_1|x) \\ T(\theta_2|x) \\ \ldots \\ T(\theta_n|x) \end{bmatrix}. \quad (4)$$

Using a transition probability function $P(y_j|x)$, we can make the probability prediction $P(x|y_j)$ by

$$P(x|y_j) = P(x)P(y_j|x)/P(y_j), \quad (5)$$

which is the classical Bayes' formula. Using a truth function $T(\theta_j|x)$, we can also make a probability prediction or produce a likelihood function by

$$P(x|\theta_j) = P(x)T(\theta_j|x)/T(\theta_j), \qquad (6)$$

where $T(\theta_j)$ is the logical probability of y_j. There is

$$T(\theta_j) = \sum_i P(x_i)T(\theta_j|x_i). \qquad (7)$$

Equation (6) is called the semantic Bayes' formula [17]. The likelihood function is subjective; it may be regarded as the hybrid of logical probability and statistical probability.

When the source $P(x)$ is changed, the above formulas for predictions still work. It is easy to prove that $P(x|\theta_j) = P(x|y_j)$ as $T(\theta_j|x) \propto P(y_j|x)$. Since the maximum of $T(\theta_j|x)$ is 1, letting $P(x|\theta_j) = P(x|y_j)$, we can obtain the optimized truth function [17]:

$$T^*(\theta_j|x) = [P(x|y_j)/P(x)]/\max[P(x|y_j)/P(x)] = P(y_j|x)/\max[P(y_j|x)], \qquad (8)$$

where x is a variable and $\max(.)$ is the maximum of the function in brackets (.).

2.2. To Review Popular Confirmation Measures

We use h_1 to denote a hypothesis, h_0 to denote its negation, and h to denote one of them. We use e_1 as another hypothesis as the evidence of h_1, e_0 as its negation, and e as one of them. We use $c(e, h)$ to represent a confirmation measure, which means the degree of inductive support. Note that $c(e, h)$ here is used as in [8], where e is on the left, and h is on the right.

In the existing studies of confirmation, logical probability and statistical probability are not definitely distinguished. We still use P for both in introducing popular confirmation measures.

The popular confirmation measures include:

- $D(e_1, h_1) = P(h_1|e_1) - P(h_1)$ (Carnap, 1962 [1]),
- $M(e_1, h_1) = P(e_1|h_1) - P(e_1)$ (Mortimer, 1988 [5]),
- $R(e_1, h_1) = \log[P(h_1|e_1)/P(h_1)]$ (Horwich, 1982 [6]),
- $C(e_1, h_1) = P(h_1, e_1) - P(e_1)P(h_1)$ (Carnap, 1962 [1]),
- $Z(h_1, e_1) = \begin{cases} [P(h_1|e_1) - P(h_1)]/P(h_0), & \text{as } P(h_1|e_1) \geq P(h_1), \\ [P(h_1|e_1) - P(h_1)]/P(h_1), & \text{otherwise,} \end{cases}$ (Shortliffe and Buchanan, 1975 [7], Crupi et al., 2007 [8]),
- $S(e_1, h_1) = P(h_1|e_1) - P(h_1|e_0)$ (Christensen, 1999 [9]),
- $N(e_1, h_1) = P(e_1|h_1) - P(e_1|h_0)$ (Nozik, 1981 [10]),
- $L(e_1, h_1) = \log[P(e_1|h_1)/P(e_1|h_0)]$ (Good, 1984 [11]), and
- $F(e_1, h_1) = [P(e_1|h_1) - P(e_1|h_0)]/[P(e_1|h_1) + P(e_1|h_0)]$ (Kemeny and Oppenheim, 1952 [12]).

Two measures D and C proposed by Carnap are for incremental confirmation and absolute confirmation respectively. There are more confirmation measures in [8,24]. Measure F is also denoted by l^* [13], L [8], or k [24]. Most authors explain that probabilities they use, such as $P(h_1)$ and $P(h_1|e_1)$ in D, R, and C, are logical probabilities. Some authors explain that probabilities they use, such as $P(e_1|h_1)$ in F, are statistical probabilities.

Firstly, we need to clarify that confirmation is to assess what kind of evidence supports what kind of hypotheses. Let us have a look at the following three hypotheses:

- Hypothesis 1: $h_1(x) =$ "x is elderly", where x is a variable for an age and $h_1(x)$ is a predicate. An instance $x=70$ may be the evidence, and the truth value $T(\theta_1|70)$ of proposition $h_1(70)$ should be 1. If $x=50$, the (uncertain) truth value should be less, such as 0.5. Let $e_1 =$ "$x \geq 60$", true e_1 may also be the evidence that supports h_1 so that $T(\theta_1|e_1) > T(\theta_1)$.

- Hypothesis 2: $h_1(x)$ = "If age $x \geq 60$, then x is elderly", which is a hypothetical judgment, a major premise, or a rule. Note that $x = 70$ or $x \geq 60$ is only the evidence of the consequent "x is elderly" instead of the evidence of the rule. The rule's evidence should be a sample with many examples.
- Hypothesis 3: $e_1 \to h_1$ = "If age $x \geq 60$, then x is elderly", which is the same as Hypothesis 2. The difference is that e_1 = "$x \geq 60$"; h_1 = "x is elderly". The evidence is a sample with many examples like $\{(e_1, h_1), (e_1, h_0), \ldots\}$, or a sampling distribution $P(e, h)$, where P means statistical probability.

Hypothesis 1 has a (uncertain) truth function or a conditional logic probability function between 0 and 1, which is ascertained by our definition or usage. Hypothesis 1 need not be confirmed. Hypothesis 2 or Hypothesis 3 is what we need to confirm. The degree of confirmation is between −1 and 1.

There exist two different understandings about $c(e, h)$:

- Understanding 1: The h is the major premise to be confirmed, and e is the evidence that supports h; h and e are so used by Elles and Fitelson [14].
- Understanding 2: The e and h are those in rule $e \to h$ as used by Kemeny and Oppenheim [12]. The e is only the evidence that supports consequent h instead of the major premise $e \to h$ (see Section 2.3 for further analysis).

Fortunately, although researchers understand $c(e, h)$ in different ways, most researchers agree to use a sample including four types of examples $(e_1, h_1), (e_0, h_1), (e_1, h_0),$ and (e_0, h_0) as the evidence to confirm a rule and to use the four examples' numbers $a, b, c,$ and d (see Table 1) to construct confirmation measures. The following statements are based on this common view.

Table 1. The numbers of four types of examples for confirmation measures.

	e_0	e_1
h_1	b	a
h_0	d	c

The a is the number of example (e_1, h_1). For example, e_1 = "raven" ("raven" is a label or the abbreviate of "x is a raven") and h_1 = "black"; a is the number of black ravens. Similarly, b is the number of black non-raven things; c is the number of non-black ravens; d is the number of non-black and non-raven things.

To make the confirmation task clearer, we follow Understanding 2 to treat $e \to h$ = "if e then h" as the rule to be confirmed and replace $c(e, h)$ with $c(e \to h)$. To research confirmation is to construct or select the function $c(e \to h) = f(a, b, c, d)$.

To screen reasonable confirmation measures, Elles and Fitelson [14] propose the following symmetries:

- Hypothesis Symmetry (HS): $c(e_1 \to h_1) = -c(e_1 \to h_0)$ (two consequents are opposite),
- Evidence Symmetry (ES): $c(e_1 \to h_1) = -c(e_0 \to h_1)$ (two antecedents are opposite),
- Commutativity Symmetry (CS): $c(e_1 \to h_1) = c(h_1 \to e_1)$, and
- Total Symmetry (TS): $c(e_1 \to h_1) = c(e_0 \to h_0)$.

They conclude that only HS is desirable; the other three symmetries are not desirable. We call this conclusion the symmetry/asymmetry requirement. Their conclusion is supported by most researchers. Since TS is the combination of HS and ES, we only need to check HS, ES, and CS. According to this symmetry/asymmetry requirement, only measures $L, F,$ and Z among the measures mentioned above are screened out. It is uncertain whether N can be ruled out by this requirement [15]. See [14,25,26] for more discussions about the symmetry/asymmetry requirement.

Greco et al. [15] propose monotonicity as a desirable property. If $f(a, b, c, d)$ does not decrease with a or d and does not increase with b or c, then we say that $f(a, b, c, d)$ has the monotonicity.

Measures L, F, and Z have this monotonicity, whereas measures D, M, and N do not have. If we further require that $c(e\rightarrow h)$ are normalizing (between −1 and 1) [8,12], then only F and Z are screened out. There are also other properties discussed [15,19]. One is logicality, which means $c(e\rightarrow h) = 1$ without counterexample and $c(e\rightarrow h) = -1$ without positive example. We can also screen out F and Z using the logicality requirement.

Consider the medical test, such as the test for COVID-19. Let e_1 = "positive" (e.g., "x is positive", where x is a specimen), e_0 = "negative", h_1 = "infected" (e.g.,"x is infected"), and h_0 = "uninfected". Then the positive likelihood ratio is $LR^+ = P(e_1|h_1)/P(e_1|h_0)$, which indicates the reliability of the rule $e_1\rightarrow h_1$. Measures L and F have the one-to-one correspondence with LR:

$$L(e_1\rightarrow h_1) = \log LR^+; \tag{9}$$

$$F(e_1, h_1) = (LR^+ - 1)/(LR^+ + 1). \tag{10}$$

Hence, L and F can also be used to assess the reliability of the medical test. In comparison with LR and L, F can indicate the distance between a test (any F) and the best test (F = 1) or the worst test (F = −1) better than LR and L. However, LR can be used for the probability predictions of diseases more conveniently [27].

2.3. To Distinguish a Major Premise's Evidence and Its Consequent's Evidence

The evidence for the consequent of a syllogism is the minor premise, whereas the evidence for a major premise or a rule is a sample or a sampling distribution P(e, h). In some researchers' studies, e is used sometimes as the minor premise, and sometimes as an example or a sample; h is used sometimes as a consequent, and sometimes as a major premise. Researchers use c(e, h) or c(h, e) instead of $c(e\rightarrow h)$ because they need to avoid the contradiction between the two understandings. However, if we distinguish the two types of evidence, it has no problem to use $c(e\rightarrow h)$. We only need to emphasize that the evidence for a major premise is a sampling distribution P(e, h) instead of e.

If h is used as a major premise and e is used as the evidence (such as in [14,28]), −e (the negation of e) is puzzling because there are four types of examples instead of two. Suppose $h = p\rightarrow q$ and that e is one of (p, q), (p, −q), (−p, q), and (−p, −q). If (p, −q) is the counterexample, and other three examples (p, q), (−p, q) and (−p, −q) are positive examples, which support $p\rightarrow q$, then (−p, q) and (−p, −q) should also support $p\rightarrow \neg q$ because of the same reason. However, according to HS [14], it is unreasonable that the same evidence supports both $p\rightarrow q$ and $p\rightarrow \neg q$. In addition, e is a sample with many examples in general. A sample's negation or a sample's probability is also puzzling.

Fortunately, though many researchers say that e is the evidence of a major premise h, they also treat e as the antecedent and treat h as the consequent of a major premise because, only in this way, one can calculate the probabilities or conditional probabilities of e and h for a confirmation measure. Why, then, should we replace c(e, h) with $c(e\rightarrow h)$ to make the task clearer? Section 5.3 will show that h used as a major premise will result in the misunderstanding of the symmetry/asymmetry requirement.

2.4. Incremental Confirmation or Absolute Confirmation

Confirmation is often explained as assessing the impact of evidence on hypotheses, or the impact of the premise on the consequent of a rule [14,19]. However, this paper has a different point of view that confirmation is to assess how well a sample or sampling distribution supports a major premise or a rule; the impact on the rule (e.g., the increment of degree of confirmation) may be made by newly added examples.

Since one can use one or several examples to calculate the degree of confirmation with a confirmation measure, many researchers call their confirmation incremental confirmation [14,15]. There are also researchers who claim that we need absolute confirmation [29]. This paper supports absolute confirmation.

The problem with incremental confirmation is that the degrees of confirmation calculated are often bigger than 0.5 and are irrelevant to our prior knowledge or a, b, c, and d that we knew before. It is unreasonable to ignore prior knowledge. Suppose that the logical probability of h_1 = "x is elderly" is 0.2; the evidence is one or several people with age(s) $x > 60$; the conditionally logical probability of h_1 is 0.9. With measure D, the degree of confirmation is 0.9 − 0.2 = 0.7, which is very large and irrelevant to the prior knowledge.

In confirmation function $f(a, b, c, d)$, the numbers a, b, c, and d should be those of all examples including past and current examples. A measure $f(a, b, c, d)$ should be an absolute confirmation measure. Its increment should be

$$\Delta f = f(a + \Delta a, b + \Delta b, c + \Delta c, d + \Delta d) - f(a, b, c, d). \tag{11}$$

The increment of the degree of confirmation brought about by a new example is closely related to the number of old examples. Section 5.2 will further discuss incremental confirmation and absolute confirmation.

2.5. The Semantic Channel and the Degree of Belief of Medical Tests

We now consider the Shannon channel and the semantic channel of the medical test. The relation between h and e is shown in Figure 2.

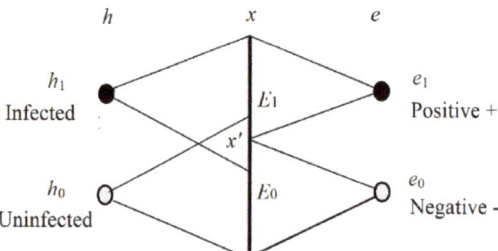

Figure 2. The relationship between Positive/Negative and Infected/Uninfected in the medical test.

In Figure 2, h_1 denotes an infected specimen (or person), h_0 denotes an uninfected specimen, e_1 is positive, and e_0 is negative. We can treat e_1 as a prediction "h is infected" and e_0 as a prediction "h is uninfected". In other word, h is a true label or true statement, and e is a prediction or selected label. The x is the observed feature of h; E_1 and E_2 are two sub-sets of the domain of x. If x is in E_1, then e_1 is selected; if x is in E_0, then e_0 is selected.

Figure 3 shows the relationship between h and x by two posterior probability distributions $P(x|h_0)$ and $P(x|h_1)$ and the magnitudes of four conditional probabilities (with four colors).

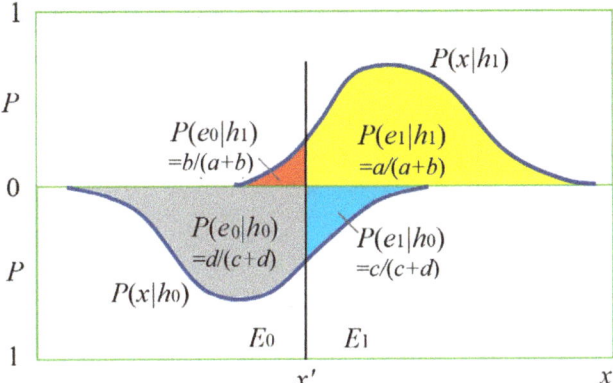

Figure 3. The relationship between two feature distributions and four conditional probabilities for the Shannon channel of the medical test.

In the medical test, $P(e_1|h_1)$ is called sensitivity [18], and $P(h_0|e_0)$ is called specificity. They ascertain a Shannon channel, which is denoted by $P(e|h)$, as shown in Table 2.

Table 2. Sensitivity and specificity ascertain a Shannon's Channel $P(e|h)$.

	Negative e_0	Positive e_1
Infected h_1	$P(e_0|h_1)$ = 1 − sensitivity	$P(e_1|h_1)$ = sensitivity
Uninfected h_0	$P(e_0|h_0)$ = specificity	$P(e_1|h_0)$ = 1 − specificity

We regard predicate $e_1(h)$ as the combination of believable and unbelievable parts (see Figure 4). The truth function of the believable part is $T(E_1|h) \in \{0,1\}$. The unbelievable part is a tautology, whose truth function is always 1. Then we have the truth functions of predicates $e_1(h)$ and $e_0(h)$:

$$T(\theta_{e1}|h) = b_1' + b_1' T(E_1|h); \quad T(\theta_{e0}|h) = b_0' + b_0' T(E_0|h). \tag{12}$$

where model parameter b_1' is the proportion of the unbelievable part, and also the truth value for the counter-instance h_0.

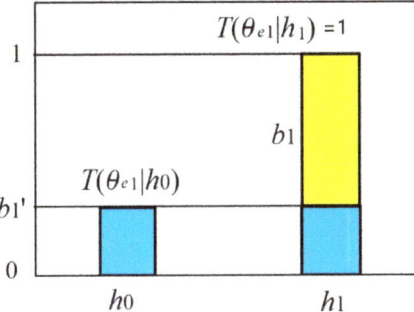

Figure 4. Truth function $T(\theta_{e1}|h)$ includes the believable part with proportion b_1 and the unbelievable part with proportion b_1' ($b_1' = 1 - |b_1|$).

The four truth values form a semantic channel, as shown in Table 3.

Table 3. The semantic channel ascertained by b_1' and b_0' for the medical test.

	e_0 (Negative)	e_1 (Positive)
h_1 (infected)	$T(\theta e_0|h_1) = b_0'$	$T(\theta_{e1}|h_1) = 1$
h_0 (uninfected)	$T(\theta_{e0}|h_0) = 1$	$T(\theta_{e1}|h_0) = b_1'$

For medical tests, the logical probability of e_1 is

$$T(\theta_{e1}) = \sum_i P(h_i)T(\theta_{e1}|h_i) = P(h_1) + b_1{}'P(h_0), \qquad (13)$$

The likelihood function is

$$P(h|\theta_{e1}) = P(h)T(\theta_{e1}|h)/T(\theta_{e1}). \qquad (14)$$

$P(h|\theta_j)$ is also the predicted probability of h according to $T(\theta_{e1}|h)$ or the semantic meaning of e_1.

To measure subjective or semantic information, we need subjective probability or logical probability [17]. To measure confirmation, we need statistical probability.

2.6. Semantic Information Formulas and the Nicod–Fisher Criterion

According to the semantic information G theory [17], the (amount of) semantic information conveyed by y_j about x_i is defined with the log-normalized-likelihood:

$$I(x_i;\theta_j) = \log \frac{P(x_i|\theta_j)}{P(x_i)} = \log \frac{T(\theta_j|x_i)}{T(\theta_j)}, \qquad (15)$$

where $T(\theta_j|x_i)$ is the truth value of proposition $y_j(x_i)$ and $T(\theta_j)$ is the logical probability of y_j. If $T(\theta_j|x)$ is always 1, then this semantic information formula becomes Carnap and Bar-Hillel's semantic information formula [30].

In semantic communication, we often see hypotheses or predictions, such as "The temperature is about 10 °C", "The time is about seven o'clock", or "The stock index will go up about 10% next month". Each one of them may be represented by y_j = "x is about x_j." We can express the truth functions of y_j by

$$T(\theta_j|x) = \exp[-(x-x_j)^2/(2\sigma^2)]. \qquad (16)$$

Introducing Equation (16) into Equation (15), we have

$$I(x_i;\theta_j) = \log[1/T(\theta_j)] - (x_i - x_j)^2/(2\sigma^2), \qquad (17)$$

by which we can explain that this semantic information is equal to the Carnap–Bar-Hillel's semantic information minus the squared relative deviation. This formula is illustrated in Figure 5.

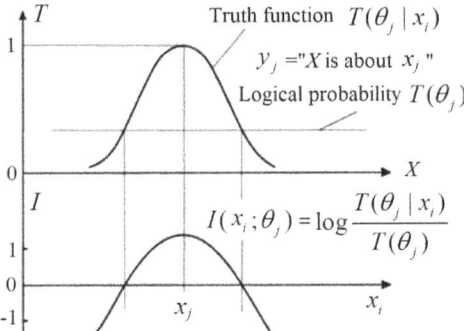

Figure 5. The semantic information conveyed by y_j about x_i.

Figure 5 indicates that the smaller the logical probability is, the more information there is; and the larger the deviation is, the less information there is. Thus, a wrong hypothesis will convey negative information. These conclusions accord with Popper's thought (see [2], p. 294).

To average $I(x_i; \theta_j)$, we have generalized Kullback–Leibler information or relative cross-entropy:

$$I(X;\theta_j) = \sum_i P(x_i|y_j) \log \frac{P(x_i|\theta_j)}{P(x_i)} = \sum_i P(x_i|y_j) \log \frac{T(\theta_j|x_i)}{T(\theta_j)}, \qquad (18)$$

where $P(x|y_j)$ is the sampling distribution, and $P(x|\theta_j)$ is the likelihood function. If $P(x|\theta_j)$ is equal to $P(x|y_j)$, then $I(X; \theta_j)$ reaches its maximum and becomes the relative entropy or the Kullback–Leibler divergence.

Consider medical tests, the semantic information conveyed by e_1 about h becomes

$$I(h_i; \theta_{e1}) = \log \frac{P(h_i|\theta_{e1})}{P(h_i)} = \log \frac{T(\theta_{e1}|h)}{T(\theta_{e1})}. \qquad (19)$$

The average semantic information is:

$$I(h; \theta_{e1}) = \sum_{i=0}^{1} P(h_i|e_1) \log \frac{P(h_i|\theta_{e1})}{P(h_i)} = \sum_{i=0}^{1} P(h_i|e_1) \log \frac{T(\theta_{e1}|h_i)}{T(\theta_{e1})} \qquad (20)$$

where $P(h_i|e_1)$ is the conditional probability from a sample.

We now consider the relationship between the likelihood and the average semantic information.

Let **D** be a sample $\{(h(t), e(t))|t = 1$ to $N; h(t)\in\{h_0, h_1\}; e(t)\in\{e_0, e_1\}\}$, which includes two sub-samples or conditional samples $\mathbf{H}_0$ with label e_0 and $\mathbf{H}_1$ with label e_1. When N data points in **D** come from Independent and Identically Distributed random variables, we have the log-likelihood

$$L(\theta_{e1}) = \log P(\mathbf{H}_1|\theta_{e1}) = \log P(h(1), h(2), \ldots, h(N)|\theta_{e1}) = \log \prod_{i=0}^{1} P(h_i|\theta_{e1})^{N_{1i}}$$
$$= N_1 \sum_{i=0}^{1} P(h_i|e_1) \log P(h_i|\theta_{ej}) = -N_1 H(h|\theta_{e1}). \qquad (21)$$

where N_{1i} is the number of example (h_i, e_1) in **D**; N_1 is the size of $\mathbf{H}_1$. $H(h|\theta_{e1})$ is the cross-entropy. If $P(h|\theta_{e1}) = P(h|e_1)$, then the cross-entropy becomes the Shannon entropy. Meanwhile, the cross-entropy reaches its minimum, and the likelihood reaches its maximum.

Comparing the above two equations, we have

$$I(h; \theta_{e1}) = L(\theta_{e1})/N_1 - \sum_{i=0}^{1} P(h_i|e_1) \log P(h_i) \quad (22)$$

which indicates the relationship between the average semantic information and the likelihood. Since the second term on the right side is constant, the maximum likelihood criterion is equivalent to the maximum average semantic information criterion. It is easy to find that a positive example (e_1, h_1) increases the average log-likelihood $L(\theta_{e1})/N_1$; a counterexample (e_1, h_0) decreases it; examples (e_0, h_0) and (e_0, h_1) with e_0 are irrelevant to it.

The Nicod criterion about confirmation is that a positive example (e_1, h_1) supports rule $e_1 \to h_1$; a counterexample (e_1, h_0) undermines $e_1 \to h_1$. No reference exactly indicates if Nicod affirmed that (e_0, h_1) and (e_0, h_1) are irrelevant to $e_1 \to h_1$. If Nicod did not affirm, we can add this affirmation to the criterion, then call the corresponding criterion the Nicod–Fisher criterion, since Fisher proposed the maximum likelihood estimation. From now on, we use the Nicod–Fisher criterion to replace the Nicod criterion.

2.7. Selecting Hypotheses and Confirming Rules: Two Tasks from the View of Statistical Learning

Researchers have noted the similarity between most confirmation measures and information measures. One explanation [31] is that information is the average of confirmatory impact. However, this paper gives a different explanation as follows.

There are three tasks in statistical learning: label learning, classification, and reliability analysis. There are similar tasks in inductive reasoning:

- Induction. It is similar to label learning. For uncertain hypotheses, label learning is to train a likelihood function $P(x|\theta_j)$ or a truth function $T(\theta_j|x)$ by a sampling distribution [17]. The Logistic function often used for binary classifications may be treated as a truth function.
- Hypothesis selection. It is like classification according to different criteria.
- Confirmation. It is similar to reliability analysis. The classical methods are to provide likelihood ratios and correct rates (including false rates, as those in Table 8).

Classification and reliability analysis are two different tasks. Similarly, hypothesis selection and confirmation are two different tasks.

In statistical learning, classification depends on the criterion. The often-used criteria are the maximum posterior probability criterion (which is equivalent to the maximum correctness criterion) and the maximum likelihood criterion (which is equivalent to the maximum semantic information criterion [17]). The classifier for binary classifications is

$$e(x) = \begin{cases} e_1, & \text{if } P(\theta_1|x) \geq P(\theta_0|x), \ P(x|\theta_1) \geq P(x|\theta_0), \text{ or } I(x; \theta_1) \geq I(x; \theta_0); \\ e_0, & \text{otherwise}. \end{cases} \quad (23)$$

After the above classification, we may use information criterion to assess how well e_j is used to predict h_j:

$$\begin{aligned} I^*(h_j; \theta_{ej}) = I(h_j; e_j) &= \log \frac{P(h_j|e_j)}{P(h_j)} = \log \frac{P(e_j|h_j)}{P(e_j)} \\ &= \log P(h_j|e_j) - \log P(h_j) = \log P(e_j|h_j) - \log P(e_j) \\ &= \log P(h_j, e_j) - \log[P(h_j)P(e_j)], \end{aligned} \quad (24)$$

where I^* means optimized semantic information. With information amounts $I(h_i; \theta_{ej})$ ($i, j = 0, 1$), we can optimize the classifier [17]:

$$e_j^* = f(x) = \underset{e_j}{\operatorname{argmax}}[P(h_0|x)I(h_0; \theta_{ej}) + P(h_1|x)I(h_1; \theta_{ej})]. \quad (25)$$

The new classifier will provide the new Shannon's channel $P(e|h)$. The maximum mutual information classification can be achieved by repeating Equations (23) and (25) [17,32].

With the above classifiers, we can make prediction e_j = "x is h_j" according to x. To tell information receivers how reliable the rule $e_j \to h_j$ is, we need the likelihood ratio LR to indicate how good the channel is or need the correct rate to indicate how good the probability prediction is. Confirmation is similar. We need to provide a confirmation measure similar to LR, such as F, and a confirmation measure similar to the correct rate. The difference is that the confirmation measures should change between −1 and 1.

According to above analyses, it is easy to find that confirmation measures D, N, R, and C are more like information measures for assessing and selecting predictions instead of confirming rules. Z is their normalization [8]; it seems between an information measure and a confirmation measure. However, confirming rules is different from measuring predictions' information; it needs the proportions of positive examples and counterexamples.

3. Two Novel Confirmation Measures

3.1. To Derive Channel Confirmation Measure b^*

We use the maximum semantic information criterion, which is consistent with the maximum likelihood criterion, to derive the channel confirmation measure. According to Equations (13) and (18), the average semantic information conveyed by e_1 about h is

$$I(h;\theta_{e1}) = P(h_0|e_1) \log \frac{b'_1}{P(h_1 + b'_1 P(h_0)} + P(h_1|e_1) \log \frac{1}{P(h_1 + b'_1 P(h_0)} \tag{26}$$

Letting $dI(h;\theta_{e1})/db_1' = 0$, we can obtain the optimized b_1':

$$b_1'^* = \frac{P(h_0|e_1)}{P(h_0)} / \frac{P(h_1|e_1)}{P(h_1)}, \tag{27}$$

where $P(h_1|e_1)/P(h_1) \geq P(h_0|e_1)/P(h_0)$. The b'^* can be called a disconfirmation measure. Letting both the numerator and the denominator multiply by $P(e_1)$, the above formula becomes:

$$b_1'^* = P(e_1|h_0)/P(e_1|h_1) = (1 - \text{specificity})/\text{sensibility} = 1/LR^+. \tag{28}$$

According to the semantic information G theory [17], when a truth function is proportional to the corresponding transition probability function, e.g., $T^*(\theta_{e1}|h) \propto P(e_1|h)$, the average semantic information reaches its maximum. Using $T^*(\theta_{e1}|h) \propto P(e_1|h)$, we can directly obtain

$$\frac{b_1'^*}{P(e_1|h_0)} = \frac{1}{P(e_1|h_1)} \tag{29}$$

and Equation (28). We call

$$b_1^* = 1 - b_1'^* = [P(e_1|h_1) - P(e_1|h_0)]/P(e_1|h_1) \tag{30}$$

the degree of confirmation of the rule $e_1 \to h_1$. Considering $P(e_1|h_1) < P(e_1|h_0)$, we have

$$b_1^* = b_1'^* - 1 = [P(e_1|h_0) - P(e_1|h_1)]/P(e_1|h_0). \tag{31}$$

Combining the above two formulas, we obtain

$$b_1^* = b^*(e_1 \to h_1) = \frac{P(e_1|h_1) - P(e_1|h_0)}{\max[P(e_1|h_1), P(e_1|h_0)]} = \frac{LR^+ - 1}{\max(LR^+, 1)}. \tag{32}$$

Since
$$b_1^* = b^*(e_1 \rightarrow h_0) = \frac{P(e_1|h_0) - P(e_1|h_1)}{\max[P(e_1|h_0), P(e_1|h_1)]} = -b^*(e_1 \rightarrow h_1), \tag{33}$$

the b_1^* possesses HS or Consequent Symmetry.

In the same way, we obtain

$$b_0^* = b^*(e_0 \rightarrow h_0) = \frac{P(e_0|h_0) - P(e_0|h_1)}{\max[P(e_0|h_0), P(e_0|h_1)]} = \frac{LR^- - 1}{\max(LR^-, 1)}. \tag{34}$$

Using Consequent Symmetry, we can obtain $b^*(e_1 \rightarrow h_0) = -b^*(e_1 \rightarrow h_1)$ and $b^*(e_0 \rightarrow h_1) = -b^*(e_0 \rightarrow h_0)$.

Using measure b^* or F, we can answer the question: if the result of NAT is negative and the result of CT is positive, which should we believe? Section 4.2 will provide the answer that is consistent with the improved diagnosis of COVID-19 in Wuhan.

Compared with F, b^* is better for probability predictions. For example, from $b_1^* > 0$ and $P(h)$, we obtain

$$P(h_1|\theta_{e1}) = P(h_1)/[\ P(h_1) + b_1'^* P(h_0)] = P(h_1)/[1 - b_1^* P(h_0)]. \tag{35}$$

This formula is much simpler than the classical Bayes' formula (see Equation (5)).

If $b_1^* = 0$, then $P(h_1|\theta_{e1}) = P(h_1)$. If $b_1^* < 0$, then we can make use of HS or Consequent Symmetry to obtain $b_{10}^* = b_1^*(e_1 \rightarrow h_0) = |b_1^*(e_1 \rightarrow h_1)| = |b_1^*|$. Then we have

$$P(h_0|\theta_{e1}) = P(h_0)/[\ P(h_0) + b_{10}'^* P(h_1)] = P(h_0)/[1 - b_{10}^* P(h_1)]. \tag{36}$$

We can also obtain $b_1^* = 2F_1/(1 + F_1)$ from $F_1 = F(e_1 \rightarrow h_1)$ for the probability prediction $P(h_1|\theta_{e1})$, but the calculation of probability predictions with F_1 is a little complicated.

So far, it is still problematic to use b^*, F, or another measure to handle the Raven Paradox. For example, as shown in Table 13, the increment of $F(e_1 \rightarrow h_1)$ caused by $\Delta d = 1$ is 0.348 − 0.333, whereas the increment caused by $\Delta a = 1$ is 0.340 − 0.333. The former is greater than the latter, which means that a piece of white chalk can support "Ravens are black" better than a black raven. Hence measure F does not accord with the Nicod–Fisher criterion. Measures b^* and Z do not either.

Why does not measure b^* and F accord with the Nicod–Fisher criterion? The reason is that the likelihood $L(\theta_{e1})$ is related to prior probability $P(h)$, whereas b^* and F are irrelevant to $P(h)$.

3.2. To Derive Prediction Confirmation Measure c^*

Statistics not only uses the likelihood ratio to indicate how reliable a testing means (as a channel) is but also uses the correct rate to indicate how reliable a probability prediction is. Measure F and b^* like LR cannot indicate the quality of a probability prediction. Most other measures have similar problems.

For example, we assume that an NAT for COVID-19 [33] has sensitivity $P(e_1|h_1) = 0.5$ and specificity $P(e_0|h_0) = 0.95$. We can calculate $b_1'^* = 0.1$ and $b_1^* = 0.9$. When the prior probability $P(h_1)$ of the infection changes, predicted probability $P(h_1|\theta_{e1})$ (see Equation (35)) changes with the prior probability, as shown in Table 4. We can obtain the same results using the classical Bayes' formula (see Equation (5)).

Table 4. Predictive probability $P(h_1|\theta_{e1})$ changes with prior probability $P(h_1)$ as $b_1^* = 0.9$.

	Common People	Risky Group	High-Risky Group
$P(h_1)$	0.001	0.1	0.25
$P(h_1|\theta_{e1})$	0.002	0.19	0.77

Data in Table 4 show that measure b^* cannot indicate the quality of probability predictions. Therefore, we need to use $P(h)$ to construct a confirmation measure that can reflect the correct rate.

We now treat probability prediction $P(h|\theta_{e1})$ as the combination of a believable part with proportion c_1 and an unbelievable part with proportion c_1', as shown in Figure 6. We call c_1 the degree of belief of the rule $e_1 \to h_1$ as a prediction.

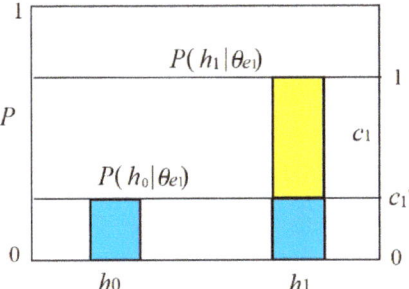

Figure 6. Likelihood function $P(h|\theta_{e1})$ may be regarded as a believable part plus an unbelievable part.

When the prediction accords with the fact, e.g., $P(h|\theta_{e1}) = P(h|e_1)$, c_1 becomes c_1^*. The degree of disconfirmation for predictions is

$$c'^*(e_1 \to h_1) = P(h_0|e_1)/P(h_1|e_1), \text{ if } P(h_0|e_1) \le P(h_1|e_1);$$
$$c'^*(e_1 \to h_1) = P(h_1|e_1)/P(h_0|e_1), \text{ if } P(h_1|e_1) \le P(h_0|e_1).$$
(37)

Further, we have the prediction confirmation measure

$$c_1^* = c^*(e_1 \to h_1) = \frac{P(h_1|e_1) - P(h_0|e_1)}{\max(P(h_1|e_1), P(h_0|e_1))}$$
$$= \frac{2P(h_1|e_1) - 1}{\max(P(h_1|e_1), 1 - P(h_1|e_1))} = \frac{2CR_1 - 1}{\max(CR_1, 1 - CR_1)}.$$
(38)

where $CR_1 = P(h_1|\theta_{e1}) = P(h_1|e_1)$ is the correct rate of rule $e_1 \to h_1$. This correct rate means that the probability of h_1 we predict as $x \in E_1$ is CR_1. Letting both the numerator and denominator of Equation (38) multiply by $P(e_1)$, we obtain

$$c_1^* = c^*(e_1 \to h_1) = \frac{P(h_1, e_1) - P(h_0, e_1)}{\max(P(h_1, e_1), P(h_0, e_1))} = \frac{a - c}{\max(a, c)}.$$
(39)

The sizes of four areas covered by two curves in Figure 7 may represent a, b, c, and d.

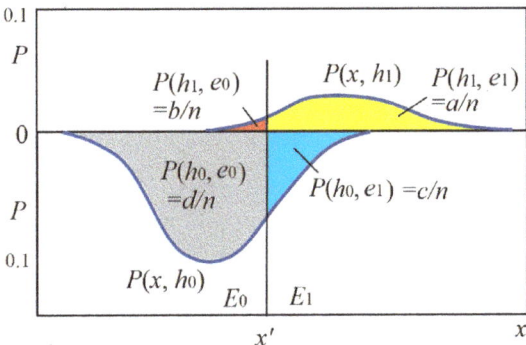

Figure 7. The numbers of positive examples and counterexamples for $c^*(e_0 \to h_0)$ (see the left side) and $c^*(e_1 \to h_1)$ (see the right side).

In like manner, we obtain

$$c_0^* = c^*(e_0 \to h_0) = \frac{P(h_0, e_0) - P(h_1, e_0)}{\max(P(h_0, e_0), P(h_1, e_0))} = \frac{d-b}{\max(d,b)}. \tag{40}$$

Making use of Consequent Symmetry, we can obtain $c^*(e_1 \to h_0) = -c^*(e_1 \to h_1)$ and $c^*(e_0 \to h_1) = -c^*(e_0 \to h_0)$.

In Figure 7, the sizes of the two areas covered by two curves are $P(h_0)$ and $P(h_1)$, which are different. If $P(h_0) = P(h_1) = 0.5$, then prediction confirmation measure c^* is equal to channel confirmation measure b^*.

Using measure c^*, we can directly assess the quality of the probability predictions. For $P(h_1|\theta_{e1}) = 0.77$ in Table 4, we have $c_1^* = (0.77 - 0.23)/0.77 = 0.701$. We can also use c^* for probability predictions. When $c_1^* > 0$, according to Equation (39), we have the correct rate of rule $e_1 \to h_1$:

$$CR_1 = P(h_1|\theta_{e1}) = 1/(1 + c_1'^*) = 1/(2 - c_1^*) \tag{41}$$

For example, if $c_1^* = 0.701$, then $CR_1 = 1/(2-0.701) = 0.77$. If $c^*(e_1 \to h_1) = 0$, then $CR_1 = 0.5$. If $c^*(e_1 \to h_1) < 0$, we may make use of HS to have $c_{10}^* = c^*(e_1 \to h_0) = |c^*_1|$, and then make probability prediction:

$$P(h_0|\theta_{e1}) = 1/(2 - c_{10}^*),$$
$$P(h_1|\theta_{e1}) = 1 - P(h_0|\theta_{e1}) = (1 - c_{10}^*)/(2 - c_{10}^*). \tag{42}$$

We may define another prediction confirmation measure by replacing operation max( ) with +:

$$c_{F1} = c_F^*(e_1 \to h_1) = \frac{P(h_1|e_1) - P(h_0|e_1)}{P(h_1|e_1) + P(h_0|e_1)} = P(h_1|e_1) - P(h_0|e_1)$$
$$= \frac{P(h_1, e_1) - P(h_0, e_1)}{P(e_1)} = \frac{a-c}{a+c}. \tag{43}$$

The c_F^* is also convenient for probability predictions when $P(h)$ is certain. There is

$$P(h_1|\theta_{e1}) = CR_1 = (1 + c_{F1}^*)/2;$$
$$P(h_0|\theta_{e1}) = 1 - CR_1 = (1 - c_{F1}^*)/2. \tag{44}$$

However, when $P(h)$ is variable, we should still use b^* with $P(h)$ for probability predictions.

It is easy to prove that $c^*(e_1 \to h_1)$ and $c_F^*(e_1 \to h_1)$ possess all the above-mentioned desirable properties.

3.3. Converse Channel/Prediction Confirmation Measures $b^*(h \to e)$ and $c^*(h \to e)$

Greco et al. [19] divide confirmation measures into

- Bayesian confirmation measures with $P(h|e)$ for $e \to h$,
- Likelihoodist confirmation measures with $P(e|h)$ for $e \to h$,
- converse Bayesian confirmation measures with $P(h|e)$ for $h \to e$, and
- converse Likelihoodist confirmation measures with $P(e|h)$ for $h \to e$.

Similarly, this paper divides confirmation measures into

- channel confirmation measure $b^*(e \to h)$,
- prediction confirmation measure $c^*(e \to h)$,
- converse channel confirmation measure $b^*(h \to e)$, and
- converse prediction confirmation measure $c^*(h \to e)$.

We now consider $c^*(h_1 \to e_1)$. The positive examples' proportion and the counterexamples' proportion can be found in the upside of Figure 7. Then we have

$$c^*(h_1 \to e_1) = \frac{P(e_1|h_1) - P(e_0|h_1)}{\max(P(e_1|h_1), P(e_0|h_1))} = \frac{a-b}{\max(a,b)}. \quad (45)$$

The correct rate reflected by $c^*(h_1 \to e_1)$ is sensitivity or true positive rate $P(h_1|e_1)$. The correct rate reflected by $c^*(h_0 \to e_0)$ is specificity or true negative rate $P(h_0|e_0)$.

Consider the converse channel confirmation measure $b^*(h_1 \to e_1)$. Now the source is $P(e)$ instead of $P(h)$. We may swap e_1 with h_1 in $b^*(e_1 \to h_1)$ or swap a with d and b with c in $f(a, b, c, d)$ to obtain

$$b^*(h_1 \to e_1) = \frac{P(h_1|e_1) - P(h_1|e_0)}{P(h_1|e_1) \vee P(h_1|e_0)} = \frac{ad - bc}{a(b+d) \vee b(a+c)} \quad (46)$$

where $\vee$ is the operator for the maximum of two numbers and is used to replace max( ). There are also four types of converse channel/prediction confirmation formulas with a, b, c, and d (see Table 7). Due to Consequent Symmetry, there are the eight types of converse channel/prediction confirmation formulas altogether.

3.4. Eight Confirmation Formulas for Different Antecedents and Consequents

Table 5 shows the positive examples' and counterexamples' proportions needed by measures b^* and c^*.

Table 5. Eight proportions for calculating $b^*(e \to h)$ and $c^*(e \to h)$.

	e_0 (Negative)	e_1 (Positive)
h_1 (infected)	$P(e_0|h_1) = b/(a+b)$	$P(e_1|h_1) = a/(a+b)$
h_0 (uninfected)	$P(e_0|h_0) = d/(c+d)$	$P(e_1|h_0) = c/(c+d)$
h_1 (infected)	$P(h_1|e_0) = b/(b+d)$	$P(h_1|e_1) = a/(a+c)$
h_0 (uninfected)	$P(h_0|e_0) = d/(b+d)$	$P(h_0|e_1) = c/(a+c)$

Table 6 provides four types of confirmation formulas with a, b, c, and d for rule $e \to h$, where function max( ) is replaced with the operator $\vee$.

Table 6. Channel/prediction confirmation measures expressed by a, b, c, and d.

	$b^*(e \to h)$ (for Channels, Refer to Figure 3)	$c^*(e \to h)$ (for Predictions, Refer to Figure 7)
$e_1 \to h_1$	$\frac{P(e_1|h_1) - P(e_1|h_0)}{P(e_1|h_1) \vee P(e_1|h_0)} = \frac{ad-bc}{a(c+d) \vee c(a+b)}$	$\frac{P(h_1|e_1) - P(h_0|e_1)}{P(h_1|e_1) \vee P(h_0|e_1)} = \frac{a-c}{a \vee c}$
$e_0 \to h_0$	$\frac{P(e_0|h_0) - P(e_0|h_1)}{P(e_0|h_0) \vee P(e_0|h_1)} = \frac{ad-bc}{d(a+b) \vee b(c+d)}$	$\frac{P(h_0|e_0) - P(h_1|e_0)}{P(h_0|e_0) \vee P(h_1|e_0)} = \frac{d-b}{d \vee b}$

These confirmation measures are related to the misreporting rates of the rule $e \to h$. For example, smaller $b^*(e_1 \to h_1)$ or $c^*(e_1 \to h_1)$ means that the test shows positive for more uninfected people.

Table 7 includes four types of confirmation measures for $h \to e$.

Table 7. Converse channel/prediction confirmation measures expressed by a, b, c, and d.

	$b^*(h \to e)$ (for Converse Channels)	$c^*(h \to e)$ (for Converse Predictions, Refer to Figure 7)
$h_1 \to e_1$	$\frac{P(h_1|e_1) - P(h_1|e_0)}{P(h_1|e_1) \vee P(h_1|e_0)} = \frac{ad-bc}{a(b+d) \vee b(a+c)}$	$\frac{P(e_1|h_1) - P(e_0|h_1)}{P(e_1|h_1) \vee P(e_0|h_1)} = \frac{a-b}{a \vee b}$
$h_0 \to e_0$	$\frac{P(h_0|e_0) - P(h_0|e_1)}{P(h_0|e_0) \vee P(h_0|e_1)} = \frac{ad-bc}{d(a+c) \vee c(b+d)}$	$\frac{P(e_0|h_0) - P(e_1|h_0)}{P(e_0|h_0) \vee P(e_1|h_0)} = \frac{d-c}{d \vee c}$

These confirmation measures are related to the underreporting rates of the rule $h \to e$. For example, smaller $b^*(h_1 \to e_1)$ or $c^*(h_1 \to e_1)$ means that the test shows negative for more infected people. Underreports are more serious problems.

Each of the eight types of confirmation measures in Tables 6 and 7 has its consequent-symmetrical form. Therefore, there are 16 types of function $f(a, b, c, d)$ altogether for confirmation.

In a prediction and converse prediction confirmation formula, the conditions of two conditional probabilities are the same; they are the antecedents of rules so that a confirmation measure c^* only depends on the two numbers of positive examples and counterexamples. Therefore, these measures accord with the Nicod–Fisher criterion.

If we change "∨" into "+" in $f(a, b, c, d)$, then measure b^* becomes measure $b_F^* = F$, and measure c^* becomes measure c_F^*. For example,

$$c_F^*(e_1 \to h_1) = (a - c)/(a + c). \tag{47}$$

3.5. Relationship Between Measures b^* and F

Measure b^* is like measure F. The two measures changes with likelihood ratio LR, as shown in Figure 8.

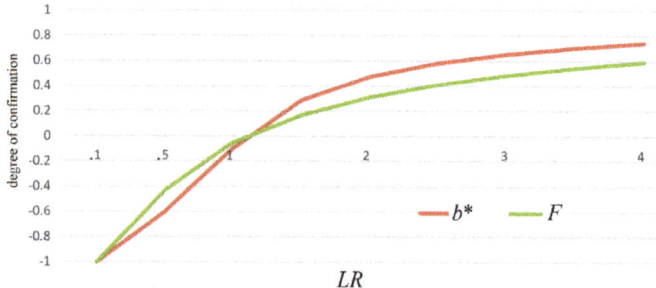

Figure 8. Measures b^* and F change with likelihood ratio LR.

Measure F has four confirmation formulas for different antecedents and consequents [8], which are related to measure b_F^* as follows:

$$F(e_1 \to h_1) = \frac{P(e_1|h_1) - P(e_1|h_0)}{P(e_1|h_1) + P(e_1|h_0)} = \frac{ad - bc}{ad + bc + 2ac} = b_F^*(e_1 \to h_1) \tag{48}$$

$$F(h_1 \to e_1) = \frac{P(h_1|e_1) - P(h_1|e_0)}{P(h_1|e_1) + P(h_1|e_0)} = \frac{ad - bc}{ad + bc + 2ab} = b_F^*(h_1 \to e_1) \tag{49}$$

$$F(e_0 \to h_0) = \frac{P(e_0|h_0) - P(e_0|h_1)}{P(e_0|h_0) + P(e_0|h_1)} = \frac{ad - bc}{ad + bc + 2bd} = b_F^*(e_0 \to h_0) \tag{50}$$

$$F(h_0 \to e_0) = \frac{P(h_0|e_0) - P(h_0|e_1)}{P(h_0|e_0) + P(h_0|e_1)} = \frac{ad - bc}{ad + bc + 2cd} = b_F^*(h_0 \to e_0) \tag{51}$$

F is equivalent to b_F^*. Measure b^* has all the above-mentioned desirable properties as well as measure F. The differences are that measure b^* has a greater absolute value than measure F; measure b^* can be used for probability predictions more conveniently (see Equation (35)).

3.6. Relationships between Prediction Confirmation Measures and Some Medical Test's Indexes

Channel confirmation measures are related to likelihood ratios, whereas Prediction Confirmation Measures (PCMs) including converse PCMs are related to correct rates and false rates in the medical test.

To help us understand the significances of PCMs in the medical test, Table 8 shows that each PCM is related to which correct rate and which false rate.

Table 8. PCMs (Prediction Confirmation Measures) are related to different correct rates and false rates in the medical test [18].

PCM	Correct Rate Positively Related to c^*	False Rate Negatively Related to c^*
$c^*(e_1 \rightarrow h_1)$	$P(h_1\|e_1)$: PPV (Positive Predictive Value)	$P(h_0\|e_1)$: FDR (False Discovery Rate)
$c^*(e_0 \rightarrow h_0)$	$P(h_0\|e_0)$: NPV (Negative Predictive Value)	$P(h_1\|e_0)$: FOR (False Omission Rate)
$c^*(h_1 \rightarrow e_1)$	$P(e_1\|h_1)$: Sensitivity or TPR (True Positive Rate)	$P(e_0\|h_1)$: FNR (False Negative Rate)
$c^*(h_0 \rightarrow e_0)$	$P(e_0\|h_0)$: Specificity or TNR (True Negative Rate)	$P(e_1\|h_0)$: FPR (False Positive Rate)

The false rates related to PCMs are the misreporting rates of the rule $e \rightarrow h$, whereas the false rates related to converse PCMs are the underreporting rates of the rule $h \rightarrow e$. For example, False Discovery Rate $P(h_0|e_1)$ is also the misreporting rate of rule $e_1 \rightarrow h_1$; False Negative Rate $P(e_0|h_1)$ is also the underreporting rate of rule $h_1 \rightarrow e_1$.

4. Results

4.1. Using Three Examples to Compare Various Confirmation Measures

In China's war against COVID-19, people often ask the question: since the true positive rate, e.g., sensitivity, of NAT is so low (less than 0.5), why do we still believe it? Medical experts explain that though NAT has low sensitivity, it has high specificity, and hence its positive is very believable.

We use the following two extreme examples (see Figure 9) to explain why a test with very low sensitivity can provide more believable positive than another test with very high sensitivity, and whether popular confirmation measures support this conclusion.

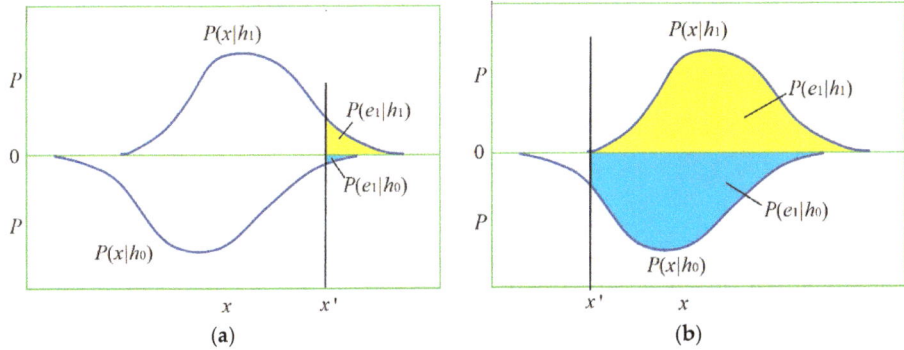

Figure 9. How the proportions of positive examples and counterexamples affect $b^*(e_1 \rightarrow h_1)$. (**a**) Example 1: positive examples' proportion is $P(e_1|h_1) = 0.1$, and counterexamples' proportion is $P(e_1|h_0) = 0.01$. (**b**) Example 2: positive examples' proportion is $P(e_1|h_1) = 1$, and counterexamples' proportion is $P(e_1|h_0) = 0.9$.

In Example 1, $b^*(e_1 \rightarrow h_1) = (0.1 - 0.01)/0.1 = 0.9$, which is very large. In Example 2, $b^*(e_1 \rightarrow h_1) = (1 - 0.9)/1 = 0.1$, which is very small. The two examples indicate that fewer counterexamples' existence is more important to b^* than more positive examples' existence. Measures F, c^*, and c_F^* also possess this characteristic, which is compatible with the Logicality requirement [15]. However, most confirmation measures do not possess this characteristic.

We supposed $P(h_1) = 0.2$ and $n = 1000$ and then calculated the degrees of confirmation with different confirmation measures for the above two examples, as shown in Table 9, where the base of log for R and L is 2. Table 9 also includes Example 3 (e.g., Ex. 3), in which $P(h_1)$ is 0.01. Example 3 reveals the difference between Z and b^* (or F).

Table 9. Three examples to show the differences between different confirmation measures.

Ex.	a, b, c, d	D	M	R	C	Z	S	N	L	F	b^*	c^*
1	20, 180, 8, 792	0.514	0.072	1.84	0.014	0.643	0.529	0.09	**3.32**	**0.818**	**0.9**	0.8
2	200, 0, 720, 80	0.017	0.08	0.12	0.016	0.022	0.217	0.1	**0.152**	**0.053**	**0.1**	−0.722
3	10, 0, 90, 900	0.09	0.9	3.32	0.009	0.091	0.1	0.091	3.46	0.833	0.91	−0.9

Data for Examples 1 and 2 show that L, F and b^* give Example 1 a much higher rating than Example 2, whereas M, C, and N give Example 2 a higher rating than Example 1 (see red numbers). The excel file for Table 9, Tables 12 and 13 can be find in Supplementary Material.

In Examples 2 and 3, where $c > a$ (counterexamples are more than positive examples), only the values of $c^*(e_1 \to h_1)$ are negative. The negative values should be reasonable for assessing probability predictions when counterexamples are more than positive examples.

The data for Example 3 show that when $P(h_0) = 0.99 >> P(h_1) = 0.01$, measure Z is very different from measures F and b^* (see blue numbers) because F and b^* are independent of P(h) unlike Z.

Although measure L (log-likelihood ratio) is compatible with F and b^*, its values, such as 3.32 and 0.152, are not intuitionistic as well as the values of F or b^*, which are normalizing.

4.2. Using Measures b^* to Explain Why And How CT is also Used to Test COVID-19

The COVID-19 outbreak in Wuhan of China in 2019 and 2020 has infected many people. In the early stage, only NAT was used to diagnose the infection. Later, many doctors found that NAT often failed to report the viral infection. Because this test has low sensitivity (which may be less than 0.5) and high specificity, we can confirm the infection when NAT is positive, but it is not good for confirming the non-infection when NAT is negative. That means that NAT-negative is not believable. To reduce the underreports of the infection, CT gained more attention because CT had higher sensitivity than NAT.

When both NAT and CT were used in Wuhan, doctors improved the diagnosis, as shown in Figure 10 and Table 11. If we diagnose the infection according to confirmation measure b^*, will the diagnosis be the same as the improved diagnosis? Besides NAT and CT, patients' symptoms, such as fever and cough, were also used for the diagnosis. To simplify the problem, we assumed that all patients had the same symptoms so that we could diagnose only according to the results of NAT and CT.

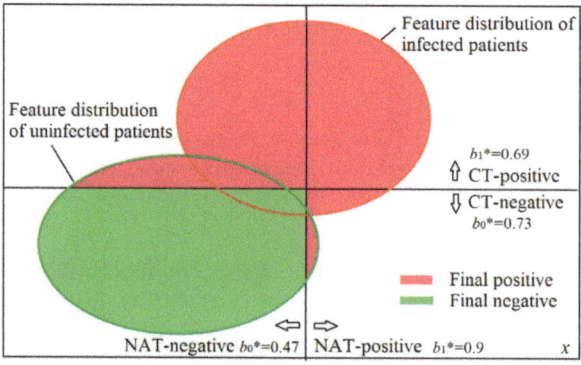

Figure 10. Using both NAT and CT to diagnose the infection of COVID-19 with the help of confirmation measure b^*.

Reference [34] introduces the sensitivity and specificity of CT that the authors achieved. According to [33,34] and other reports on the internet, the author of this paper estimated the sensitivities and specificities, as shown in Table 10.

Table 10. Sensitivities and specificities of NAT (Nucleic Acid Test) and CT for COVID-19.

	Sensitivity	Specificity
NAT	0.5	0.95
CT	0.8	0.75

Figure 10 was drawn according to Table 10. Figure 10 also shows sensitivities and specificities. For example, the half of the red circle on the right side indicates that the sensitivity of NAT is 0.5.

We use c(NAT+) to denote the degree of confirmation of NAT-positive with any measure c, and used c(NAT−), c(CT+), and c(CT−) in like manner. Then we have

$$b^*(\text{NAT+}) = [P(e_1|h_1) - P(e_1|h_0)]/P(e_1|h_1) = [0.5 - (1 - 0.95)]/0.5 = 0.9;$$

$$b^*(\text{NAT}-) = [P(e_0|h_0) - P(e_0|h_1)]/P(e_0|h_0) = [0.95 - (1 - 0.5)]/0.95 = 0.47.$$

We can also obtain b^*(CT+) = 0.69 and b^*(CT−) = 0.73 in like manner (see Table 11).

Table 11. Improved diagnosis (for final positive or negative) according to NAT and CT.

	NAT-Negative, b_0^* = 0.47	NAT-Positive, b_1^* = 0.9
CT-positive, b_1^* = 0.69	Final positive (changed)	Final positive
CT-negative, b_0^* = 0.73	Final negative	Final positive

If we only use the positive or negative of NAT as the final positive or negative, we confirm the non-infection as NAT shows negative. According to measure b^*, if we use both results of NAT and CT, when NAT shows a negative whereas CT shows positive, the final diagnosis should be positive (see blue words in Table 11) because b^*(CT+) = 0.69 is higher than b^*(NAT−) = 0.47. This diagnosis is the same as the improved diagnosis in Wuhan.

Assuming the prior probability of the infection $P(e_1)$ = 0.25, the author calculated the various degrees of confirmation with different confirmation measures for the same sensitivities and specificities, as shown in Table 12.

Table 12. Various confirmation measures for assessing the results of NAT and CT.

	D	M	Z	S	C	N	F	b^*	c^*
c(NAT−)	0.10	0.11	0.40	0.62	0.08	0.45	0.31	0.47	0.83
c(NAT+)	0.52	0.34	0.69	0.62	0.08	0.45	0.82	0.90	0.70
c(CT−)	0.17	0.14	0.67	0.43	0.10	0.55	0.58	0.73	0.91
c(CT+)	0.27	0.41	0.36	0.43	0.10	0.55	0.52	0.69	0.06
c(CT+) > c(NAT−)			No	No					No
c(NAT+) > c(CT−)					No	No			No

If there is a "No" under a measure, this measure will result in a different diagnosis from the improved diagnosis. The red numbers mean that c(CT+) < c(NAT−) or c(NAT+)<c(CT−). Measures D, M, and F, as well as b^*, are consistent with the improved diagnosis. If we change $P(h_1)$ from 0.1 to 0.6, we will find that measure M is also not consistent with the improved diagnosis. If we believe a test-positive or test-negative when its degree of confirmation is greater than 0.2, then D is also undesirable, and only measures F and b^* satisfy our requirements.

The above sensitivities and specificities in Table 10 were not specially selected. When NAT-sensitivity changed between 0.3 and 0.7, or CT-sensitivity changed between 0.6 and 0.9, it was the same that only measures D, F, and b^* were consistent with the improved diagnosis.

Measure c^* is also not suitable for the diagnosis because it reflects correctness and cannot reduce the underreports of the infection. Yet, the underreports of the infection will cause greater loss than the misreports of the infection.

4.3. How Various Confirmation Measures are Affected by Increments Δa and Δd

The following example is used to check if we can use popular confirmation measures to explain that a black raven can confirm "Ravens are black" more strongly than a piece of white chalk.

Table 13 shows the degrees of confirmation calculated with nine different measures. First, we supposed $a = d = 20$ and $b = c = 10$ to calculate the nine degrees of confirmation. Next, we only replaced a with $a + 1$ to calculate the nine degrees. Last, we only replaced d with $d + 1$ to calculate them.

Table 13. How confirmation measures are affected by $\Delta a = 1$ and $\Delta d = 1$.

	$f(a, b, c, d)$	$a = d = 20$ $b = c = 10$	$\Delta a = 1$ $\Delta d = 0$	$\Delta d = 1$ $\Delta a = 0$	$\Delta f/\Delta a - \Delta f/\Delta d$
$D(e_1 \to h_1)$	$a/(a+c) - (a+b)/n$	0.167	0.169	0.175	−0.006
$M(e_1 \to h_1)$	$a/(a+b) - (a+c)/n$	0.167	0.169	0.175	−0.006
$C(e_1 \to h_1)$	$a/n - (a+c)(a+b)/n^2$	0.083	0.086	0.086	0
$Z(e_1 \to h_1)$	$D(e_1 \to h_1)/[(c+d)/n]$	0.333	0.344	0.344	0
$S(e_1 \to h_1)$	$a/(a+c) - b/(b+d)$	0.333	0.334	0.344	0
$N(e_1 \to h_1)$	$a/(a+b) - c/(c+d)$	0.333	0.334	0.344	0
$F(e_1 \to h_1)$	$(ad-bc)/(ad+bc+2ac)$	0.333	0.340	0.348	−0.007
LR^+	$[a/(a+b)]/[c/(c+d)]$	2	2.03	2.07	−0.034
$c^*(e_1 \to h_1)$	$(a-c)/\max(a, c)$	0.5	0.524	0.5	$0.024 > 0$

The results must have exceeded many researchers' expectations. Table 13 indicates that all measures except c^* (see blue numbers) cannot ensure that $\Delta a = 1$ increases $f(a, b, c, d)$ more than $\Delta d = 1$. If we change b and c between 1 and 19, all measures except c^*, S, and N cannot ensure $\Delta f/\Delta a \geq \Delta f/\Delta d$. When $b > c$, measures S and N also cannot ensure $\Delta f/\Delta a \geq \Delta f/\Delta d$. The cause for measures D and M is that $\Delta d = 1$ decreases $P(h_1)$ and $P(e_1)$ more than increasing $P((h_1|e_1)$ and $P(e_1|h_1)$. The causes for other measures except c^* are similar.

5. Discussions

5.1. To Clarify the Raven Paradox

To clarify the Raven Paradox, some researchers including Hemple [3] affirm the Equivalence Condition and deny the Nicod–Fisher criterion; some researchers, such as Scheffler and Goodman [35], affirm the Nicod–Fisher criterion and deny the Equivalence Condition. There are also some researchers who do not fully affirm the Equivalence Condition or the Nicod–Fisher criterion.

First, we consider measure F to see if we can use it to eliminate the Raven Paradox. The difference between $F(e_1 \to h_1)$ and $F(h_0 \to e_0)$ is that their counterexamples are the same, yet, their positive examples are different. When d increases to $d + \Delta d$, $F(e_1 \to h_1)$ and $F(h_0 \to e_0)$ unequally increase. Therefore,

- though measure F denies the Equivalence Condition, it still affirms that Δd affects both $F(e_1 \rightarrow h_1)$ and $F(h_0 \rightarrow e_0)$;
- measure F does not accord the Nicod–Fisher criterion.

Measure b^* is like F. The conclusion is that measures F and b^* cannot eliminate our confusion about the Raven Paradox.

After inspecting many different confirmation measures from the perspective of the rough set theory, Greco et al. [15] conclude that Nicod criterion (e.g., the Nicod–Fisher criterion) is right, but it is difficult to find a suitable measure that accords with the Nicod criterion. However, many researchers still think that the Nicod criterion is incorrect; it accords with our intuition only because a confirmation measure $c(e_1 \rightarrow h_1)$ can evidently increase with a and slightly increase with d. After comparing different confirmation measures, Fitelson and Hawthorne [28] believe that the likelihood ratio may be used to explain that a black raven can confirm "Ravens are black" more strongly than a non-black non-raven thing.

Unfortunately, Table 13 shows that the increments of all measures except c^* caused by $\Delta d = 1$ are greater than or equal to those caused by $\Delta a = 1$. That means that these measures support the conclusion that a piece of white chalk can confirm "Ravens are black" more strongly than (or as well as) a black raven. Therefore, these measures cannot be used to clarify the Raven Paradox.

However, measure c^* is different. Since $c^*(e_1 \rightarrow h_1) = (a - c)/(a \vee c)$ and $c^*(h_0 \rightarrow e_0) = (d - c)/(d \vee c)$, the Equivalence Condition does not hold, and measure c^* accords with the Nicod–Fisher criterion very well. Hence, the Raven Paradox does not exist anymore according to measure c^*.

5.2. About Incremental Confirmation and Absolute Confirmation

In Table 13, if the initial numbers are $a = d = 200$ and $b = c = 100$, the increments of all measures caused by $\Delta a = 1$ will be much less than those in Table 13. For example, $D(e_1 \rightarrow h_1)$ increases from 0.1667 to 0.1669; $c^*(e_1 \rightarrow h_1)$ increase from 0.5 to 0.5025. The increments are about 1/10 of those in Table 13. Therefore, the increment of the degree of confirmation brought about by a new example is closely related to the number of old examples or our prior knowledge.

The absolute confirmation requires that

- the sample size n is big enough;
- each example is selected independently;
- examples are representative.

Otherwise, the degree of confirmation calculated is unreliable. We need to replace the degree of confirmation with the degree interval of confirmation, such as [0.5, 1] instead of 1.

5.3. Is Hypothesis Symmetry or Consequent Symmetry desirable?

Elles and Fitelson defined HS by $c(e, h) = -c(e, -h)$. Actually, it means $c(x, y) = -c(x, -y)$ for any x and y. Similarly, ES is Antecedent Symmetry, which means $c(x, y) = -c(-x, y)$ for any x and y. Since e and h are not the antecedent and the consequent of a major premise from their point of view, they cannot say Antecedent Symmetry and Consequent Symmetry. Consider that $c(e, h)$ becomes $c(h, e)$. According the literal meaning of HS (Hypothesis Symmetry), one may misunderstand HS as shown in Table 14.

Table 14. Misunderstood HS (Hypothesis Symmetry) and ES (Evidence Symmetry).

	HS or Consequent Symmetry	ES or Antecedent Symmetry
Misunderstood HS	$c(e, h) = -c(e, -h)$	$c(h, e) = -c(-h, e)$
Misunderstood ES	$c(h, e) = -c(h, -e)$	$c(e, h) = -c(-e, h)$

For example, the misunderstanding happens in [8,19], where the authors call $c(h, e) = -c(h, -e)$ ES. However, it is in fact HS or Consequent Symmetry. In [19], the authors think that $F(H, E)$ (where the right side is evidence) should have HS: $F(H, E) = -F(-H, E)$, whereas $F(E, H)$ should have ES: $F(E, H) = -F(-E, H)$. However, this "ES" does not accord with the original meaning of ES in [14]. Both $F(H, E)$ and $F(E, H)$ possess HS instead of ES. The more serious thing because of the misunderstanding is that [19] concludes that ES and EHS (e.g., $c(H, E) = c(-H, -E)$), as well as HS, are desirable, and hence, measures S, N, and C are particularly valuable.

The author of this paper approves the conclusion of Elles and Fitelson that only HS (e.g., Consequent Symmetry) is desirable. Therefore, it is necessary to make clear that e and h in $c(e, h)$ are the antecedent and the consequent of the rule $e \rightarrow h$. To avoid the misunderstanding, we had better replace $c(e, h)$ with $c(e \rightarrow h)$ and use "Antecedent Symmetry" and "Consequent Symmetry" instead of "Evidence Symmetry" and "Hypothesis Symmetry".

5.4. About Bayesian Confirmation and Likelihoodist Confirmation

Measure D proposed by Carnap is often referred to as the standard Bayesian confirmation measure. The above analyses, however, show that D is only suitable as a measure for selecting hypotheses instead of a measure for confirming major premises. Carnap opened the direction of Bayesian confirmation, but his explanation about D easily lets us confuse a major premise's evidence (a sample) and a consequent's evidence (a minor premise).

Greco et al. [19] call confirmation measures with conditional probability $p(h|e)$ as Bayesian confirmation measures, those with $P(e|h)$ as Likelihoodist confirmation measures, and those for $h \rightarrow e$ as converse Bayesian/Likelihoodist confirmation measures. This division is very enlightening. However, the division of confirmation measures in this paper does not depend on symbols, but on methods. The optimized proportion of the believable part in the truth function is the channel confirmation measure b^*, which is similar to the likelihood ratio, reflecting how good the channel is. The optimized proportion of the believable part in the likelihood function is the prediction confirmation measure c^*, which is similar to the correct rate, reflecting how good the probability prediction is. The b^* may be called the logical Bayesian confirmation measure because it is derived with Logical Bayesian Inference [17], although $P(e|h)$ may be used for b^*. The c^* may be regarded as the likelihoodist confirmation measure, although $P(h|e)$ may be used for c^*.

This paper also provides converse channel/prediction confirmation measures for rule $h \rightarrow e$. Confirmation measures $b^*(e \rightarrow h)$ and $c^*(e \rightarrow h)$ are related to misreporting rates, whereas converse confirmation measures $b^*(h \rightarrow e)$ and $c^*(h \rightarrow e)$ are related to underreporting rates.

5.5. About the Certainty Factor for Probabilistic Expert Systems

The Certainty Factor, which is denoted by CF, was proposed by Shortliffe and Buchanan for a backward chaining expert system [7]. It indicates how true an uncertain inference $h \rightarrow e$ is. The relationship between measures CF and Z is $CF(h \rightarrow e) = Z(e \rightarrow h)$ [36].

As pointed out by Heckerman and Shortliffe [36], the Certainty Factor method has been widely adopted in rule-based expert systems, it also has its theoretical and practical limitations. The main reason is that the Certainty Factor method is not compatible with statistical probability theory. They believe that the belief-network representation can overcome many of the limitations of the Certainty Factor model; however, the Certainty Factor model is simpler than the belief-network representation; it is possible to combine both to develop simpler probabilistic expert systems.

Measure $b^*(e_1 \rightarrow h_1)$ is related to the believable part of the truth function of predicate $e_1(h)$. It is similar to $CF(h_1 \rightarrow e_1)$. The differences are that $b^*(e_1 \rightarrow h_1)$ is independent of $P(h)$ whereas $CF(h_1 \rightarrow e_1)$ is related to $P(h)$; $b^*(e_1 \rightarrow h_1)$ is compatible with statistical probability theory whereas $CF(h_1 \rightarrow e_1)$ is not.

Is it possible to use measure b^* or c^* as the Certainty Factor to simplify belief-networks or probabilistic expert systems? This issue is worth exploring.

5.6. How Confirmation Measures F, b, and c* are Compatible with Popper's Falsification Thought*

Popper affirms that a counterexample can falsify a universal hypothesis or a major premise. However, for an uncertain major premise, how do counterexamples affect its degree of confirmation? Confirmation measures F, b^*, and c^* can reflect the importance of counterexamples. In Example 1 of Table 9, the proportion of positive examples is small, and the proportion of counterexamples is smaller still, so that the degree of confirmation is large. This example shows that to improve the degree of confirmation, it is not necessary to increase the conditional probability $P(e_1|h_1)$ (for b^*) or $P(h_1|e_1)$ (for c^*). In Example 2 of Table 9, although the proportion of positive examples is large, the proportion of counterexamples is not small so that the degree of confirmation is very small. This example shows that to raise degree of confirmation, it is not sufficient to increase the posterior probability. It is necessary and sufficient to decrease the relative proportion of counterexamples.

Popper affirms that a counterexample can falsify a universal hypothesis, which can be explained by that for the falsification of a strict universal hypothesis, it is important to have no counterexample. Now for the confirmation of a universal hypothesis that is not strict or uncertain, we can explain that it is important to have fewer counterexamples. Therefore, confirmation measures F, b^*, and c^* are compatible with Popper's falsification thought.

Scheffler and Goodman [35] proposed selective confirmation based on Popper's falsification thought. They believe that black ravens support "Ravens are black" because black ravens undermine "Ravens are not black". Their reason why non-black ravens support "Ravens are not black" is that non-black ravens undermine the opposite hypothesis "Ravens are black". Their explanation is very meaningful. However, they did not provide the corresponding confirmation measure. Measure $c^*(e_1 \to h_1)$ is what they need.

6. Conclusions

Using the semantic information and statistical learning methods and taking the medical test as an example, this paper has derived two confirmation measures $b^*(e \to h)$ and $c^*(e \to h)$. The measure b^* is similar to the measure F proposed by Kemeny and Oppenheim; it can reflect the channel characteristics of the medical test like the likelihood ratio, indicating how good a testing means is. Measure $c^*(e \to h)$ is similar to the correct rate but varies between −1 and 1. Both b^* and c^* can be used for probability predictions. The b^* is suitable for predicting the probability of disease when the prior probability of disease is changed. Measures b^* and c^* possess symmetry/asymmetry proposed by Elles and Fitelson [14], monotonicity proposed by Greco et al. [16], normalizing property (between −1 and 1) suggested by many researchers. The new confirmation measures support absolute confirmation instead of incremental confirmation.

This paper has shown that most popular confirmation measures cannot help us diagnose the infection of COVID-19, but measures F and b^* and the like, which are the functions of likelihood ratio, can. It has also proved that popular confirmation measures did not support the conclusion that a black raven could confirm more strongly than a non-black non-raven thing, such as a piece of chalk. It has shown that measure c^* could definitely deny the Equivalence Condition and exactly reflect Nicod–Fisher Criterion, and hence, could be used to eliminate the Raven Paradox. The new confirmation measures b^* and c^* as well as F indicates that fewer counterexamples' existence is more important than more positive examples' existence; therefore, measures F, b^*, and c^* are compatible with Popper's falsification thought.

When the sample is small, the degree of confirmation calculated by any confirmation measure is not reliable, and hence, the degree of confirmation should be replaced with the degree interval of confirmation. We need further studies combining the theory of hypothesis testing. It is also worth conducting further studies ensuring that the new confirmation measures are used as the Certainty Factors for belief-networks.

Supplementary Materials: The Excel File for Data in Tables 9, 12 and 13 is available online at http://survivor99.com/lcg/Table9-12-13NAT.zip. We can test different confirmation measures by changing *a*, *b*, *c*, and *d*.

Funding: This research received no external funding.

Acknowledgments: The author thanks Zhilin Zhang of Fudan University and Jianyong Zhou of Changsha University because this study benefited from communication with them. The author thanks Peizhuang Wang of Liaoning Technical University for his long-term support and encouragement. The author also thanks the anonymous reviewers for their comments and suggestions, which evidently improved this paper.

Conflicts of Interest: The author declares no conflict of interest.

References

1. Carnap, R. *Logical Foundations of Probability*, 2nd ed.; University of Chicago Press: Chicago, IL, USA, 1962.
2. Popper, K. *Conjectures and Refutations*, 1st ed.; Routledge: London, UK; New York, NY, USA, 2002.
3. Hempel, C.G. Studies in the Logic of Confirmation. *Mind* **1945**, *54*, 1–26, 97–121. [CrossRef]
4. Nicod, J. *Le Problème Logique De L'induction*; Alcan: Paris, France, 1924; p. 219, (Engl. Transl. The logical problem of induction. In *Foundations of Geometry and Induction*; Routledge: London, UK, 2000.).
5. Mortimer, H. *The Logic of Induction*; Prentice Hall: Paramus, NJ, USA, 1988.
6. Horwich, P. *Probability and Evidence*; Cambridge University Press: Cambridge, UK, 1982.
7. Shortliffe, E.H.; Buchanan, B.G. A model of inexact reasoning in medicine. *Math. Biosci.* **1975**, *23*, 351–379. [CrossRef]
8. Crupi, V.; Tentori, K.; Gonzalez, M. On Bayesian measures of evidential support: Theoretical and empirical issues. *Philos. Sci.* **2007**, *74*, 229–252. [CrossRef]
9. Christensen, D. Measuring confirmation. *J. Philos.* **1999**, *96*, 437–461. [CrossRef]
10. Nozick, R. *Philosophical Explanations*; Clarendon: Oxford, UK, 1981.
11. Good, I.J. The best explicatum for weight of evidence. *J. Stat. Comput. Simul.* **1984**, *19*, 294–299. [CrossRef]
12. Kemeny, J.; Oppenheim, P. Degrees of factual support. *Philos. Sci.* **1952**, *19*, 307–324. [CrossRef]
13. Fitelson, B. Studies in Bayesian Confirmation Theory. Ph.D. Thesis, University of Wisconsin, Madison, WI, USA, 2001.
14. Eells, E.; Fitelson, B. Symmetries and asymmetries in evidential support. *Philos. Stud.* **2002**, *107*, 129–142. [CrossRef]
15. Greco, S.; Slowiński, R.; Szczęch, I. Properties of rule interestingness measures and alternative approaches to normalization of measures. *Inf. Sci.* **2012**, *216*, 1–16. [CrossRef]
16. Greco, S.; Pawlak, Z.; Slowiński, R. Can Bayesian confirmation measures be useful for rough set decision rules? *Eng. Appl. Artif. Intell.* **2004**, *17*, 345–361. [CrossRef]
17. Lu, C. Semantic information G theory and Logical Bayesian Inference for machine learning. *Information* **2019**, *10*, 261. [CrossRef]
18. Sensitivity and specificity. Wikipedia the Free Encyclopedia. Available online: https://en.wikipedia.org/wiki/Sensitivity_and_specificity (accessed on 27 February 2020).
19. Greco, S.; Slowiński, R.; Szczech, I. Measures of rule interestingness in various perspectives of confirmation. *Inf. Sci.* **2016**, *346–347*, 216–235. [CrossRef]
20. Lu, C. A generalization of Shannon's information theory. *Int. J. Gen. Syst.* **1999**, *28*, 453–490. [CrossRef]
21. Shannon, C.E. A mathematical theory of communication. *Bell Syst. Tech. J.* **1948**, *27*, 379–429, 623–656. [CrossRef]
22. Tarski, A. The semantic conception of truth and the foundations of semantics. *Philos. Phenomenol. Res.* **1994**, *4*, 341–376. [CrossRef]
23. Davidson, D. Truth and meaning. *Synthese* **1967**, *17*, 304–323. [CrossRef]
24. Tentori, K.; Crupi, V.; Bonini, N.; Osherson, D. Comparison of confirmation measures. *Cognition* **2007**, *103*, 107–119. [CrossRef]
25. Glass, D.H. Entailment and symmetry in confirmation measures of interestingness. *Inf. Sci.* **2014**, *279*, 552–559. [CrossRef]
26. Susmaga, R.; Szczęch, I. Selected group-theoretic aspects of confirmation measure symmetries. *Inf. Sci.* **2016**, *346–347*, 424–441. [CrossRef]

27. Thornbury, I.R.; Fryback, D.G.; Edwards, W. Likelihood ratios as a measure of the diagnostic usefulness of excretory urogram information. *Radiology* **1975**, *114*, 561–565. [CrossRef]
28. Fitelson, B.; Hawthorne, J. How Bayesian confirmation theory handles the paradox of the ravens. In *The Place of Probability in Science*; Eells, E., Fetzer, J., Eds.; Springer: Dordrecht, Germany, 2010; pp. 247–276.
29. Huber, F. What Is the Point of Confirmation? *Philos. Sci.* **2005**, *72*, 1146–1159. [CrossRef]
30. Carnap, R.; Bar-Hillel, Y. *An Outline of a Theory of Semantic Information*; Technical Report No. 247; Research Lab. of Electronics, MIT: Cambridge, MA, USA, 1952.
31. Crupi, V.; Tentori, K. State of the field: Measuring information and confirmation. *Stud. Hist. Philos. Sci.* **2014**, *47*, 81–90. [CrossRef]
32. Lu, C. Semantic channel and Shannon channel mutually match and iterate for tests and estimations with maximum mutual information and maximum likelihood. In Proceedings of the 2018 IEEE International Conference on Big Data and Smart Computing, Shanghai, China, 15 January 2018; IEEE Computer Society Press Room: Washington, DC, USA, 2018; pp. 15–18.
33. Available online: http://news.cctv.com/2020/02/13/ARTIHIHFAHyTYO6NEovYRMNh200213.shtml (accessed on 13 February 2020).
34. Wang, S.; Kang, B.; Ma, J.; Zeng, X.; Xiao, M.; Guo, J.; Cai, M.; Yang, J.; Li, Y.; Meng, X.; et al. A deep learning algorithm using CT images to screen for Corona Virus Disease (COVID-19). *medRxiv* **2020**. [CrossRef]
35. Scheffler, I.; Goodman, N.J. Selective confirmation and the ravens: A reply to Foster. *J. Philos.* **1972**, *69*, 78–83. [CrossRef]
36. Heckerman, D.E.; Shortliffe, E.H. From certainty factors to belief networks. *Artif. Intell. Med.* **1992**, *4*, 35–52. [CrossRef]

© 2020 by the author. Licensee MDPI, Basel, Switzerland. This article is an open access article distributed under the terms and conditions of the Creative Commons Attribution (CC BY) license (http://creativecommons.org/licenses/by/4.0/).

Article
On a Class of Tensor Markov Fields

Enrique Hernández-Lemus [1,2]

[1] Computational Genomics Division, National Institute of Genomic Medicine, 14610 Mexico City, Mexico; ehernandez@inmegen.gob.mx; Tel.: +52-55-5350-1970
[2] Centro de Ciencias de la Complejidad, Universidad Nacional Autónoma de México, 04510 Mexico City, Mexico

Received: 6 March 2020; Accepted: 9 April 2020; Published: 16 April 2020

Abstract: Here, we introduce a class of Tensor Markov Fields intended as probabilistic graphical models from random variables spanned over multiplexed contexts. These fields are an extension of Markov Random Fields for tensor-valued random variables. By extending the results of Dobruschin, Hammersley and Clifford to such tensor valued fields, we proved that tensor Markov fields are indeed Gibbs fields, whenever strictly positive probability measures are considered. Hence, there is a direct relationship with many results from theoretical statistical mechanics. We showed how this class of Markov fields it can be built based on a statistical dependency structures inferred on information theoretical grounds over empirical data. Thus, aside from purely theoretical interest, the Tensor Markov Fields described here may be useful for mathematical modeling and data analysis due to their intrinsic simplicity and generality.

Keywords: Markov random fields; probabilistic graphical models; multilayer networks

1. General Definitions

Here, we introduce Tensor Markov Fields, i.e., Markov random fields [1,2] over tensor spaces. Tensor Markov Fields (TMFs) represent the joint probability distribution for a set of tensor-valued random variables.

Let $X = X_\alpha^\beta$ be one of such tensor-valued random variables. Here $X_i^j \in X$ may represent either a variable $i \in \alpha$, that may exist in a given context or layer $j \in \beta$ (giving rise to a class of so-called multilayer graphical models or multilayer networks) or a single tensor-valued quantity X_i^j. A TMF will be an undirected multilayer graph representing the statistical dependency structure of X as given by the joint probability distribution $\mathbb{P}(X)$.

As an extension of the case of Markov random fields, a TMF is a multilayer graph $\hat{G} = (V, E)$ formed by a set V of vertices or nodes (the X_i^j's) and a set $E \subseteq V \times V$ of edges connecting the nodes, either on the same *layer* or through different layers (Figure 1). The set of edges represents a neighborhood law N stating which vertex is connected (dependent) to which other vertex in the multilayer graph. With this in mind, a TMF can be also represented (slightly abusing notation) as $\hat{G} = (V, N)$. The set of neighbors of a given point X_i^j will be denoted $N_{X_i^j}$.

Entropy 2020, 22, 451

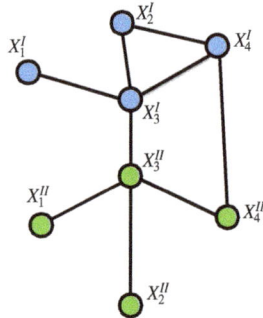

Figure 1. A Tensor Markov Field: represented as a multilayer graph spanning over X_i^j with $i = \{1, 2, 3, 4\}$ and $j = \{I, II\}$. To illustrate, layer I is colored in blue and layer II is colored green.

1.1. Configuration

It is possible to assign to each point in the multilayer graph, one of a finite set S of labels. Such assignment will be called a *configuration*. We will assign probability measures to the set Ω of all possible configurations ω. Hence, ω_A represents the configuration ω restricted to the subset A of V. It is possible to think of ω_A as a configuration on the smaller multilayer graph $\hat{G}_A$ restricting V to points of A (Figure 2).

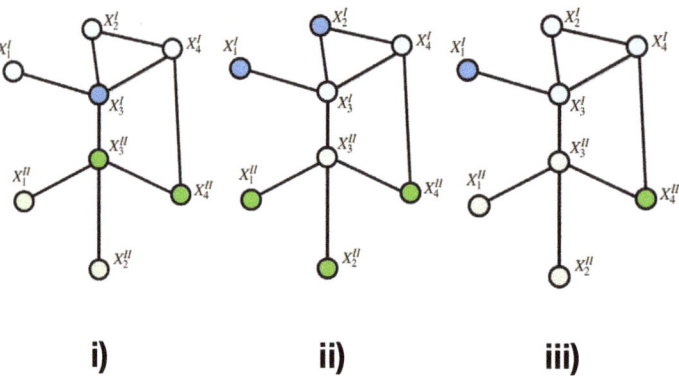

Figure 2. Three different configurations of a Tensor Markov Fieldpanels (i), (ii) and (iii) present different configurations or *states* of the TMF. Labels are represented by color intensity.

1.2. Local Characteristics

It is also possible to extend the notion of *local characteristics* from MRFs. The local characteristics of a probability measure $\mathbb{P}$ defined on Ω are the conditional probabilities of the form:

$$\mathbb{P}(\omega_t \,|\, \omega_{T \setminus t}) = \mathbb{P}(\omega_t \,|\, \omega_{N_t}) \tag{1}$$

i.e., the probability that the point t is assigned the value ω_t, given the values at all other points of the multilayer graph. In order to make explicit the tensorial nature of the multilayer graph $\hat{G}$, let us re-write Equation (1). Let us also recall the fact that the probability measure will define a tensor Markov random field (a TMF) if the local characteristics depend only of the knowledge of the outcomes at neighboring points, i.e., if for every ω

$$\mathbb{P}(\omega_{X_i^j} \mid \omega_{\hat{G}\setminus X_i^j}) = \mathbb{P}(\omega_{X_i^j} \mid \omega_{N_{X_i^j}}) \tag{2}$$

1.3. Cliques

Given an arbitrary graph (or in the present case a multilayer graph), we shall say that a set of points C is a *clique* if every pair of points in C are neighbors (see Figure 3). This definition includes the empty set as a clique. A clique is thus a set whose *induced subgraph* is complete, for this reason cliques are also called *complete induced subgraphs* or *maximal subgraphs* (although these latter term may be ambiguous).

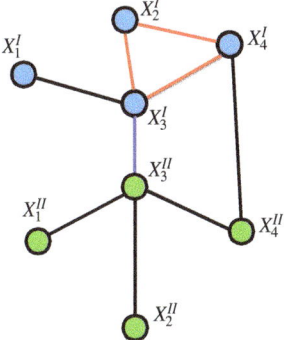

Figure 3. Cliques on a Tensor Markov Field: The set $\{X_2^I, X_3^I, X_4^I\}$ forms an intra-layer 2-clique (as marked by the red edges, all on layer I), the set $\{X_3^I, X_3^{II}\}$ forms an inter-layer 1-clique (marked by the blue edge connecting layers I and II). However, the set $\{X_3^I, X_3^{II}, X_4^I, X_4^{II},\}$ is not a clique since there are no edges between X_3^I and X_4^{II} nor between X_3^I and X_4^{II}.

1.4. Configuration Potentials

A *potential* η is a way to assign a number $\eta_A(\omega)$ to every subconfiguration ω_A of a configuration ω in the multilayer graph $\hat{G}$. Given a potential, we shall say that it defines (or better, induces) a *dimensionless energy* $U(\omega)$ on the set of all configurations ω by

$$U(\omega) = \sum_A \eta_A(\omega) \tag{3}$$

In the preceeding expression, for fixed ω, the sum is taken over all subsets $A \subseteq V$ including the empty set. We can define a probability measure, called the *Gibbs measure induced* by U as

$$\mathbb{P}(\omega) = \frac{e^{-U(\omega)}}{Z} \tag{4}$$

with Z a normalization constant called the *partition function*.

$$Z = \sum_\omega e^{-U(\omega)} \tag{5}$$

In physics, the term *potential* is often used in connection with the so-called potential energies. Physicists often call η_A a *dimensionless potential energy*, and they call $\phi_A = e^{-\eta_A}$ a potential.

Equations (4) and (5) can be thus rewritten as:

$$\mathbb{P}(\omega) = \frac{\prod_A \phi_A(\omega)}{Z} \tag{6}$$

$$Z = \sum_\omega \prod_A \phi_A(\omega) \tag{7}$$

Since this latter use is more common in probability and graph theory, we will refer to Equations (6) and (7) as the definitions of Gibbs measure and partition function (respectively) unless otherwise stated.

1.5. Gibbs Fields

A potential is called a nearest neighbor Gibbs potential if $\phi_A(\omega) = 1$ whenever A is not a clique. It is customary to refer as a *Gibbs measure* to a measure induced by a nearest neighbor Gibbs potential. However, it is possible to define more general Gibbs measures by considering other types of potentials.

The inclusion of all cliques in the calculation of the Gibbs measure is necessary to establish the equivalence between Gibbs random fields and Markov random fields. Let us see how a nearest neighbor Gibbs measure on a multilayer graph determines a TMF.

Let $\mathbb{P}(\omega)$ be a probability measure determined on Ω by a nearest neighbor Gibbs potential ϕ:

$$\mathbb{P}(\omega) = \frac{\prod_C \phi_C(\omega)}{Z} \tag{8}$$

With the product taken over all cliques C on the multilayer graph $\hat{G}$. Then,

$$\mathbb{P}(\omega_{X_i^j} | \omega_{\hat{G} \setminus X_i^j}) = \frac{\mathbb{P}(\omega)}{\sum_{\omega'} \mathbb{P}(\omega')} \tag{9}$$

Here ω' is any configuration which agrees with ω at all points except X_i^j.

$$\mathbb{P}(\omega_{X_i^j} | \omega_{\hat{G} \setminus X_i^j}) = \frac{\prod_C \phi_C(\omega)}{\sum_{\omega'} \prod_C \phi_C(\omega')} \tag{10}$$

For any clique C that does not contain X_i^j, $\phi_C(\omega) = \phi_C(\omega')$, So that all the terms that correspond to cliques that do not contain the point X_i^j cancel both from the numerator and the denominator in Equation (10), therefore this probability depends only on the values x_i^j at X_i^j and its neighbors. $\mathbb{P}$ defines thus a TMF.

A more general proof of this equivalence is given by Hammersley-Clifford theorem that will be presented in the following section.

2. Extended Hammersley Clifford Theorem

Here we will outline a proof for an extension of Hammersley-Clifford theorem for Tensor Markov Fields (i.e., we will show that a Tensor Markov Field is equivalent to a Tensor Gibbs Field).

Let $\hat{G} = (V, N)$ be a multilayer graph representing a TMF as defined in the previous section. With $V = X_\alpha^\beta = \{X_i^j\}$, a set of vertices over a tensor field and N a neighborhood law that connects vertices over this tensor field. The field $\hat{G}$ obeys the following neighborhood law given its Markov property (see Equation (2))

$$\mathbb{P}(X_i^j | X_{\hat{G} \setminus X_i^j}) = \mathbb{P}(X_i^j | X_{N_i^j}) \tag{11}$$

Here $X_{N_i^j}$ is any neighbor of X_i^j. The Hammersley-Clifford theorem states that a MRF is also a local Gibbs field. In the case of a TMF we have the following expression:

$$\mathbb{P}(X) = \frac{1}{Z} \prod_{c \in C_{\hat{G}}} \phi_c(X_c) \tag{12}$$

In order to prove the equivalence of Equations (11) and (12), we will first bult a deductive (backward direction) part of the proof to be complemented with a constructive (forward direction) part as presented in the following subsections.

2.1. Backward Direction

Let us consider Equation (11) at the light of Bayes' theorem:

$$\mathbb{P}(X_i^j | X_{\hat{G} \setminus X_i^j}) = \frac{\mathbb{P}(X_i^j, X_{N_i^j})}{\mathbb{P}(X_{N_i^j})} \tag{13}$$

Using a clique-approach to calculate the joint and marginal probabilities (see next subsection to support the following statement):

$$\mathbb{P}(X_i^j | X_{\hat{G} \setminus X_i^j}) = \frac{\sum_{\hat{G} \setminus D_i^j} \prod_{c \in C_{\hat{G}}} \phi_c(X_c)}{\sum_{X_i^j} \sum_{\hat{G} \setminus D_i^j} \prod_{c \in C_{\hat{G}}} \phi_c(X_c)} \tag{14}$$

Let us split the product $\prod_{c \in C_{\hat{G}}} \phi_c(X_c)$ into two products, one over the set of cliques that contain X_i^j (let us call it C_i^j) and another set formed by cliques not containing X_i^j (let us call it R_i^j):

$$\mathbb{P}(X_i^j | X_{\hat{G} \setminus X_i^j}) = \frac{\sum_{\hat{G} \setminus D_i^j} \prod_{c \in C_i^j} \phi_c(X_c) \prod_{c \in R_i^j} \phi_c(X_c)}{\sum_{X_i^j} \sum_{\hat{G} \setminus D_i^j} \prod_{c \in C_i^j} \phi_c(X_c) \prod_{c \in R_i^j} \phi_c(X_c)} \tag{15}$$

Factoring out the terms depending on X_i^j (that do not contribute to cliques in the domain $\hat{G} \setminus X_i^j$):

$$\mathbb{P}(X_i^j | X_{\hat{G} \setminus X_i^j}) = \frac{\prod_{c \in C_i^j} \phi_c(X_c) \sum_{\hat{G} \setminus D_i^j} \prod_{c \in R_i^j} \phi_c(X_c)}{\sum_{X_i^j} \prod_{c \in C_i^j} \phi_c(X_c) \sum_{\hat{G} \setminus D_i^j} \prod_{c \in R_i^j} \phi_c(X_c)} \tag{16}$$

The term $\sum_{\hat{G} \setminus D_i^j} \prod_{c \in R_i^j} \phi_c(X_c)$ does not involve X_i^j (by construction) so, it can be factored out from the summation over X_i^j in the denominator.

$$\mathbb{P}(X_i^j | X_{\hat{G} \setminus X_i^j}) = \frac{\prod_{c \in C_i^j} \phi_c(X_c) \sum_{\hat{G} \setminus D_i^j} \prod_{c \in R_i^j} \phi_c(X_c)}{\sum_{\hat{G} \setminus D_i^j} \prod_{c \in R_i^j} \phi_c(X_c) \sum_{X_i^j} \prod_{c \in C_i^j} \phi_c(X_c)} \tag{17}$$

We can cancel the term in the numerator and denominator:

$$\mathbb{P}(X_i^j | X_{\hat{G} \setminus X_i^j}) = \frac{\prod_{c \in C_i^j} \phi_c(X_c)}{\sum_{X_i^j} \prod_{c \in C_i^j} \phi_c(X_c)} \tag{18}$$

Then we multiply by $\frac{\prod_{c \in R_i^j} \phi_c(X_c)}{\prod_{c \in R_i^j} \phi_c(X_c)}$

$$\mathbb{P}(X_i^j | X_{\hat{G} \setminus X_i^j}) = \frac{\prod_{c \in C_i^j} \phi_c(X_c) \prod_{c \in R_i^j} \phi_c(X_c)}{\sum_{X_i^j} \prod_{c \in C_i^j} \phi_c(X_c) \prod_{c \in R_i^j} \phi_c(X_c)} \qquad (19)$$

Remembering that $C_i^j \cup R_i^j = C_{\hat{G}}$,

$$\mathbb{P}(X_i^j | X_{\hat{G} \setminus X_i^j}) = \frac{\prod_{c \in \hat{G}} \phi_c(X_c)}{\sum_{X_i^j} \prod_{c \in \hat{G}} \phi_c(X_c)} \qquad (20)$$

Equation (20) is nothing but the definition of a local Gibbs Tensor Field (Equation (12)).

2.2. Forward Direction

In this subsection we will show how to express the clique potential functions $\phi_c(X_c)$, given the joint probability distribution over the tensor field and the Markov property.

Consider any subset $\sigma \subset \hat{G}$ of the multilayer graph $\hat{G}$. We define a candidate potential function (following Möbius inversion lemma) [3] as follows:

$$f_\sigma(X_\sigma = x_\sigma) = \prod_{\zeta \subset \sigma} \mathbb{P}(X_\zeta = x_\zeta, X_{\hat{G} \setminus \zeta} = 0)^{-1^{|\sigma| - |\zeta|}} \qquad (21)$$

In order for f_σ to be a proper clique potential, it must satisfy the following two conditions:

(i) $\prod_{\sigma \subset \hat{G}} f_\sigma(X_\sigma) = \mathbb{P}(X)$
(ii) $f_\sigma(X_\sigma) = 1$ whenever σ is not a clique

To prove (i), we need to show that all factors in $f_\sigma(X_\sigma = x_\sigma)$ cancel out, except for $\mathbb{P}(X)$. To do this, it will be useful to consider the following *combinatorial expansion of zero*:

$$0 = (1-1)^K = C_0^K - C_1^K + C_2^K + \cdots + (-1)^K C_K^K \qquad (22)$$

Here, of course C_B^A is the number of combinations of B elements from an A-element set.

Let us consider any subset ζ of $\hat{G}$. Let us consider a factor $\Delta = \mathbb{P}(X_\zeta = x_\zeta, X_{\hat{G} \setminus \zeta} = 0)$. For the case of $f\zeta(X_\zeta)$ it occurs as $\Delta^{-1^0} = \Delta$. Such factor also occurs in subsets contaning ζ and other additional elements. If it includes ζ and one additional element, there are $C_1^{|\hat{G}| - |\zeta|}$ such functions. The additional element creates an inverse factor $\Delta^{-1^1} = \Delta^{-1}$. The functions over subsets contaning ζ and two additional elements contributes with a factor $\Delta^{-1^2} = \Delta^1 = \Delta$. If we continue this process and consider Equation (22), it is evident that all odd cardinality difference terms cancel out with all even cardinality difference terms so that the only remaining factor corresponds to $\zeta = \hat{G}$ equal to $\mathbb{P}(X)$ thus fulfilling condition (i).

In order to show how condition (ii) is fulfilled, we will need to use the Markov property of TMFs. Let us consider $\sigma* \subset \hat{G}$ that is not a clique. Then it will be possible to find two nodes X_i^h and X_j^k in $\sigma*$ that are not connected to each other. Let us recall Equation (21):

$$f_\sigma(X_{\sigma*} = x_{\sigma*}) = \prod_{\zeta \subset \sigma*} \mathbb{P}(X_\zeta = x_\zeta, X_{\hat{G} \setminus \zeta} = 0)^{-1^{|\sigma*| - |\zeta|}} \qquad (23)$$

An arbitrary subset ζ may belong to any of the following classes: (i) $\zeta = \omega$ a generic subset of σ; (ii) $\zeta = \omega \cup \{X_i^h\}$; (iii) $\zeta = \omega \cup \{X_j^k\}$ or iv) $\zeta = \omega \cup \{X_i^h, X_j^k\}$. If we write down Equation (23) factored down to these contributions we get:

$$f_\sigma(X_{\sigma*} = x_{\sigma*}) = \prod_{\omega \subset \sigma* \setminus \{X_i^h, X_j^k\}} \left[\frac{\mathbb{P}(X_\omega, X_{\hat{G} \setminus \omega} = 0) \mathbb{P}(X_{\omega \cup \{X_i^h, X_j^k\}}, X_{\hat{G} \setminus \omega \cup \{X_i^h, X_j^k\}} = 0)}{\mathbb{P}(X_{\omega \cup \{X_i^h\}}, X_{\hat{G} \setminus \omega \cup \{X_i^h\}} = 0) \mathbb{P}(X_{\omega \cup \{X_j^k\}}, X_{\hat{G} \setminus \omega \cup \{X_j^k\}} = 0)} \right]^{-1^{|\sigma*|-|\xi|}} \tag{24}$$

Let us consider two of the factors in Equation (24) at the light of Bayes' theorem:

$$\frac{\mathbb{P}(X_\omega, X_{\hat{G} \setminus \omega} = 0)}{\mathbb{P}(X_{\omega \cup \{X_i^h\}}, X_{\hat{G} \setminus \omega \cup \{X_i^h\}} = 0)} = \frac{\mathbb{P}(X_{\{X_i^h\}} = 0 | X_{\{X_j^k\}} = 0, X_\omega, X_{\hat{G} \setminus \omega \cup \{X_i^h, X_j^k\}} = 0) \mathbb{P}(X_{\{X_j^k\}} = 0, X_\omega, X_{\hat{G} \setminus \omega \cup \{X_i^h, X_j^k\}} = 0)}{\mathbb{P}(X_{\{X_i^h\}} | X_{\{X_j^k\}} = 0, X_\omega, X_{\hat{G} \setminus \omega \cup \{X_i^h, X_j^k\}} = 0) \mathbb{P}(X_{\{X_j^k\}} = 0, X_\omega, X_{\hat{G} \setminus \omega \cup \{X_i^h, X_j^k\}} = 0)} \tag{25}$$

We can notice that the priors in the numerator and denominator of Equation (25) are the same. We can then cancel them out. Since by definition X_i^h and X_j^k are conditionally independent given the rest of the multilayer graph, we can also replace the default value $X_j^k = 0$ for X_j^k instead.

$$\frac{\mathbb{P}(X_\omega, X_{\hat{G} \setminus \omega} = 0)}{\mathbb{P}(X_{\omega \cup \{X_i^h\}}, X_{\hat{G} \setminus \omega \cup \{X_i^h\}} = 0)} = \frac{\mathbb{P}(X_{\{X_i^h\}} = 0 | X_{\{X_j^k\}}, X_\omega, X_{\hat{G} \setminus \omega \cup \{X_i^h, X_j^k\}} = 0) \mathbb{P}(X_{\{X_j^k\}}, X_\omega, X_{\hat{G} \setminus \omega \cup \{X_i^h, X_j^k\}} = 0)}{\mathbb{P}(X_{\{X_i^h\}} | X_{\{X_j^k\}}, X_\omega, X_{\hat{G} \setminus \omega \cup \{X_i^h, X_j^k\}} = 0) \mathbb{P}(X_{\{X_j^k\}}, X_\omega, X_{\hat{G} \setminus \omega \cup \{X_i^h, X_j^k\}} = 0)} \tag{26}$$

Since X_i^h and X_j^k are conditionally independent given the rest of the multilayer graph, we can also replace the condition for X_j^k with any other, without affecting X_i^h. By adjusting this prior *conveniently*, we can write out:

$$\frac{\mathbb{P}(X_\omega, X_{\hat{G} \setminus \omega} = 0)}{\mathbb{P}(X_{\omega \cup \{X_i^h\}}, X_{\hat{G} \setminus \omega \cup \{X_i^h\}} = 0)} = \frac{\mathbb{P}(X_{\omega \cup \{X_j^k\}}, X_{\hat{G} \setminus \omega \cup \{X_j^k\}} = 0)}{\mathbb{P}(X_{\omega \cup \{X_i^h, X_j^k\}}, X_{\hat{G} \setminus \omega \cup \{X_i^h, X_j^k\}} = 0)} \tag{27}$$

By substituting Equation (27) into Equation (24) we get (condition (ii)):

$$f_\sigma * (X_{\sigma*}) = 1 \tag{28}$$

3. An Information-Theoretical Class of Tensor Markov Fields

Let us consider again the set of tensor-valued random variables $X = X_\alpha^\beta$. It is possible to calculate, for every duplex in X, the mutual information function $I(\cdot, \cdot)$ [4]:

$$I(X_i^h, X_j^k) = \sum_\Omega \sum_{\Omega'} p(X_i^h, X_j^k) \log \frac{p(X_i^h, X_j^k)}{p(X_i^h) p(X_j^k)} \tag{29}$$

Let us consider a multilayer graph scenario. From now on, the indices i, j will refer to the random variables, whereas h, k will be indices for the layers. Ω and Ω' are the respective sampling spaces (that may, of course, be equal). In order to discard self-information, let us define the *off-diagonal mutual information* as follows:

$$I^\dagger(X_i^h, X_j^k) = I(X_i^h, X_j^k) \times \left(1 - \delta_{X_i^h X_j^k}\right) \tag{30}$$

With the bi-delta function $\delta_{X_i^h X_j^k}$ defined as:

$$\delta_{X_i^h X_j^k} = \begin{cases} 1, & \text{if } i = j \text{ and } h = k \\ 0, & \text{otherwise} \end{cases} \tag{31}$$

By having the complete set of off-diagonal mutual information functions for all the random variables and layers, it is possible to define the following hyper-matrix elements:

$$A_{ij}^{hk} = \Theta\left[I^\dagger(X_i^h, X_j^k) - I_0\right] \tag{32}$$

as well as:

$$W_{ij}^{hk} = A_{ij}^{hk} \circ I^\dagger(X_i^h, X_j^k) \tag{33}$$

Here $\Theta[\cdot]$ is Heavyside's function and I_0 is a lower bound for mutual information (a threshold) to be considered *significant*.

We call A_{ij}^{hk} and W_{ij}^{hk} the *adjacency hypermatrix* and the *strength hypermatrix* respectively (notice that $\circ$ in Equation (33) represents the product of a scalar times a hypermatrix). The adjacency hyper-matrix and the strength hyper-matrix define the (unweighted and weighted, respectively) neighborhood law of the associated TMF, hence the statistical dependency structure for the set of random variables and contexts (layers).

Although the adjacency and strength hypermatrices are indeed, proper representations of the undirected (unweighted and weighted) dependency structure of $\mathbb{P}(X)$, it has been considered advantageous to embed them into a tensor linear structure, in order to be able to work out some of the mathematical properties of such fields relying on the methods of tensor algebra. One relevant proposal in this regard, has been advanced by De Domenico and collaborators, in the context of multilayer networks.

Following the ideas of De Domenico and co-workers [5], we introduce the unweighted and weighted adjacency 4-tensors (respectively) as follows:

$$\mathbb{A} = \sum_{h,k=1}^{L} \sum_{i,j=1}^{N} A_{ij}^{hk} \otimes \zeta_{\beta\delta}^{\alpha\gamma} \tag{34}$$

$$\mathbb{W} = \sum_{h,k=1}^{L} \sum_{i,j=1}^{N} W_{ij}^{hk} \otimes \zeta_{\beta\delta}^{\alpha\gamma} \tag{35}$$

Here, $\zeta_{\beta\delta}^{\alpha\gamma} = \zeta_{\beta\delta}^{\alpha\gamma}[ijhk]$ is a unit four-tensor whose role is to provide the hypermatrices with the desired linear properties (projections, contractions, etc.). Square brackets indicate that the indices i, j, h and k belong to the α, β, γ and δ dimensions and $\otimes$ represents a form of a *tensor matricization* product (i.e., the one producing a 4-tensor out of a 4-index hypermatrix times a unitary 4-tensor).

3.1. Conditional Independence in Tensor Markov Fields

In order to discuss the conditional independence structure induced by the present class of TMFs, let us analyze Equation (32). As already mentioned, the hyper-adjacency matrix A_{ij}^{hk} represents the neigborhood law (as given by the Markov property) on the multilayer graph $\hat{G}$ (i.e., the TMF). Every non-zero entry on this hypermatrix represents a statistical dependence relation between two elements on X. The conditional dependence structure on TMFs inferred from mutual information measures via Equation (32) are related not only to the statistical independence conditions (as given by a zero mutual information measure between two elements), but also to the lower bound I_0 and in general to the dependency structure of the whole multilayer graph.

The definition of conditional independence (CI) for tensor random variables is as follows:

$$(X_i^h \perp\!\!\!\perp X_j^k) | X_l^m \iff \mathbb{F}_{X_i^h, X_j^k | X_l^m = X_l^m*}(X_i^h*, X_j^k*) = \mathbb{F}_{X_i^h | X_l^m = X_l^m*}(X_i^h*) \cdot \mathbb{F}_{X_j^k | X_l^m = X_l^m*}(X_j^k*) \tag{36}$$

$\forall \; X_i^h, X_j^k, X_l^m \in X.$

Here $\perp\!\!\!\perp$ represents conditional independence between two random variables, were $\mathbb{F}_{X_i^h, X_j^k | X_l^m = X_l^m*}(X_i^h*, X_j^k*) = Pr(X_i^h \leq X_i^h*, X_j^k \leq X_j^k * | X_l^m = X_l^m*)$ is the joint conditional cumulative distribution of X_i^h and X_j^k given X_l^m and X_i^h*, X_j^k* and X_l^m* are realization events of the corresponding random variables.

In the case of MRFs (and by extension TMFs), CI is defined by means of (multi)graph separation: in this sense we say that $X_i^h \perp\!\!\!\perp_{\hat{G}} X_j^k | X_l^m$ iff X_l^m separates X_i^h from X_j^k in the multilayer graph $\hat{G}$. This means that if we remove node X_l^m there are no undirected paths from X_i^h to X_j^k in $\hat{G}$.

Conditional independence in random fields is often considered in terms of subsets of V. Let A, B and C be three subsets of V. The statement $X_A \perp\!\!\!\perp_{\hat{G}} X_B | X_C$, which holds only iff C separates A from B in the multilayer graph $\hat{G}$, meaning that if we remove all vertices in C there will be no paths connecting any vertex in A to any vertex in B is called the *global Markov property* of TMFs.

The smallest set of vertices that renders a vertex X_i^h conditionally independent of all other vertices in the multilayer graph is called its *Markov blanket*, denoted $mb(X_i^h)$. If we define the *closure* of a node X_i^h as $\mathcal{C}(X_i^h)$ then $X_i^h \perp\!\!\!\perp \hat{G} \setminus \mathcal{C}(X_i^h) | mb(X_i^h)$.

It is possible to show that in a TMF, the Markov blanket of a vertex is its set of first neighbors. This is called the *undirected local Markov property*. Starting from the local Markov property it is possible to show that two vertices X_i^h and X_j^k are conditionally independent given the rest if there is no direct edge between them. This has been called the *pairwise Markov property*.

If we denote by $\hat{G}_{X_i^h \to X_j^k}$ the set of undirected paths in the multilayer graph $\hat{G}$ connecting vertices X_i^h and X_j^k, then the pairwise Markov property of a TMF can be stated as:

$$X_i^h \perp\!\!\!\perp X_j^k | \hat{G} \setminus \{X_i^h, X_j^k\} \iff \hat{G}_{X_i^h \to X_j^k} = \emptyset \tag{37}$$

It is clear that the global Markov property implies the local Markov property which in turn implies the pairwise Markov property. For systems with positive definity probability densities, it has been probed (in the case of MRFs) that pairwise Markov actually implied global Markov (See [6] p. 119 for a proof). For the present extension this is important since it is easier to assess pairwise conditional independence statements.

3.2. Indepence Maps

Let $I_{\hat{G}}$ denote the set of all conditional independence relations encoded by the multilayer graph $\hat{G}$ (i.e., those CI relations given by the Global Markov property). Let $I_\mathbb{P}$ be the set of all CI relations implied by the probability distribution $\mathbb{P}(X_i^j)$. A multilayer graph $\hat{G}$ will be called an *independence map* (I-map) for a probability distribution $\mathbb{P}(X_i^j)$, if all CI relations implied by $\hat{G}$ hold for $\mathbb{P}(X_i^j)$, i.e., $I_{\hat{G}} \subseteq I_\mathbb{P}$ [6].

The converse statement is not necessarily true, i.e., there may be some CI relations implied by $\mathbb{P}(X_i^j)$ that are not encoded in the multilayer graph $\hat{G}$. We may be usually interested in *minimal I-maps*, i.e., I-maps from which none of the edges could be removed without destroying its CI properties.

Every distribution has a unique minimal I-map (and a given graph representation). Let $\mathbb{P}(X_i^j) > 0$. Let $\hat{G}^\dagger$ be the multilayer graph obtained by introducing edges between all pairs of vertices X_i^h, X_j^k such that $X_i^h \perp\!\!\!\perp X_j^k | X \setminus \{X_i^h, X_j^k\}$, then $\hat{G}^\dagger$ is the unique minimal I-map. We call $\hat{G}$ a *perfect map* of $\mathbb{P}$ when there is no dependencies $\hat{G}$ which are not indicated by $\mathbb{P}$, i.e., $I_{\hat{G}} = I_\mathbb{P}$ [6].

3.3. Conditional Independence Tests

Conditional independence tests are useful to evaluate whether CI conditions apply either exactly or in the case of applications under a certain bounded error. In order to be able to write down expressions for C.I. tests let us introduce the following *conditional kernels* [7]:

$$\mathbb{C}_A(B) = \mathbb{P}(B|A) = \frac{\mathbb{P}(AB)}{\mathbb{P}(A)} \tag{38}$$

as well as their generalized recursive relations:

$$\mathbb{C}_{ABC}(D) = \mathbb{C}_{AB}(D|C) = \frac{\mathbb{C}_{AB}(CD)}{\mathbb{C}_{AB}(C)} \tag{39}$$

The conditional probability of X_h^k given X_i^j can be thus written as:

$$\mathbb{C}_{X_i^j}(X_h^k) = \mathbb{P}(X_h^k|X_i^j) = \frac{\mathbb{P}(X_h^k, X_i^j)}{\mathbb{P}(X_i^j)} \qquad (40)$$

We can then write down expressions for Markov conditional independence as follows:

$$X_i^j \perp\!\!\!\perp X_h^k | X_l^m \Rightarrow \mathbb{P}(X_i^j, X_h^k | X_l^m) = \mathbb{P}(X_i^j | X_l^m) \times \mathbb{P}(X_h^k | X_l^m) \qquad (41)$$

Following Bayes' theorem, CI conditions –in this case– will be of the form:

$$\mathbb{P}(X_i^j, X_h^k | X_l^m) = \frac{\mathbb{P}(X_i^j, X_l^m)}{\mathbb{P}(X_l^m)} \times \frac{\mathbb{P}(X_h^k, X_l^m)}{\mathbb{P}(X_l^m)} = \frac{\mathbb{P}(X_i^j, X_l^m) \times \mathbb{P}(X_h^k, X_l^m)}{\mathbb{P}(X_l^m)^2} \qquad (42)$$

Equation (42) is useful since in large scale data applications is computationally cheaper to work with joint and marginal probabilities rather than conditionals.

Now let us consider the case of conditional independence given several conditional variables. The case for CI given two variables could be written—using conditional kernels—as follows:

$$X_i^j \perp\!\!\!\perp X_h^k | X_l^m, X_n^o \Rightarrow \mathbb{P}(X_i^j, X_h^k | X_l^m, X_n^o) = \mathbb{P}(X_i^j | X_l^m, X_n^o) \times \mathbb{P}(X_h^k | X_l^m, X_n^o) \qquad (43)$$

Hence,

$$\mathbb{P}(X_i^j, X_h^k | X_l^m, X_n^o) = \mathbb{C}_{X_l^m, X_n^o}(X_i^j) \times \mathbb{C}_{X_l^m, X_n^o}(X_h^k) \qquad (44)$$

Using Bayes' theorem,

$$\mathbb{P}(X_i^j, X_h^k | X_l^m, X_n^o) = \frac{\mathbb{P}(X_i^j, X_l^m, X_n^o)}{\mathbb{P}(X_l^m, X_n^o)} \times \frac{\mathbb{P}(X_h^k, X_l^m, X_n^o)}{\mathbb{P}(X_l^m, X_n^o)} \qquad (45)$$

$$\mathbb{P}(X_i^j, X_h^k | X_l^m, X_n^o) = \frac{\mathbb{P}(X_i^j, X_l^m, X_n^o) \times \mathbb{P}(X_h^k, X_l^m, X_n^o)}{\mathbb{P}(X_l^m, X_n^o)^2} \qquad (46)$$

In order to generalize the previous results to CI relations given an arbitrary set of conditionals, let us consider the following *sigma-algebraic* approach:

Let Σ_{ih}^{jk} be the σ-algebra of all subsets of X that do not contain X_i^j or X_h^k. If we consider the contravariant index $i \in \alpha$ with $i = 1, 2, \ldots, N$ and the covariant index $j \in \beta$ with $j = 1, 2, \ldots, L$, then there are $\mathcal{M} = \frac{NL}{2}$ such σ-algebras in X (let us recall that TMFs are *undirected* graphical models).

A relevant problem for network reconstruction is that of establishing the more general Markov pairwise CI conditions, i.e., the CI relations for every edge not drawn in the graph. Two arbitrary nodes X_i^j and X_h^k are conditionally independent given the rest of the graph iff:

$$X_i^j \perp\!\!\!\perp X_h^k | \Sigma_{ih}^{jk} \Rightarrow \mathbb{P}(X_i^j, X_h^k | \Sigma_{ih}^{jk}) = \mathbb{P}(X_i^j | \Sigma_{ih}^{jk}) \times \mathbb{P}(X_h^k | \Sigma_{ih}^{jk}) \qquad (47)$$

By using conditional kernels, the recursive relations and Bayes' theorem it is possible to write down $\mathcal{M}$ expressions of the form:

$$\mathbb{P}(X_i^j, X_h^k | \Sigma_{ih}^{jk}) = \frac{\mathbb{P}(X_i^j, \Sigma_{ih}^{jk}) \times \mathbb{P}(X_h^k, \Sigma_{ih}^{jk})}{\mathbb{P}(\Sigma_{ih}^{jk})^2} \qquad (48)$$

The family of Equations (48) represent the CI relations for all the non-existing edges in the hypergraph $\hat{G}$, i.e., every pair of nodes X_i^j and X_h^k not-connected in $\hat{G}$ must be conditionally independent given the rest of the nodes in the graph. These expression may serve to implement exact tests or optimization strategies for *graph reconstruction* and/or *graph sparsification* in applications considering a mutual information threshold I_0 as in Equation (32).

In brief, for every node pair with a mutual information value lesser than I_0, the presented graph reconstruction approach will not draw an edge, hence implying CI between the two nodes given the rest. Such CI condition may be tested on the data to see whether it holds or the threshold itself can be determined by resorting to optimization schemes (e.g. error bounds) in Equation (48).

4. Graph Theoretical Features and Multilinear Structure

Once the probabilistic properties of TMFs have been set, it may be fit to briefly present some of their graph theoretical features, as well as some preliminaries as to the reasons to embed hyperadjecency matrices into multilayer adjacency tensors. Given that TMFs are indeed PGMs, some of their graph characteristics will result relevant here.

Since the work by De Domenico and coworkers [5] covers in great detail how the multilinear structure of the multilayer adjacency tensor allows the calculation of these quantities—usually as projection operations—we will only mention connectivity degree vectors since these are related with the size of the TMF dependency neighborhoods.

Let us recall multilayer adjacency tensors, as defined in Equations (34) and (35). To ease presentation, we will work with the unweighted tensor $\mathbb{A}^{\alpha\gamma}_{\beta\delta}$ (Equation (34)). The *multidegree centrality vector* K^α which contains the connectivity degrees of the nodes spanning different layers can be written as follows:

$$K^\alpha = \mathbb{A}^{\alpha\gamma}_{\beta\delta} U^\delta_\gamma u^\beta \qquad (49)$$

Here U^δ_γ is a rank 2 tensor that contains a 1 in every component and u^β is a rank 1 tensor that contains a 1 in every component—these quantities are called 1—tensors by De Domenico and coworkers [5]. It can be shown that K^α is indeed given by the sums of the *connectivity degree vectors* k^α corresponding to all different layers:

$$K^\alpha = \sum_{h=1}^{L} \sum_{k=1}^{L} k^\alpha(hk) \qquad (50)$$

$k^\alpha(hk)$ is the vector of connections that nodes in the set $\alpha = 1, 2, \ldots, N$ in layer h have to any other nodes in layer k. Whereas K^α is the vector with connections in all the layers. Appropriate projections will yield measures such as the size of the neighborhood to a given vertex $|N_{X^j_i}|$, the size of its Markov blanket $|mb(X^h_i)|$, or other similar quantities.

5. Specific Applications

After having considered some of the properties of this class of Tensor Markov Fields, it may become evident that aside from purely theoretical importance, there is a number of important applications that may arise as probabilistic graphical models in tensor valued problems, among the ones that are somewhat evident are the following:

- The analysis of multidimensional biomolecular networks such as the ones arising from multi-omic experiments (For a real-life example, see Figure 4) [8–10];
- Probabilistic graphical models in computer vision (especially 3D reconstructions and 4D [3D+time] rendering) [11];
- The study of fracture mechanics in continuous deformable media [12];
- Probabilistic network models for seismic dynamics [13];
- Boolean networks in control theory [14].

Some of these problems are being treated indeed as multiple instances of Markov fields or as multipartite graphs or hypergraphs. However, it may become evident that when random variables *across layers* are interdependent (which is often the case), the definitions of potentials, cliques and partition functions, as well as the conditional statistical independence features become manageable (and in some cases even meaningful) under the presented formalism of Tensor Markov Fields.

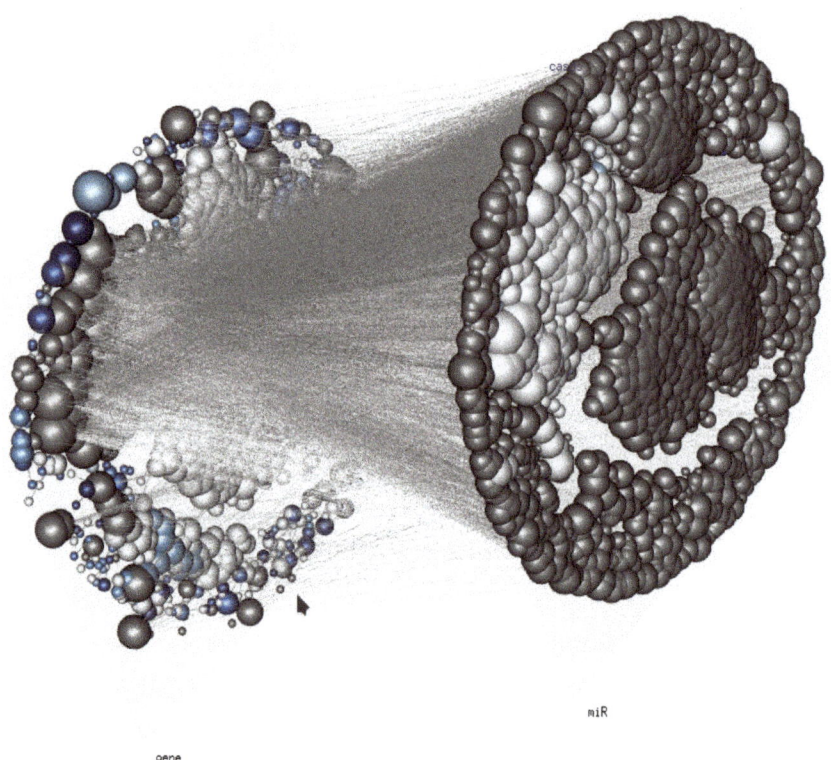

Figure 4. Gene and microRNA regulatory network: A Tensor Markov Field depicting the statistical dependence of genome wide gene and microRNA (miR) on a human phenotype. Edge width is given by the mutual information $I^\dagger(X_i^j, X_h^k)$ between expression levels of genes (layer j) and miRs (layer k) in a very large corpus of RNASeq samples, vertex size is proportional to the *degree*, i.e., the size of the node's neighborhood, $N_{X_i^j}$.

6. Conclusions

Here we have presented the definitions and fundamental properties of Tensor Markov Fields, i.e., random Markov fields over tensor spaces. We have proved –by extending the results of Dobruschin, Hammersley and Clifford to such tensor valued fields– that tensor Markov fields are indeed Gibbs fields whenever strictly positive probability measures are considered. We also introduced a class of tensor Markov fields obtained by using information theoretical statistical dependence measures inducing local and global Markov properties, and show how these can be used as probabilistic graphical models in multi-context environments much in the spirit of the so-called multilayer network approach. Finally, we discuss the convenience of embedding tensor Markov fields in the multilinear tensor representation of multilayer networks.

Funding: This research received no external funding.

Conflicts of Interest: The author declares no conflict of interest.

References

1. Dobruschin, R.L. The description of a random field by means of conditional probabilities and conditions of its regularity. *Theory Probab. Appl.* **1968**, *13*, 197–224. [CrossRef]
2. Grimmett, G.R. A theorem about random fields. *Bull. Lond. Math. Soc.* **1973**, *5*, 81–84. [CrossRef]
3. Rota, G.C. On the foundations of combinatorial theory I: Theory of Möbius functions. *Probab. Theory Relat. Fields* **1964**, *2*, 340–368. [CrossRef]
4. Cover, T.M.; Thomas, J.A. *Elements of Information Theory*; John Wiley & Sons: Hoboken, NJ, USA, 2012.
5. De Domenico, M.; Solé-Ribalta, A.; Cozzo, E.; Kivelä, M.; Moreno, Y.; Porter, M.A.; Gómez, S.; Arenas, A. Mathematical formulation of multilayer networks. *Phys. Rev. X* **2013**, *3*, 041022. [CrossRef]
6. Koller, D.; Friedman, N. Probabilistic Graphical Models: Principles and Techniques (Adaptive Computation and Machine Learning series). *Mit Press. Aug* **2009**, *31*, 2009.
7. Williams, D. *Probability with Martingales*; Cambridge University Press: Cambridge, UK, 1991.
8. Hernández-Lemus, E.; Espinal-Enríquez, J.; de Anda-Jáuregui, G. Probabilistic multilayer networks. *arXiv* **2018**, arXiv:1808.07857.
9. de Anda-Jauregui, G.; Hernandez-Lemus, E. Computational Oncology in the Multi-Omics Era: State of the Art. *Front. Oncol.* **2020**, *10*, 423. [CrossRef]
10. Hernández-Lemus, E.; Reyes-Gopar, H.; Espinal-Enríquez, J.; Ochoa, S. The Many Faces of Gene Regulation in Cancer: A Computational Oncogenomics Outlook. *Genes* **2019**, *10*, 865. [CrossRef] [PubMed]
11. McGee, F.; Ghoniem, M.; Melançon, G.; Otjacques, B.; Pinaud, B. The state of the art in multilayer network visualization. In *Computer Graphics Forum*; John Wiley & Sons: Hoboken, NJ, USA, 2019; Volume 38, pp. 125–149.
12. Krejsa, M.; Koubová, L.; Flodr, J.; Protivínský, J.; Nguyen, Q.T. Probabilistic prediction of fatigue damage based on linear fracture mechanics. *Fract. Struct. Integr.* **2017**, *39*, 143–159. [CrossRef]
13. Abe, S.; Suzuki, N. Complex network of earthquakes. In *International Conference on Computational Science*; Springer: Berlin/Heidelberg, Germany, 2004; pp. 1046–1053.
14. Liu, F.; Cui, Y.; Wang, J.; Ji, D. Observability of probabilistic Boolean multiplex networks. *Asian J. Control.* **2020**, *1*, 1–8. doi:10.1002/asjc.2290. [CrossRef]

© 2020 by the authors. Licensee MDPI, Basel, Switzerland. This article is an open access article distributed under the terms and conditions of the Creative Commons Attribution (CC BY) license (http://creativecommons.org/licenses/by/4.0/).

Article

Objective Bayesian Inference in Probit Models with Intrinsic Priors Using Variational Approximations

Ang Li [†,‡], Luis Pericchi [*,†,‡] and Kun Wang [†,‡]

Río Piedras Campus, University of Puerto Rico, 00925 San Juan, Puerto Rico; ang.li@upr.edu (A.L.); kun.wang@upr.edu (K.W.)
* Correspondence: luis.pericchi@upr.edu
† 14 Ave. Universidad Ste. 1401, San Juan, PR 00925, USA.
‡ These authors contributed equally to this work.

Received: 3 March 2020; Accepted: 26 April 2020; Published: 30 April 2020

Abstract: There is not much literature on objective Bayesian analysis for binary classification problems, especially for intrinsic prior related methods. On the other hand, variational inference methods have been employed to solve classification problems using probit regression and logistic regression with normal priors. In this article, we propose to apply the variational approximation on probit regression models with intrinsic prior. We review the mean-field variational method and the procedure of developing intrinsic prior for the probit regression model. We then present our work on implementing the variational Bayesian probit regression model using intrinsic prior. Publicly available data from the world's largest peer-to-peer lending platform, LendingClub, will be used to illustrate how model output uncertainties are addressed through the framework we proposed. With LendingClub data, the target variable is the final status of a loan, either charged-off or fully paid. Investors may very well be interested in how predictive features like FICO, amount financed, income, etc. may affect the final loan status.

Keywords: objective Bayesian inference; intrinsic prior; variational inference; binary probit regression; mean-field approximation

1. Introduction

There is not much literature on objective Bayesian analysis for binary classification problems, especially for intrinsic prior related methods. By far, only two articles have explored intrinsic prior related methods on classification problems. Reference [1] implements integral priors into the generalized linear models with various link functions. In addition, reference [2] considers intrinsic priors for probit models. On the other hand, variational inference methods have been employed to solve classification problem with logistic regression ([3]) and probit regression ([4,5]) with normal priors. Variational approximation methods have been reviewed in [6,7], and more recently [8].

In this article, we propose to apply variational approximations on probit regression models with intrinsic priors. In Section 4, we review the mean-field variational method that will be used in this article. In Section 3, procedures for developing intrinsic priors for probit models will be introduced following [2]. Our work is presented in Section 5. Our motivations for combining intrinsic prior methodology and variational inference is as following

- Avoiding manually set ad hoc plugin priors by automatically generating a family of non-informative priors that are less sensible.
- Reference [1,2] do not consider inference of posterior distributions of parameters. Their focus is on model comparison. Although the development of intrinsic priors itself comes from a model selection background, we thought it would be interesting to apply intrinsic priors on inference

problems. In fact, some recently developed priors that proposed to solve inference or estimation problems turned out to be also intrinsic priors. For example, the Scaled Beta2 prior [9] and the Matrix-F prior [10].
- Intrinsic priors concentrate probability near the null hypothesis, a condition that is widely accepted and should be required of a prior for testing a hypothesis.
- Also, intrinsic priors have flat tails that prevents finite sample inconsistency [11].
- For inference problems with large data set, variational approximation methods are much faster than MCMC-based methods.

As for model comparison, due to the fact that the output of variational inference methods cannot be employed directly to compare models, we propose in Section 5.3 to simply make use of the variational approximation of the posterior distribution as an importance function and get the Monte Carlo estimated marginal likelihood by importance sampling for model comparison.

2. Background and Development of Intrinsic Prior Methodology

2.1. Bayes Factor

The Bayesian framework of model selection coherently involves the use of probability to express all uncertainty in the choice of model, including uncertainty about the unknown parameters of a model. Suppose that models $M_1, M_2, ..., M_q$ are under consideration. We shall assume that the observed data $\mathbf{x} = (x_1, x_2, ..., x_n)$ is generated from one of these models but we do not know which one it is. We express our uncertainty through prior probability $P(M_j), j = 1, 2, ..., q$. Under model M_i, $\mathbf{x}$ has density $f_i(\mathbf{x}|\boldsymbol{\theta}_i, M_i)$, where $\boldsymbol{\theta}_i$ are unknown model parameters, and the prior distribution for $\boldsymbol{\theta}_i$ is $\pi_i(\boldsymbol{\theta}_i|M_i)$. Given observed data and prior probabilities, we can then evaluate the posterior probability of M_i using Bayes' rule

$$P(M_i|\mathbf{x}) = \frac{p_i(\mathbf{x}|M_i)P(M_i)}{\sum_{j=1}^{q} p_j(\mathbf{x}|M_j)P(M_j)}, \tag{1}$$

where

$$p_i(\mathbf{x}|M_i) = \int f_i(\mathbf{x}|\boldsymbol{\theta}_i, M_i)\pi_i(\boldsymbol{\theta}_i|M_i)d\boldsymbol{\theta}_i \tag{2}$$

is the marginal likelihood of $\mathbf{x}$ under M_i, also called the evidence for M_i [12]. A common choice of prior model probabilities is $P(M_j) = \frac{1}{q}$, so that each model has the same initial probability. However, there are other alternatives of assigning probabilities to correct for multiple comparison (See [13]). From (1), the posterior odds are therefore the prior odds multiplied by the Bayes factor

$$\frac{P(M_j|\mathbf{x})}{P(M_i|\mathbf{x})} = \frac{P(M_j)p_j(\mathbf{x})}{P(M_i)p_i(\mathbf{x})} = \frac{P(M_j)}{P(M_i)} \times B_{ji}. \tag{3}$$

where the Bayes factor of M_j to M_i is defined by

$$B_{ji} = \frac{p_j(\mathbf{x})}{p_i(\mathbf{x})} = \frac{\int f_j(\mathbf{x}|\boldsymbol{\theta}_j)\pi_j(\boldsymbol{\theta}_j)d\boldsymbol{\theta}_j}{\int f_i(\mathbf{x}|\boldsymbol{\theta}_i)\pi_i(\boldsymbol{\theta}_i)d\boldsymbol{\theta}_i}. \tag{4}$$

Here we omit the dependence on models M_j, M_i to keep the notation simple. The marginal likelihood, $p_i(\mathbf{x})$ expresses the preference shown by the observed data for different models. When $B_{ji} > 1$, the data favor M_j over M_i, and when $B_{ji} < 1$ the data favor M_i over M_j. A scale for interpretation of B_{ji} is given by [14].

2.2. Motivation and Development of Intrinsic Prior

Computing B_{ji} requires specification of $\pi_i(\theta_i)$ and $\pi_j(\theta_j)$. Often in Bayesian analysis, when prior information is weak, one can use non-informative (or default) priors $\pi_i^N(\theta_i)$. Common choices for non-informative priors are the uniform prior, $\pi_i^U(\theta_i) \propto 1$; the Jeffreys prior, $\pi_i^J(\theta_i) \propto \left[\det(\mathbf{I}_i(\theta_i))\right]^{1/2}$ where $\mathbf{I}_i(\theta_i)$ is the expected Fisher information matrix corresponding to M_i.

Using any of the π_i^N in (4) would yield

$$B_{ji}^N = \frac{p_j^N(\mathbf{x})}{p_i^N(\mathbf{x})} = \frac{\int f_j(\mathbf{x}|\theta_j)\pi_j^N(\theta_j)d\theta_j}{\int f_i(\mathbf{x}|\theta_i)\pi_i^N(\theta_i)d\theta_i}. \tag{5}$$

The difficulty with (5) is that π_i^N are typically improper and hence are defined only up to an unspecified constant c_i. So B_{ji}^N is defined only up to the ratio c_j/c_i of two unspecified constants.

An attempt to circumvent the ill definition of the Bayes factors for improper non-informative priors is the intrinsic Bayes factor introduced by [15], which is a modification of a partial Bayes factor [16]. To define the intrinsic Bayes factor we consider the set of subsamples $\mathbf{x}(l)$ of the data $\mathbf{x}$ of minimal size l such that $0 < p_i^N(\mathbf{x}(l)) < \infty$. These subsamples are called training samples (not to be confused with training sample in machine learning). In addition, there is a total number of L such subsamples.

The main idea here is that training sample $\mathbf{x}(l)$ will be used to convert the improper $\pi_i^N(\theta_i)$ to proper posterior

$$\pi_i^N(\theta_i|\mathbf{x}(l)) = \frac{f_i(\mathbf{x}(l)|\theta_i)\pi_i^N(\theta_i)}{p_i^N(\mathbf{x}(l))} \tag{6}$$

where $p_i^N(\mathbf{x}(l)) = \int f_i(\mathbf{x}(l)|\theta_i)\pi_i^N(\theta_i)d\theta_i$. Then, the Bayes factor for the remaining of the data $\mathbf{x}(n-l)$, where $\mathbf{x}(l) \cup \mathbf{x}(n-l) = \mathbf{x}$, using $\pi_i^N(\theta_i|\mathbf{x}(l))$ as prior is called a "partial" Bayes factor,

$$B_{ji}^N(\mathbf{x}(n-l)|\mathbf{x}(l)) = \frac{\int f_j(\mathbf{x}(n-l)|\theta_j)\pi_j^N(\theta_j|\mathbf{x}(l))d\theta_j}{\int f_i(\mathbf{x}(n-l)|\theta_i)\pi_i^N(\theta_i|\mathbf{x}(l))d\theta_i} \tag{7}$$

This partial Bayes factor is a well-defined Bayes factor, and can be written as $B_{ji}^N(\mathbf{x}(n-l)|\mathbf{x}(l)) = B_{ji}^N(\mathbf{x})B_{ij}(\mathbf{x}(l))$, where $B_{ji}^N(\mathbf{x}) = \frac{p_j^N(\mathbf{x})}{p_i^N(\mathbf{x})}$ and $B_{ij}(\mathbf{x}(l)) = \frac{p_i^N(\mathbf{x}(l))}{p_j^N(\mathbf{x}(l))}$. Clearly, $B_{ji}^N(\mathbf{x}(n-l)|\mathbf{x}(l))$ will depend on the choice of the training samples $\mathbf{x}(l)$. To eliminate this arbitrariness and increase stability, reference [15] suggests averaging over all training samples and obtained the arithmetic intrinsic Bayes factor (AIBF)

$$B_{ji}^{AIBF}(\mathbf{x}) = B_{ji}^N(\mathbf{x})\frac{1}{L}\sum_{l=1}^{L}B_{ij}^N(\mathbf{x}(l)). \tag{8}$$

The strongest justification of the arithmetic IBF is its asymptotic equivalence with a proper Bayes factor arising from *Intrinsic priors*. These intrinsic priors were identified through an asymptotic analysis (see [15]). For the case where M_i is nested in M_j, it can be shown that the intrinsic priors are given by

$$\pi_i^I(\theta_i) = \pi_i^N(\theta_i) \text{ and } \pi_j^I(\theta_j) = \pi_j^N(\theta_j)E_{M_j}\left[\frac{m_i^N(\mathbf{x}(l))}{m_j^N(\mathbf{x}(l))}\bigg|\theta_j\right]. \tag{9}$$

3. Objective Bayesian Probit Regression Models

3.1. Bayesian Probit Model and the Use of Auxiliary Variables

Consider a sample $\mathbf{y} = (y_1, ..., y_n)$, where $Y_i, i = 1, ..., n$, is a $0-1$ random variable such that under model M_j, it follows a probit regression model with a $j+1$-dimensional vector of covariates x_i, where $j \leq p$. Here, p is the total number of covariate variables under our consideration. In addition, this probit model M_j has the form

$$Y_i | \beta_0, ..., \beta_j, M_j \sim \text{Bernoulli}(\Phi(\beta_0 x_{0i} + \beta_1 x_{1i} + ... + \beta_j x_{ji})), \quad 1 \leq i \leq n, \tag{10}$$

where Φ denotes the standard normal cumulative distribution function and $\beta_j = (\beta_0, ..., \beta_j)$ is a vector of dimension $j+1$. The first component of the vector x_i is set equal to 1 so that when considering models of the form (10), the intercept is in any submodel. The maximum length of the vector of covariates is $p+1$. Let $\pi(\beta)$, proper or improper, summarize our prior information about β. Then the posterior density of β is given by

$$\pi(\beta|y) = \frac{\pi(\beta) \prod_{i=1}^n \Phi(x_i'\beta)^{y_i}(1 - \Phi(x_i'\beta))^{1-y_i}}{\int \pi(\beta) \prod_{i=1}^n \Phi(x_i'\beta)^{y_i}(1 - \Phi(x_i'\beta))^{1-y_i} d\beta},$$

which is largely intractable.

As shown by [17], the Bayesian probit regression model becomes tractable when a particular set of auxiliary variables is introduced. Based on the data augmentation approach [18], introducing n latent variables $Z_1, ..., Z_n$, where

$$Z_i | \beta \sim N(x_i'\beta, 1).$$

The probit model (10) can be thought of as a regression model with incomplete sampling information by considering that only the sign of z_i is observed. More specifically, define $Y_i = 1$ if $Z_i > 0$ and $Y_i = 0$ otherwise. This allows us to write the probability density of y_i given z_i

$$p(y_i | z_i) = \mathbb{I}(z_i > 0)\mathbb{I}(y_i = 1) + \mathbb{I}(z_i \leq 0)\mathbb{I}(y_i = 0).$$

Expansion of the parameter set from $\{\beta\}$ to $\{\beta, \mathbf{Z}\}$ is the key to achieving a tractable solution for variational approximation.

3.2. Development of Intrinsic Prior for Probit Models

For the sample $\mathbf{z} = (z_1, ..., z_n)'$, the null normal model is

$$M_1 : \{N_n(\mathbf{z}|\alpha \mathbf{1}_n, \mathbf{I}_n), \pi(\alpha)\}.$$

For a generic model M_j with $j+1$ regressors, the alternative model is

$$M_j : \{N_n(\mathbf{z}|\mathbf{X}_j \beta_j, \mathbf{I}_n), \pi(\beta_j)\},$$

where the design matrix $\mathbf{X}_j$ has dimensions $n \times (j+1)$. Intrinsic prior methodology for the linear model was first developed by [19], and was further developed in [20] by using the methods of [21]. This intrinsic methodology gives us an automatic specification of the priors $\pi(\alpha)$ and $\pi(\beta)$, starting with the non-informative priors $\pi^N(\alpha)$ and $\pi^N(\beta)$ for α and β, which are both improper and proportional to 1.

The marginal distributions for the sample **z** under the null model, and under the alternative model with intrinsic prior, are formally written as

$$p_1(\mathbf{z}) = \int N_n(\mathbf{z}|\alpha \mathbf{1}_n, \mathbf{I}_n) \pi^N(\alpha) d\alpha,$$
$$p_j(\mathbf{z}) = \int \int N_n(\mathbf{z}|\mathbf{X}_j \boldsymbol{\beta}_j, \mathbf{I}_n) \pi^I(\boldsymbol{\beta}|\alpha) \pi^N(\alpha) d\alpha d\boldsymbol{\beta}. \quad (11)$$

However, these are marginals of the sample **z**, but our selection procedure requires us to compute the Bayes factor of model M_j versus the reference model M_1 for the sample $\mathbf{y} = (y_1, ..., y_n)$. To solve this problem, reference [2] proposed to transform the marginal $p_j(\mathbf{z})$ into the marginal $p_j(\mathbf{y})$ by using the probit transformations $y_i = 1(z_i > 0), i = 1, ..., n$. These latter marginals are given by

$$p_j(\mathbf{y}) = \int_{A_1 \times ... \times A_n} p_j(\mathbf{z}) d\mathbf{z} \quad (12)$$

where

$$A_i = \begin{cases} (0, \infty) & \text{if } y_i = 1, \\ (-\infty, 0) & \text{if } y_i = 0. \end{cases} \quad (13)$$

4. Variational Inference

4.1. Overview of Variational Methods

Variational methods have their origins in the 18th century with the work of Euler, Lagrange, and others on the calculus of variations (The derivation in this section is standard in the literature on variational approximation and will at times follow the arguments in [22,23]). Variational inference is a body of deterministic techniques for making approximate inference for parameters in complex statistical models. Variational approximations are a much faster alternative to Markov Chain Monte Carlo (MCMC), especially for large models, and are a richer class of methods than the Laplace approximation [6].

Suppose we have a Bayesian model and a prior distribution for the parameters. The model may also have latent variables, here we shall denote the set of all latent variables and parameters by $\boldsymbol{\theta}$. In addition, we denote the set of all observed variables by **X**. Given a set of n independent, identically distributed data, for which $\mathbf{X} = \{\mathbf{x}_1, ..., \mathbf{x}_n\}$ and $\boldsymbol{\theta} = \{\boldsymbol{\theta}_1, ..., \boldsymbol{\theta}_n\}$, our probabilistic model (e.g., probit regression model) specifies the joint distribution $p(\mathbf{X}, \boldsymbol{\theta})$, and our goal is to find an approximation for the posterior distribution $p(\boldsymbol{\theta}|\mathbf{X})$ as well as for the marginal likelihood $p(\mathbf{X})$. For any probability distribution $q(\boldsymbol{\theta})$, we have the following decomposition of the log marginal likelihood

$$\ln p(\mathbf{X}) = \mathcal{L}(q) + \text{KL}(q||p)$$

where we have defined

$$\mathcal{L}(q) = \int q(\boldsymbol{\theta}) \ln \left\{ \frac{p(\mathbf{X}, \boldsymbol{\theta})}{q(\boldsymbol{\theta})} \right\} d\boldsymbol{\theta} \quad (14)$$

$$\text{KL}(q||p) = -\int q(\boldsymbol{\theta}) \ln \left\{ \frac{p(\boldsymbol{\theta}|\mathbf{X})}{q(\boldsymbol{\theta})} \right\} d\boldsymbol{\theta} \quad (15)$$

We refer to (14) as the lower bound of the log marginal likelihood with respect to the density q, and (15) is by definition the Kullback–Leibler divergence of the posterior $q(\boldsymbol{\theta}|\mathbf{X})$ from the density q. Based on this decomposition, we can maximize the lower bound $\mathcal{L}(q)$ by optimization with respect to the distribution $q(\boldsymbol{\theta})$, which is equivalent to minimizing the KL divergence. In addition, the lower bound is attained when the KL divergence is zero, which happens when $q(\boldsymbol{\theta})$ equals the posterior distribution $p(\boldsymbol{\theta}|\mathbf{X})$. It would be hard to find such a density since the true posterior distribution is intractable.

4.2. Factorized Distributions

The essence of the variational inference approach is approximation to the posterior distribution $p(\boldsymbol{\theta}|\mathbf{X})$ by $q(\boldsymbol{\theta})$ for which the q dependent lower bound $\mathcal{L}(q)$ is more tractable than the original model evidence. In addition, tractability is achieved by restricting q to a more manageable class of distributions, and then maximizing $\mathcal{L}(q)$ over that class.

Suppose we partition elements of $\boldsymbol{\theta}$ into disjoint groups $\{\boldsymbol{\theta}_i\}$ where $i = 1, ..., M$. We then assume that the q density factorizes with respect to this partition, i.e.,

$$q(\boldsymbol{\theta}) = \prod_{i=1}^{M} q_i(\boldsymbol{\theta}_i). \tag{16}$$

The product form is the only assumption we made about the distribution. Restriction (16) is also known as *mean-field* approximation and has its root in Physics [24].

For all distributions $q(\boldsymbol{\theta})$ with the form (16), we need to find the distribution for which the lower bound $\mathcal{L}(q)$ is largest. Restriction of q to a subclass of product densities like (16) gives rise to explicit solutions for each product component in terms of the others. This fact, in turn, leads to an iterative scheme for obtaining the solutions. To achieve this, we first substitute (16) into (14) and then separate out the dependence on one of the factors $q_j(\boldsymbol{\theta}_j)$. Denoting $q_j(\boldsymbol{\theta}_j)$ by q_j to keep the notation clear, we obtain

$$\begin{aligned}\mathcal{L}(q) &= \int \prod_{i=1}^{M} q_i \left\{ \ln p(\mathbf{X}, \boldsymbol{\theta}) - \sum_{i=1}^{M} \ln q_i \right\} d\boldsymbol{\theta} \\ &= \int q_j \left\{ \int \ln p(\mathbf{X}, \boldsymbol{\theta}) \prod_{i \neq j} q_i d\boldsymbol{\theta}_i \right\} d\boldsymbol{\theta}_j - \int q_j \ln q_j d\boldsymbol{\theta}_j + \text{constant} \\ &= \int q_j \ln \tilde{p}(\mathbf{X}, \boldsymbol{\theta}_j) d\boldsymbol{\theta}_j - \int q_j \ln q_j d\boldsymbol{\theta}_j + \text{constant}\end{aligned} \tag{17}$$

where $\tilde{p}(\mathbf{X}, \boldsymbol{\theta}_j)$ is given by

$$\ln \tilde{p}(\mathbf{X}, \boldsymbol{\theta}_j) = \mathbb{E}_{i \neq j}[\ln p(\mathbf{X}, \boldsymbol{\theta})] + \text{constant}. \tag{18}$$

The notation $\mathbb{E}_{i \neq j}[\cdot]$ denotes an expectation with respect to the q distributions over all variables $\mathbf{z}_i$ for $i \neq j$, so that

$$\mathbb{E}_{i \neq j}[\ln p(\mathbf{X}, \boldsymbol{\theta})] = \int \ln p(\mathbf{X}, \boldsymbol{\theta}) \prod_{i \neq j} q_i d\boldsymbol{\theta}_i.$$

Now suppose we keep the $\{q_{i \neq j}\}$ fixed and maximize $\mathcal{L}(q)$ in (17) with respect to all possible forms for the density $q_j(\boldsymbol{\theta}_j)$. By recognizing that (17) is the negative KL divergence between $\tilde{p}(\mathbf{X}, \boldsymbol{\theta}_j)$ and $q_j(\boldsymbol{\theta}_j)$, we notice that maximizing (17) is equivalent to minimize the KL divergence, and the minimum occurs when $q_j(\boldsymbol{\theta}_j) = \tilde{p}(\mathbf{X}, \boldsymbol{\theta}_j)$. The optimal $q_j^*(\boldsymbol{\theta}_j)$ is then

$$\ln q_j^*(\boldsymbol{\theta}_j) = \mathbb{E}_{i \neq j}[\ln p(\mathbf{X}, \boldsymbol{\theta})] + \text{constant}. \tag{19}$$

The above solution says that the log of the optimal q_j is obtained simply by considering the log of the joint distribution of all parameter, latent and observable variables and then taking the expectation with respect to all the other factors q_i for $i \neq j$. Normalizing the exponential of (19), we have

$$q_j^*(\theta_j) = \frac{\exp(\mathbb{E}_{i\neq j}[\ln p(\mathbf{X},\theta)])}{\int \exp(\mathbb{E}_{i\neq j}[\ln p(\mathbf{X},\theta)])d\theta_j}.$$

The set of equations in (19) for $j = 1, ..., M$ are not an explicit solution because the expression on the right hand side of (19) for the optimal q_j^* depends on expectations taken with respect to the other factors q_i for $i \neq j$. We will need to first initialize all of the factors $q_i(\theta_i)$ and then cycle through the factors one by one and replace each in turn with an updated estimate given by the right hand side of (19) evaluated using the current estimates for all of the other factors. Convexity properties can be used to show that convergence to at least local optima is guaranteed [25]. The iterative procedure is described in Algorithm 1.

Algorithm 1 Iterative procedure for obtaining the optimal densities under factorized density restriction (16). The updates are based on the solutions given by (19).

1: Initialize $q_2^*(\theta_2), ..., q_M^*(\theta_M)$.
2: Cycle through

$$q_1^*(\theta_1) \leftarrow \frac{\exp(\mathbb{E}_{i\neq 1}[\ln p(\mathbf{X},\theta)])}{\int \exp(\mathbb{E}_{i\neq 1}[\ln p(\mathbf{X},\theta)])d\theta_1}$$

$$\vdots$$

$$q_M^*(\theta_M) \leftarrow \frac{\exp(\mathbb{E}_{i\neq M}[\ln p(\mathbf{X},\theta)])}{\int \exp(\mathbb{E}_{i\neq M}[\ln p(\mathbf{X},\theta)])d\theta_M}$$

until the increase in $\mathcal{L}(q)$ is negligible.

---

5. Incorporate Intrinsic Prior with Variational Approximation to Bayesian Probit Models

5.1. Derivation of Intrinsic Prior to Be Used in Variational Inference

Let $\mathbf{X}_l$ be the design matrix of a minimal training sample (mTS) of a normal regression model M_j for the variable $\mathbf{Z} \sim N(\mathbf{X}_j\boldsymbol{\beta}_j, \mathbf{I}_{j+1})$. We have, for the $j+1$-dimensional parameter $\boldsymbol{\beta}_j$,

$$\int N_{j+1}(\mathbf{z}_l|\mathbf{X}_l\boldsymbol{\beta}_j, \mathbf{I}_{j+1})d\boldsymbol{\beta}_j = \begin{cases} |\mathbf{X}_l'\mathbf{X}_l|^{-1/2} & \text{if rank of } \mathbf{X}_l \geq j+1 \\ \infty & \text{otherwise} \end{cases}.$$

Therefore, it follows that the mTS size is $j + 1$ [2]. Given that priors for α and β are proportional to 1, the intrinsic prior for β conditional on α could be derived. Let $\boldsymbol{\beta}_0$ denote the vector with the first component equal to α and the others equal to zero. Based on Formula (9), we have

$$\pi^I(\boldsymbol{\beta}|\alpha) = \pi_j^N(\boldsymbol{\beta})\mathbb{E}_{\mathbf{z}_l|\boldsymbol{\beta}}^{M_j}\left[\frac{p_1(\mathbf{z}_l|\alpha)}{\int p_j(\mathbf{z}_l|\boldsymbol{\beta})\pi_j^N(\boldsymbol{\beta})d\boldsymbol{\beta}}\right]$$

$$= \mathbb{E}_{\mathbf{z}_l|\boldsymbol{\beta}}^{M_j}\left[\frac{\exp\{-\frac{1}{2}(\mathbf{z}_l - \mathbf{X}_l\boldsymbol{\beta}_0)'(\mathbf{z}_l - \mathbf{X}_l\boldsymbol{\beta}_0)\}}{\int \exp\{-\frac{1}{2}(\mathbf{z}_l - \mathbf{X}_l\boldsymbol{\beta})'(\mathbf{z}_l - \mathbf{X}_l\boldsymbol{\beta})\}d\boldsymbol{\beta}}\right]$$

$$= (2\pi)^{-\frac{(j+1)}{2}}|(\mathbf{X}_l'\mathbf{X}_l)^{-1}|^{-\frac{1}{2}} \times \mathbb{E}_{\mathbf{z}_l|\boldsymbol{\beta}}^{M_j}\left[\exp\{-\frac{1}{2}(\mathbf{z}_l - \mathbf{X}_l\boldsymbol{\beta}_0)'(\mathbf{z}_l - \mathbf{X}_l\boldsymbol{\beta}_0)\}\right]$$

$$= (2\pi)^{-\frac{(j+1)}{2}}|2(\mathbf{X}_l'\mathbf{X}_l)^{-1}|^{-\frac{1}{2}}\exp\{-\frac{1}{2}[(\boldsymbol{\beta} - \boldsymbol{\beta}_0)'\frac{\mathbf{X}_l'\mathbf{X}_l}{2}(\boldsymbol{\beta} - \boldsymbol{\beta}_0)]\}.$$

Therefore,

$$\pi^I(\beta|\alpha) = N_{j+1}(\beta|\beta_0, 2(X_l'X_l)^{-1}), \text{ where } \beta_0 = \begin{pmatrix} \alpha \\ 0 \\ \vdots \\ 0 \end{pmatrix}_{(j+1)\times 1}.$$

Notice that $X_l'X_l$ is unknown because it is a theoretical design matrix corresponding to the training sample z_l. It can be estimated by averaging over all submatrices containing $j+1$ rows of the $n \times (j+1)$ design matrix X_j. This average is $\frac{j+1}{n}X_j'X_j$ (See [26] and Appendix A in [2]), and therefore

$$\pi^I(\beta|\alpha) = N_{j+1}(\beta|\beta_0, \frac{2n}{j+1}(X_j'X_j)^{-1}).$$

Next, based on $\pi^I(\beta|\alpha)$, the intrinsic prior for β can be obtained by

$$\pi^I(\beta) = \int \pi^I(\beta|\alpha)\pi^I(\alpha)d\alpha. \tag{20}$$

Since we assume that $\pi^I(\alpha) = \pi^N(\alpha)$ is proportional to one, set $\pi^N(\alpha) = c$ where c is an arbitrary positive constant. Denote $\frac{2n}{j+1}(X_j'X_j)^{-1}$ by $\Sigma_{\beta|\alpha}$, we obtain

$$\begin{aligned}
\pi^I(\beta) &= \int \pi^I(\beta|\alpha)\pi^I(\alpha)d\alpha \\
&= c \cdot (2\pi)^{-\frac{j+1}{2}}|\Sigma_{\beta|\alpha}|^{-\frac{1}{2}} \int \exp\{-\frac{1}{2}(\beta-\beta_0)'\Sigma_{\beta|\alpha}^{-1}(\beta-\beta_0)\}d\alpha \\
&\propto \exp\{-\frac{1}{2}\beta'\Sigma_{\beta|\alpha}^{-1}\beta\} \times \int \exp\{-\frac{1}{2}[\beta_0'\Sigma_{\beta|\alpha}^{-1}\beta_0 - 2\beta'\Sigma_{\beta|\alpha}^{-1}\beta_0]\}d\alpha \\
&\propto \exp\{-\frac{1}{2}\beta'\Sigma_{\beta|\alpha}^{-1}\beta\} \times \int \exp\{-\frac{1}{2}(\Sigma_{\beta|\alpha_{(1,1)}}^{-1}\alpha^2 - 2\beta'\Sigma_{\beta|\alpha_{(\cdot 1)}}^{-1}\alpha)\}d\alpha
\end{aligned} \tag{21}$$

where $\Sigma_{\beta|\alpha_{(1,1)}}^{-1}$ is component of $\Sigma_{\beta|\alpha}^{-1}$ at position row 1 column 1 and $\Sigma_{\beta|\alpha_{(\cdot 1)}}^{-1}$ is the first column of $\Sigma_{\beta|\alpha}^{-1}$. Denote $\Sigma_{\beta|\alpha_{(1,1)}}^{-1}$ by σ_{11} and $\Sigma_{\beta|\alpha_{(\cdot 1)}}^{-1}$ by γ_1, we then obtain

$$\begin{aligned}
\pi^I(\beta) &\propto \exp\{-\frac{1}{2}\beta'\Sigma_{\beta|\alpha}^{-1}\beta\} \times \int \exp\{-\frac{1}{2}\sigma_{11}(\alpha - \frac{\beta'\gamma_1}{\sigma_{11}})^2 + \frac{1}{2}\frac{(\beta'\gamma_1)^2}{\sigma_{11}}\}d\alpha \\
&\propto \exp\{-\frac{1}{2}(\beta'\Sigma_{\beta|\alpha}^{-1}\beta - \beta'\frac{\gamma_1\gamma_1'}{\sigma_{11}}\beta)\} \times \sqrt{2\pi}\sigma_{11}^{-1/2} \\
&\propto \exp\{-\frac{1}{2}\beta'(\Sigma_{\beta|\alpha}^{-1} - \frac{\gamma_1\gamma_1'}{\sigma_{11}})\beta\}.
\end{aligned} \tag{22}$$

Therefore, we have derived that

$$\pi^I(\beta) \propto N_{j+1}(0, (\Sigma_{\beta|\alpha}^{-1} - \frac{\gamma_1\gamma_1'}{\sigma_{11}})^{-1}). \tag{23}$$

For model comparison, the specific form of the intrinsic prior may be needed, including the constant factor. Therefore, by following (21) and (22) we have

$$\pi^I(\beta) = c \cdot (2\pi)^{-\frac{j+1}{2}} |\Sigma_{\beta|\alpha}|^{-\frac{1}{2}} (2\pi)^{\frac{j+1}{2}} |(\Sigma_{\beta|\alpha}^{-1} - \frac{\gamma_1 \gamma_1'}{\sigma_{11}})^{-1}|^{\frac{1}{2}} \sqrt{2\pi} \sigma_{11}^{-1/2} \times N_{j+1}(0, (\Sigma_{\beta|\alpha}^{-1} - \frac{\gamma_1 \gamma_1'}{\sigma_{11}})^{-1})$$

$$= c \cdot |\Sigma_{\beta|\alpha}(\Sigma_{\beta|\alpha}^{-1} - \frac{\gamma_1 \gamma_1'}{\sigma_{11}})|^{-\frac{1}{2}} \sqrt{2\pi} \sigma_{11}^{-1/2} \times N_{j+1}(0, (\Sigma_{\beta|\alpha}^{-1} - \frac{\gamma_1 \gamma_1'}{\sigma_{11}})^{-1}) \qquad (24)$$

$$= c \cdot \sqrt{2\pi} \sigma_{11}^{-1/2} |(\mathbb{I} - \frac{\gamma_1 \gamma_1'}{\sigma_{11}} \Sigma_{\beta|\alpha})|^{-\frac{1}{2}} \times N_{j+1}(0, (\Sigma_{\beta|\alpha}^{-1} - \frac{\gamma_1 \gamma_1'}{\sigma_{11}})^{-1}).$$

5.2. Variational Inference for Probit Model with Intrinsic Prior

5.2.1. Iterative Updates for Factorized Distributions

We have that

$$Z_i|\beta \sim N(x_i'\beta, 1) \quad \text{and}$$
$$p(y_i|z_i) = \mathbb{I}(z_i > 0)\mathbb{I}(y_i = 1) + \mathbb{I}(z_i \leq 0)\mathbb{I}(y_i = 0)$$

in Section 3.1. We have shown in Section 5.1 that

$$\pi^I(\beta) \propto N_{j+1}(\mu_\beta, \Sigma_\beta),$$

where $\mu_\beta = 0$ and $\Sigma_\beta = (\Sigma_{\beta|\alpha}^{-1} - \frac{\gamma_1 \gamma_1'}{\sigma_{11}})^{-1}$. Since $\mathbf{y}$ is independent of β given $\mathbf{z}$, we have

$$p(\mathbf{y}, \mathbf{z}, \beta) = p(\mathbf{y}|\mathbf{z}, \beta)p(\mathbf{z}|\beta)p(\beta) \qquad (25)$$
$$= p(\mathbf{y}|\mathbf{z})p(\mathbf{z}|\beta)p(\beta).$$

To apply the variational approximation to probit regression model, unobservable variables are considered in two separate groups, coefficient parameter β and auxiliary variable $\mathbf{Z}$. To approximate the posterior distribution of β, consider the product form

$$q(\mathbf{Z}, \beta) = q_\mathbf{Z}(\mathbf{Z})q_\beta(\beta).$$

We proceed by first describing the distribution for each factor of the approximation, $q_\mathbf{Z}(\mathbf{Z})$ and $q_\beta(\beta)$. Then variational approximation is accomplished by iteratively updating the parameters of each factor distribution.

Start with $q_\mathbf{Z}(\mathbf{Z})$, when $y_i = 1$, we have

$$\log p(\mathbf{y}, \mathbf{z}, \beta) = \log \left(\prod_i \frac{1}{\sqrt{2\pi}} \exp\{-\frac{(z_i - x_i'\beta)^2}{2}\} \times \pi^I(\beta) \right) \qquad \text{where } z_i > 0.$$

Now, according to (19) and Algorithm 1, the optimal $q_\mathbf{Z}$ is proportional to

$$\mathbb{E}_\beta[\log p(\mathbf{y}, \mathbf{z}, \beta)] = -\frac{1}{2}\mathbb{E}_\beta[\mathbf{z}'\mathbf{z} - 2\beta'\mathbf{X}\mathbf{z} + \beta'\mathbf{X}'\mathbf{X}\beta] + \mathbb{E}_\beta[\log \pi^I(\beta)]$$
$$= -\frac{1}{2}\mathbf{z}'\mathbf{z} + \mathbb{E}_\beta[\beta]'\mathbf{X}'\mathbf{z} + \underbrace{-\frac{1}{2}\mathbb{E}_\beta[\beta'\mathbf{X}'\mathbf{X}\beta]}_{\text{constant}} + \underbrace{\mathbb{E}_\beta[\log \pi^I(\beta)]}_{\text{constant}}.$$

So, we have the optimal $q_\mathbf{Z}$,

$$q_\mathbf{Z}^*(\mathbf{Z}) \propto \exp\{-\frac{1}{2}\mathbf{z}'\mathbf{z} + \mathbb{E}_\beta[\beta]'\mathbf{X}'\mathbf{z} + \text{constant}\}$$
$$\propto \exp\{-\frac{1}{2}(\mathbf{z} - \mathbf{X}\mathbb{E}_\beta[\beta])'(\mathbf{z} - \mathbf{X}\mathbb{E}_\beta[\beta])\}.$$

Similar procedure could be used to develop cases when $y_i = 0$. Therefore, we have that the optimal approximation for $q_\mathbf{Z}$ is a truncated normal distribution, where

$$q_\mathbf{Z}^*(\mathbf{Z}) = \begin{cases} N_{[0,+\infty)}(\mathbf{X}\mathbb{E}_\beta[\beta]_i, 1) & \text{if } y_i = 1, \\ N_{(-\infty,0]}(\mathbf{X}\mathbb{E}_\beta[\beta]_i, 1) & \text{if } y_i = 0. \end{cases} \quad (26)$$

Denote $\mathbf{X}\mathbb{E}_\beta[\beta]$ by $\mu_\mathbf{z}$, the location of distribution $q_\mathbf{Z}^*(\mathbf{Z})$. The expectation $\mathbb{E}_\beta$ is taken with respect to the density form of $q(\beta)$ for which we shall derive now.

For $q_\beta(\beta)$, given the joint form in (25), we have

$$\log p(\mathbf{y}, \mathbf{z}, \beta) = -\frac{1}{2}\exp\{(\mathbf{z} - \mathbf{X}\beta)'(\mathbf{z} - \mathbf{X}\beta)\} - \frac{1}{2}\exp\{(\beta - \mu_\beta)'\Sigma_\beta^{-1}(\beta - \mu_\beta)\} + \textbf{constant}.$$

Taking expectation with respect to $q_\mathbf{Z}(\mathbf{z})$, we have

$$\mathbb{E}_\mathbf{Z}[\log p(\mathbf{y}, \mathbf{z}, \beta)] = -\frac{1}{2}\underbrace{\mathbb{E}_\mathbf{Z}[\mathbf{Z}'\mathbf{Z}]}_{\text{constant}} + \mathbb{E}_\mathbf{Z}[\mathbf{Z}]'\mathbf{X}\beta - \frac{1}{2}\beta'\mathbf{X}'\mathbf{X}\beta$$
$$- \frac{1}{2}\beta'\Sigma_\beta^{-1}\beta + \mu_\beta'\Sigma_\beta^{-1}\beta + \underbrace{\mu_\beta'\Sigma_\beta^{-1}\mu_\beta}_{\text{constant}}.$$

Again, based on (19) and Algorithm 1, the optimal $q_\beta(\beta)$ is proportional to $\mathbb{E}_\mathbf{Z}[\log p(\mathbf{y}, \mathbf{z}, \beta)]$,

$$q_\beta^*(\beta) \propto -\frac{1}{2}\beta'(\mathbf{X}'\mathbf{X} + \Sigma_\beta^{-1})\beta + (\mathbb{E}_\mathbf{Z}[\mathbf{Z}]'\mathbf{X} + \mu_\beta'\Sigma_\beta^{-1})\beta.$$

First notice that any constant terms, including constant factor in the intrinsic prior, were canceled out due to the ratio form of (19). Then by noticing the quadratic form in the above formula we have

$$q_\beta^*(\beta) = N(\mu_{q_\beta}, \Sigma_{q_\beta}), \quad (27)$$

where

$$\Sigma_{q_\beta} = (\mathbf{X}'\mathbf{X} + \Sigma_\beta^{-1})^{-1},$$
$$\mu_{q_\beta} = (\mathbf{X}'\mathbf{X} + \Sigma_\beta^{-1})^{-1}(\mathbb{E}_\mathbf{Z}[\mathbf{Z}]'\mathbf{X} + \mu_\beta'\Sigma_\beta^{-1}).$$

Notice that μ_{q_β}, i.e., $\mathbb{E}_\beta[\beta]$, depends on $\mathbb{E}_\mathbf{Z}[\mathbf{Z}]$. In addition, from our previous derivation, we found that the update for $\mathbb{E}_\mathbf{Z}[\mathbf{Z}]$ depends on $\mathbb{E}_\beta[\beta]$. Given that the density form of $q_\mathbf{Z}$ is truncated normal, we have

$$\mathbb{E}_\mathbf{Z}[\mathbf{Z}_i] = \begin{cases} \mathbf{X}\mathbb{E}_\beta[\beta]_i + \frac{\phi(-\mathbf{X}\mathbb{E}_\beta[\beta]_i)}{1-\Phi(-\mathbf{X}\mathbb{E}_\beta[\beta]_i)} & \text{if } y_i = 1, \\ \mathbf{X}\mathbb{E}_\beta[\beta]_i - \frac{\phi(-\mathbf{X}\mathbb{E}_\beta[\beta]_i)}{\Phi(-\mathbf{X}\mathbb{E}_\beta[\beta]_i)} & \text{if } y_i = 0, \end{cases}$$

where ϕ is the standard normal density and Φ is the standard normal cumulative density. Denote $\mathbb{E}_\mathbf{Z}[\mathbf{Z}]$ by $\mu_{q_\mathbf{Z}}$. See properties of truncated normal distribution in Appendix A. Updating procedures for parameters μ_{q_β} and $\mu_{q_\mathbf{Z}}$ of each factor distribution are summarized in Algorithm 2.

Algorithm 2 Iterative procedure for updating parameters to reach optimal factor densities q_β^* and q_Z^* in Bayesian probit regression model. The updates are based on the solutions given by (26) and (27).

1: Initialize μ_{q_Z}.
2: Cycle through

$$\mu_{q_\beta} \leftarrow (\mathbf{X}'\mathbf{X} + \Sigma_\beta^{-1})^{-1}(\mu_{q_Z}'\mathbf{X} + \mu_\beta'\Sigma_\beta^{-1}),$$

$$\mu_{q_Z} \leftarrow \mathbf{X}\mu_{q_\beta} + \frac{\phi(\mathbf{X}\mu_{q_\beta})}{\Phi(\mathbf{X}\mu_{q_\beta})^\mathbf{y}[\Phi(\mathbf{X}\mu_{q_\beta}) - 1]^{1-\mathbf{y}}},$$

until the increase in $\mathcal{L}(q)$ is negligible.

5.2.2. Evaluation of the Lower Bound $\mathcal{L}(q)$

During the process of optimization of variational approximation densities, the lower bound for the log marginal likelihood need to be evaluated and monitored to determine when the iterative updating process converges. Based on derivations from previous section, we now have the exact form for the variational inference density,

$$q(\beta, \mathbf{Z}) = q_\beta(\beta)q_Z(\mathbf{Z}).$$

According to (14), we can write down the lower bound $\mathcal{L}(q)$ with respect to $q(\beta, \mathbf{Z})$.

$$\begin{aligned}
\mathcal{L}(q) &= \int q(\beta, \mathbf{Z}) \log\left\{\frac{p(\mathbf{Y}, \beta, \mathbf{Z})}{q(\beta, \mathbf{Z})}\right\} d\beta d\mathbf{Z} \\
&= \int q_\beta(\beta) q_Z(\mathbf{Z}) \log\left\{\frac{p(\mathbf{Y}, \beta, \mathbf{Z})}{q_\beta(\beta) q_Z(\mathbf{Z})}\right\} d\beta d\mathbf{Z} \\
&= \int q_\beta(\beta) q_Z(\mathbf{Z}) \log\{p(\mathbf{Y}, \beta, \mathbf{Z})\} d\beta d\mathbf{Z} - \int q_\beta(\beta) q_Z(\mathbf{Z}) \log\{q_\beta(\beta) q_Z(\mathbf{Z})\} d\beta d\mathbf{Z} \\
&= \mathbb{E}_{\beta,\mathbf{Z}}[\log\{p(\mathbf{Y}, \mathbf{Z}|\beta)\}] + \mathbb{E}_{\beta,\mathbf{Z}}[\pi^I(\beta)] - \mathbb{E}_{\beta,\mathbf{Z}}[\log\{q_\beta(\beta)\}] - \mathbb{E}_{\beta,\mathbf{Z}}[\log\{q_Z(\mathbf{Z})\}].
\end{aligned} \quad (28)$$

As we can see in (28), $\mathcal{L}(q)$ has been divided into four different parts with expectation taken over the variational approximation density $q(\beta, \mathbf{Z}) = q_\beta(\beta) q_Z(\mathbf{Z})$. We now find the expression of these expectations one by one.

Part 1: $\mathbb{E}_{\beta,\mathbf{Z}}[\log\{p(\mathbf{Y}, \mathbf{Z}|\beta)\}]$

$$\begin{aligned}
&= \log(2\pi)^{-\frac{n}{2}} + \int\int q_\beta(\beta) q_Z(\mathbf{Z})\{-\frac{1}{2}(\mathbf{z} - \mathbf{X}\beta)'(\mathbf{z} - \mathbf{X}\beta)\} d\beta d\mathbf{z} \\
&= \log(2\pi)^{-\frac{n}{2}} + \int q_Z(\mathbf{Z}) \int q_\beta(\beta)\{-\frac{1}{2}(\beta'\mathbf{X}'\mathbf{X}\beta - 2\mathbf{z}'\mathbf{X}\beta + \mathbf{z}'\mathbf{z})\} d\beta d\mathbf{z}
\end{aligned} \quad (29)$$

Deal with the inner integral first, we have

$$\begin{aligned}
\int q_\beta(\beta)\{-\frac{1}{2}(\beta'\mathbf{X}'\mathbf{X}\beta - 2\mathbf{z}'\mathbf{X}\beta + \mathbf{z}'\mathbf{z})\} d\beta &= -\frac{1}{2}\int q_\beta(\beta)[\beta'\mathbf{X}'\mathbf{X}\beta] d\beta + \mathbf{z}'\mathbf{X}\mathbb{E}_\beta[\beta] - \frac{1}{2}\mathbf{z}'\mathbf{z} \\
&= -\frac{1}{2}\int q_\beta(\beta)[\beta'\mathbf{X}'\mathbf{X}\beta] d\beta + \mathbf{z}'\mathbf{X}\mu_{q_\beta} - \frac{1}{2}\mathbf{z}'\mathbf{z}
\end{aligned} \quad (30)$$

where

$$-\frac{1}{2}\int q_\beta(\beta)[\beta'\mathbf{X}'\mathbf{X}\beta]d\beta = -\frac{1}{2}\int q_\beta(\beta)[(\beta-\mu_{q_\beta}+\mu_{q_\beta})'\mathbf{X}'\mathbf{X}(\beta-\mu_{q_\beta}+\mu_{q_\beta})]d\beta$$
$$= -\frac{1}{2}\text{trace}(\mathbf{X}'\mathbf{X}\mathbb{E}_\beta[(\beta-\mu_{q_\beta})(\beta-\mu_{q_\beta})']) - \frac{1}{2}\mu'_{q_\beta}\mathbf{X}'\mathbf{X}\mu_{q_\beta} \quad (31)$$
$$= -\frac{1}{2}\text{trace}(\mathbf{X}'\mathbf{X}[\mu_{q_\beta}\mu'_{q_\beta}+\Sigma_{q_\beta}]).$$

Substitute (31) into (30), we got

$$\int q_\beta(\beta)\{-\frac{1}{2}(\beta'\mathbf{X}'\mathbf{X}\beta - 2\mathbf{z}'\mathbf{X}\beta + \mathbf{z}'\mathbf{z})\}d\beta = -\frac{1}{2}\text{trace}(\mathbf{X}'\mathbf{X}[\mu_{q_\beta}\mu'_{q_\beta}+\Sigma_{q_\beta}]) + \mathbf{z}'\mathbf{X}\mu_{q_\beta} - \frac{1}{2}\mathbf{z}'\mathbf{z}. \quad (32)$$

Substituting (32) back into (29) gives

$$\mathbb{E}_{\beta,\mathbf{Z}}[\log\{p(\mathbf{Y},\mathbf{Z}|\beta)\}] = \log(2\pi)^{-\frac{n}{2}} + \int q_\mathbf{Z}(\mathbf{z})\{-\frac{1}{2}\text{trace}(\mathbf{X}'\mathbf{X}[\mu_{q_\beta}\mu'_{q_\beta}+\Sigma_{q_\beta}]) + \mathbf{z}'\mathbf{X}\mu_{q_\beta} - \frac{1}{2}\mathbf{z}'\mathbf{z}\}d\mathbf{z}$$
$$= \log(2\pi)^{-\frac{n}{2}} - \frac{1}{2}\text{trace}(\mathbf{X}'\mathbf{X}[\mu_{q_\beta}\mu'_{q_\beta}+\Sigma_{q_\beta}]) - \frac{1}{2}\mathbb{E}_\mathbf{Z}[\mathbf{z}'\mathbf{z}] + \mu'_{q_z}\mu_\mathbf{z}$$
$$= \log(2\pi)^{-\frac{n}{2}} - \frac{1}{2}\text{trace}(\mathbf{X}'\mathbf{X}[\mu_{q_\beta}\mu'_{q_\beta}+\Sigma_{q_\beta}]) + \mu'_{q_z}\mu_\mathbf{z}$$
$$- \frac{1}{2}\sum_{i=1}^{n}[1+\mu_{z_i}^2 - \mu_{z_i}\frac{\phi(-\mu_{z_i})}{\Phi(-\mu_{z_i})}]^{\mathbb{I}(y_i=0)}[1+\mu_{z_i}^2 + \mu_{z_i}\frac{\phi(-\mu_{z_i})}{1-\Phi(-\mu_{z_i})}]^{\mathbb{I}(y_i=1)} \quad (33)$$
$$= \log(2\pi)^{-\frac{n}{2}} - \frac{1}{2}\text{trace}(\mathbf{X}'\mathbf{X}[\mu_{q_\beta}\mu'_{q_\beta}+\Sigma_{q_\beta}]) + \mu'_{q_z}\mu_\mathbf{z}$$
$$- \frac{1}{2}\sum_{i=1}^{n}[1+\mu_{q_{z_i}}\mu_{z_i}]^{\mathbb{I}(y_i=0)}[1+\mu_{q_{z_i}}\mu_{z_i}]^{\mathbb{I}(y_i=1)}$$
$$= \log(2\pi)^{-\frac{n}{2}} - \frac{1}{2}\text{trace}(\mathbf{X}'\mathbf{X}[\mu_{q_\beta}\mu'_{q_\beta}+\Sigma_{q_\beta}]) + \frac{1}{2}\mu'_{q_z}\mu_\mathbf{z} - \frac{n}{2}.$$

We applied properties of truncated normal distribution in Appendix B to find the expression of the second moment $\mathbb{E}_\mathbf{Z}[\mathbf{z}'\mathbf{z}]$.

Part 2: $\mathbb{E}_{\beta,\mathbf{Z}}[\log q_\mathbf{Z}(\mathbf{z})]$

$$= \int\int q_\beta(\beta)q_\mathbf{Z}(\mathbf{z})\log q_\mathbf{Z}(\mathbf{z})d\beta d\mathbf{Z}$$
$$= \int q_\mathbf{Z}(\mathbf{z})\log q_\mathbf{Z}(\mathbf{z})d\mathbf{Z}$$
$$= -\frac{n}{2}(\log(2\pi)+1)$$
$$+ \sum_{i=1}^{n}\{[\log(\Phi(-\mu_{z_i})) + \mu_{z_i}\frac{\phi(-\mu_{z_i})}{2\Phi(-\mu_{z_i})}]^{\mathbb{I}(y_i=0)}[\log(1-\Phi(-\mu_{z_i})) - \mu_{z_i}\frac{\phi(-\mu_{z_i})}{2(1-\Phi(-\mu_{z_i}))}]^{\mathbb{I}(y_i=1)}\} \quad (34)$$
$$= -\frac{n}{2}(\log(2\pi)+1) - \frac{1}{2}\mu'_\mathbf{z}\mu_\mathbf{z} + \frac{1}{2}\mu'_{q_z}\mu_\mathbf{z} + \sum_{i=1}^{n}\{[\log(\Phi(-\mu_{z_i}))]^{\mathbb{I}(y_i=0)}[\log(1-\Phi(-\mu_{z_i}))]^{\mathbb{I}(y_i=1)}\}$$

Again, see Appendix B for well-known properties of truncated normal distribution. Now subtracting (34) from (33) we got

$$\mathbb{E}_{\beta,\mathbf{Z}}[\log\{p(\mathbf{Y},\mathbf{Z}|\beta)\}] - \mathbb{E}_{\beta,\mathbf{Z}}[\log q_\mathbf{Z}(\mathbf{z})] = -\frac{1}{2}\text{trace}(\mathbf{X}'\mathbf{X}[\mu_{q_\beta}\mu'_{q_\beta}+\Sigma_{q_\beta}]) + \frac{1}{2}\mu'_\mathbf{z}\mu_\mathbf{z} +$$
$$\sum_{i=1}^{n}\{[\log(\Phi(-\mu_{z_i}))]^{\mathbb{I}(y_i=0)}[\log(1-\Phi(-\mu_{z_i}))]^{\mathbb{I}(y_i=1)}\}. \quad (35)$$

Based on the exact expression of the intrinsic prior $\pi^I(\beta)$, denoting all constant terms by C, we have

Part 3: $\mathbb{E}_{\beta,\mathbf{Z}}[\log p_\beta(\beta)]$

$$
\begin{aligned}
&= \int\int q_\mathbf{Z}(\mathbf{z}) q_\beta(\beta) \log \pi^I(\beta) d\beta d\mathbf{z} \\
&= \log C - \frac{(j+1)}{2}\log(2\pi) - \frac{1}{2}\log|\Sigma_\beta| - \frac{1}{2}\int q_\beta(\beta)[\beta'\Sigma_\beta^{-1}\beta]d\beta
\end{aligned}
\tag{36}
$$

To find the expression for the integral, we have

$$
\begin{aligned}
\int q_\beta(\beta)[\beta'\Sigma_\beta^{-1}\beta]d\beta &= \int q_\beta(\beta)(\beta - \mu_{q_\beta} + \mu_{q_\beta})'\Sigma_\beta^{-1}(\beta - \mu_{q_\beta} + \mu_{q_\beta})d\beta \\
&= \mathbb{E}[\text{trace}(\Sigma_\beta^{-1}(\beta - \mu_{q_\beta})(\beta - \mu_{q_\beta})')] + \mu_{q_\beta}'\Sigma_\beta^{-1}\mu_{q_\beta} \\
&= \text{trace}(\Sigma_\beta^{-1}\Sigma_{q_\beta}) + \mu_{q_\beta}'\Sigma_\beta^{-1}\mu_{q_\beta}
\end{aligned}
\tag{37}
$$

Substituting (37) back into (36), we obtained

$$
\mathbb{E}_{\beta,\mathbf{Z}}[\log p_\beta(\beta)] = \log C - \frac{(j+1)}{2}\log(2\pi) - \frac{1}{2}\log|\Sigma_\beta| - \frac{1}{2}[\text{trace}(\Sigma_\beta^{-1}\Sigma_{q_\beta}) + \mu_{q_\beta}'\Sigma_\beta^{-1}\mu_{q_\beta}]. \tag{38}
$$

Part 4: $\mathbb{E}_{\beta,\mathbf{Z}}[\log q_\beta(\beta)]$

$$
\begin{aligned}
&= \int\int q_\mathbf{Z}(\mathbf{z}) q_\beta(\beta) \log q_\beta(\beta) d\beta \\
&= -\frac{j+1}{2}\log(2\pi) - \frac{1}{2}\log|\Sigma_{q_\beta}| - \frac{1}{2}\int q_\beta(\beta)(\beta - \mu_{q_\beta})'\Sigma_{q_\beta}^{-1}(\beta - \mu_{q_\beta})d\beta \\
&= -\frac{j+1}{2}\log(2\pi) - \frac{1}{2}\log|\Sigma_{q_\beta}| - \frac{1}{2}\text{trace}(\Sigma_\beta^{-1}\Sigma_\beta) \\
&= -\frac{j+1}{2}(\log(2\pi) + 1) - \frac{1}{2}\log|\Sigma_{q_\beta}|
\end{aligned}
\tag{39}
$$

Combining all four parts together, we get

$$
\begin{aligned}
\mathcal{L}(q) &= \mathbb{E}_{\beta,\mathbf{Z}}[\log\{p(\mathbf{Y},\mathbf{Z}|\beta)\}] + \mathbb{E}_{\beta,\mathbf{Z}}[\pi^I(\beta)] - \mathbb{E}_{\beta,\mathbf{Z}}[\log\{q_\beta(\beta)\}] - \mathbb{E}_{\beta,\mathbf{Z}}[\log\{q_\mathbf{Z}(\mathbf{Z})\}] \\
&= \underbrace{-\frac{1}{2}\text{trace}(\mathbf{X}'\mathbf{X}[\mu_{q_\beta}\mu_{q_\beta}' + \Sigma_{q_\beta}]) + \frac{1}{2}\mu_\mathbf{z}'\mu_\mathbf{z} + \sum_{i=1}^n\{[\log(\Phi(-\mu_{\mathbf{z}_i}))]^{\mathbb{I}(y_i=0)}[\log(1-\Phi(-\mu_{\mathbf{z}_i}))]^{\mathbb{I}(y_i=1)}\}}_{\mathbb{E}_{\beta,\mathbf{Z}}[\log\{p(\mathbf{Y},\mathbf{Z}|\beta)\}]-\mathbb{E}_{\beta,\mathbf{Z}}[\log\{q_\mathbf{Z}(\mathbf{Z})\}]} \\
&\quad + \underbrace{\log C - \frac{1}{2}\log|\Sigma_\beta| - \frac{1}{2}[\text{trace}(\Sigma_\beta^{-1}\Sigma_{q_\beta}) + \mu_{q_\beta}'\Sigma_\beta^{-1}\mu_{q_\beta}] + \frac{j+1}{2} + \frac{1}{2}\log|\Sigma_{q_\beta}|}_{\mathbb{E}_{\beta,\mathbf{Z}}[\log p_\beta(\beta)] - \mathbb{E}_{\beta,\mathbf{Z}}[\log q_\beta(\beta)]}.
\end{aligned}
\tag{40}
$$

5.3. Model Comparison Based on Variational Approximation

Suppose we want to compare two models, M_1 and M_0, where M_0 is the simpler model. An intuitive thought on comparing two models by variational approximation methods is just to compare the lower bounds $\mathcal{L}(q_1)$ and $\mathcal{L}(q_0)$. However, we should note that by comparing the lower bounds, we are assuming that the KL divergences in the two approximations are the same, so that we can use just these lower bounds as guide. Unfortunately, it is not easy to measure how tight in theory any particular bound can be, if this can be accomplished we could then more accurately estimate the log marginal likelihood from the beginning. As clarified in [27], when comparing two exact log marginal likelihood, we have

$$\log p_1(\mathbf{X}) - \log p_0(\mathbf{X}) = [\mathcal{L}(q_1) + KL(q_1 \parallel p_1)] - [\mathcal{L}(q_0) - KL(q_0 \parallel p_0)] \tag{41}$$

$$= \mathcal{L}(q_1) - \mathcal{L}(q_0) + [KL(q_1 \parallel p_1) - KL(q_0 \parallel p_0)] \tag{42}$$

$$\neq \mathcal{L}(q_1) - \mathcal{L}(q_0). \tag{43}$$

The difference in log marginal likelihood, $\log p_1(\mathbf{X}) - \log p_0(\mathbf{X})$, is the quantity we wish to estimate. However, if we base this on the lower bounds difference, we are basing our model comparison on (43) rather than (42). Therefore, there exists a systematic bias towards simpler model when comparing models if $KL(q_1 \parallel p_1) - KL(q_0 \parallel p_0)$ is not zero.

Realizing that we have a variational approximation for the posterior distribution of β, we propose the following method to estimate $p(\mathbf{X})$ based on our variational approximation $q_\beta(\beta)$ (27). First, writing the marginal likelihood as

$$p(\mathbf{x}) = \int \left[\frac{p(\mathbf{x}|\beta)\pi^I(\beta)}{q_\beta(\beta)} \right] q_\beta(\beta) d\beta,$$

we can interpret it as the conditional expectation

$$p(\mathbf{x}) = \mathbb{E}\left[\frac{p(\mathbf{x}|\beta)\pi^I(\beta)}{q_\beta(\beta)} \right]$$

with respect to $q_\beta(\beta)$. Next, draw samples $\beta^{(1)}, ..., \beta^{(n)}$ from $q_\beta(\beta)$ and obtain the estimated marginal likelihood

$$\widehat{p_\mathbf{x}(\mathbf{x})} = \frac{1}{n} \sum_{i=1}^{n} \frac{p(\mathbf{x}|\beta^{(i)})\pi^I(\beta^{(i)})}{q_\beta(\beta^{(i)})}.$$

Please note that this method proposed is equivalent to importance sampling with importance function being $q_\beta(\beta)$, for which we know the exact form and the generation of the random $\beta^{(i)}$ is easy and inexpensive.

6. Modeling Probability of Default Using Lending Club Data

6.1. Introduction

LendingClub (https://www.lendingclub.com/) is the world's largest peer-to-peer lending platform. LendingClub enables borrowers to create unsecured personal loans between $1000 and $40,000. The standard loan period is three or five years. Investors can search and browse the loan listings on LendingClub website and select loans that they want to invest in based on the information supplied about the borrower, amount of loan, loan grade, and loan purpose. Investors make money from interest. LendingClub makes money by charging borrowers an origination fee and investors a service fee. To attract lenders, LendingClub publishes most of the information available in borrowers' credit reports as well as information reported by borrowers for almost every loan issued through its website.

6.2. Modeling Probability of Default—Target Variable and Predictive Features

Publicly available LendingClub data, from 2007 June to 2018 Q4, has a total of 2,260,668 issued loans. Each loan has a status, either Paid-off, Charged-off, or Ongoing. We only adopted loans with an end status, i.e., either paid-off or charged-off. In addition, that loan status is the target variable. We then selected following loan features as our predictive covariates.

- Loan term in months (either 36 or 60)
- FICO
- Issued loan amount
- DTI (Debt to income ratio, i.e., customer's total debt divided by income)
- Number of credit lines opened in past 24 months
- Employment length in years
- Annual income
- Home ownership type (own, mortgage, of rent)

We took a sample from the original data set that has customer yearly income between $15,000 and $60,000 and end up with a data set of 520,947 rows.

6.3. Addressing Uncertainty of Estimated Probit Model Using Variational Inference with Intrinsic Prior

Using the process developed in Section 5, we can update the intrinsic prior for parameters (see Figure 1) of the probit model using variational inference, and get the posterior distribution for the estimated parameters. Based on the derived parameter distributions, questions of interest may be explored with model uncertainty being considered.

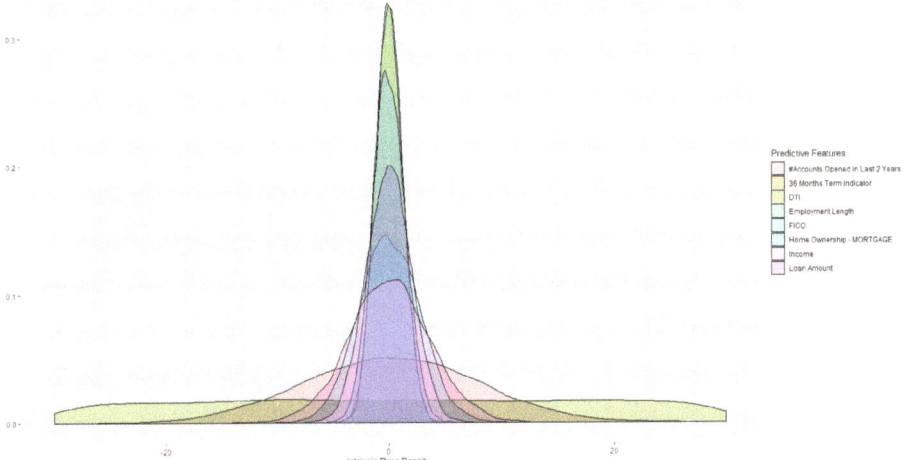

Figure 1. Intrinsic Prior.

Investors will be interested in understanding how each loan feature affect the probability of default, given a certain loan term, either 36 or 60. To answer this question, we samples 6000 cases from the original data set and draw from derived posterior distribution 100 times. We end up with 6000×100 calculated probability of default, where each one of the 6000 samples yield 100 different probit estimates based on 100 different posterior draws. We summarize some of our findings in Figure 2, where color red representing 36 months loans and green representing 60 months loans.

- In general, 60 months loans have higher risk of default.
- Given loan term months, there is a clear trend showing that high FICO means lower risk.
- Given loan term months, there is a trend showing that high DTI indicating higher risk.
- Given loan term months, there is a trend showing that more credit lines opened in past 24 months indicating higher risk.

- There is no clear pattern regarding income. This is probably because we only included customers with income between $15,000 and $60,000 in our training data, which may not representing the true income level of the whole population.

Model uncertainty could also be measured through credible intervals. Again, with the derived posterior distribution, the credible interval is just the range containing a particular percentage of estimated effect/parameter values. For instance, the 95% credible interval of the estimated parameter value of FICO is simply the central portion of the posterior distribution that contains 95% of the estimated values. Contrary to the frequentist confidence intervals, Bayesian credible interval is much more straightforward to interpret. Using the Bayesian framework created in this article, from Figure 3, we can simply state that given the observed data, the estimated effect of DTI on default has 89% probability of falling within [8.300, 8.875]. Instead of the conventional 95%, we used 89% following suggestions in [28,29], which is just as arbitrary as any of the conventions.

One of the main advantages of using variational inference over MCMC is that variational inference is much faster. Comparisons were made between the two approximation frameworks on a 64-bit Windows 10 laptop, with 32.0 GB RAM. Using the data set introduced in Section 6.2, we have that

- with a conjugate prior and following the Gibbs sampling scheme proposed by [17], it took 89.86 s to finish 100 simulations for the Gibbs sampler;
- following our method proposed in Section 5.2, it took 58.38 s to get the approximated posterior distribution and sampling 10,000 times from that posterior.

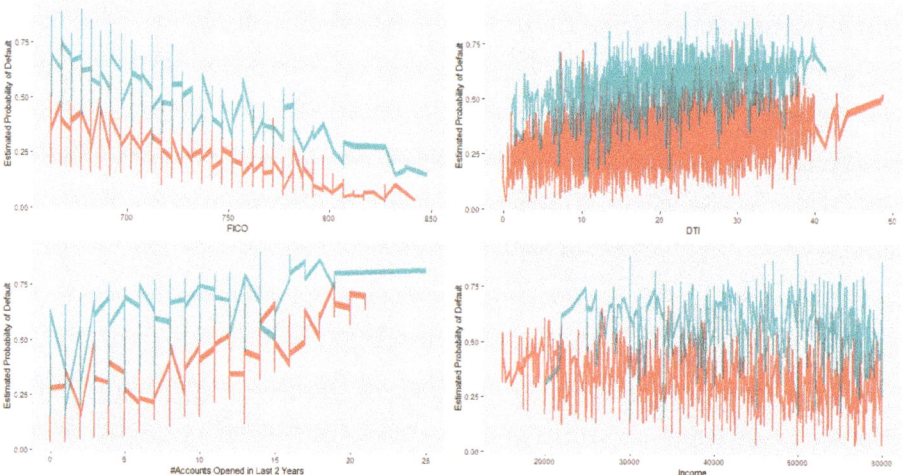

Figure 2. Effect of term months and other covariates on probability of default

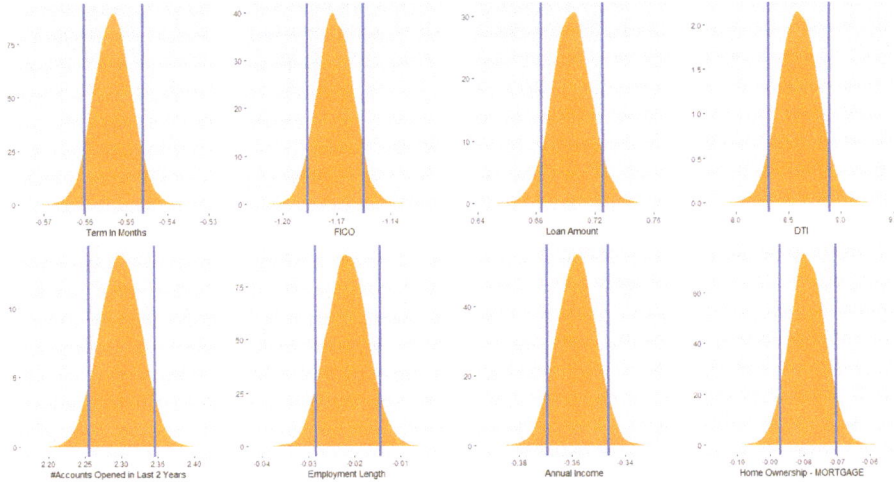

Figure 3. Credible intervals for estimated coefficients

6.4. Model Comparison

Following the procedure proposed in Section 5.3, we compare the following series of nested models. From the data set introduced in Section 6.2, 2000 records were sampled to estimate the likelihood $p(\mathbf{x}|\boldsymbol{\beta}^{(i)})$. Where $\boldsymbol{\beta}^{(i)}$ is one of the 2500 draws sampled directly from the approximated posterior distribution $q_{\boldsymbol{\beta}}(\boldsymbol{\beta})$, which serves as the importance function used to estimate the marginal likelihood $p(\mathbf{x})$.

- M_2: **FICO + Term 36 Indicator**
- M_3: **FICO + Term 36 Indicator + Loan Amount**
- M_4: **FICO + Term 36 Indicator + Loan Amount + Annual Income**
- M_5: **FICO + Term 36 Indicator + Loan Amount + Annual Income + Mortgage Indicator**

Estimated log marginal likelihood for each model is plotted in Figure 4. We can see that the model evidence has increased by adding predictive features **Loan Amount** and **Annual Income** sequentially. However, if we further adding home ownership information, i.e., **Mortgage Indicator** as a predictive feature, the model evidence decreased. We have the Bayes factor

$$BF_{45} = \frac{p(\mathbf{x}|M_4)}{p(\mathbf{x}|M_5)} = e^{-1014.78-(-1016.42)} = 5.16,$$

which suggests a substantial evidence for model M_4, indicating home ownership information may be irrelevant in predicting probability of default given that all the other predictive features are relevant.

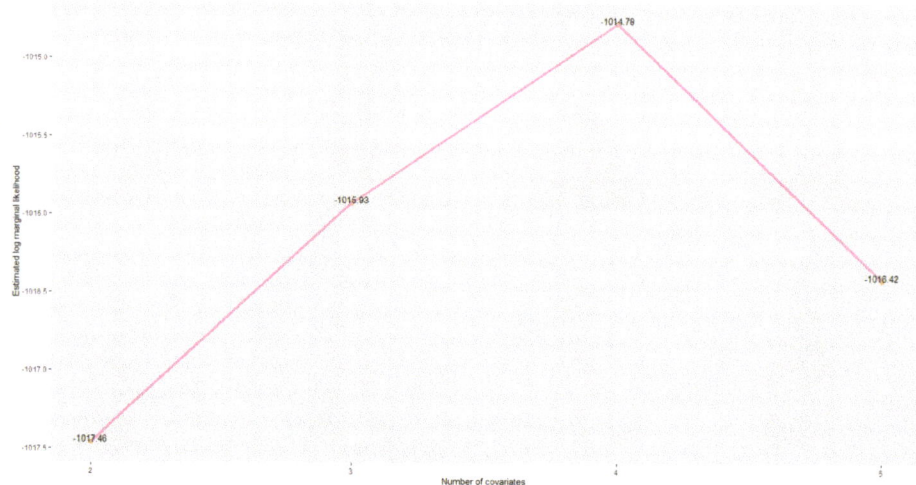

Figure 4. Log marginal likelihood comparison

7. Further Work

The authors thank the reviewers for pointing out that mean-field variational Bayes underestimates the posterior variance. This could be an interesting topic for our future research. We plan to study the *linear response variational Bayes* (LRVB) method proposed in [30] to see if it can be applied on the framework we proposed in this article. To see if we can get the approximated posterior variance close enough to the true variance using our proposed method, comparisons should be made between normal conjugate prior with the MCMC procedure, normal conjugate prior with LRVB, and intrinsic prior with LRVB.

Author Contributions: Methodology, A.L., L.P. and K.W.; software, A.L.; writing–original draft preparation, A.L., L.P. and K.W.; writing–review and editing, A.L. and L.P.; visualization, A.L. All authors have read and agreed to the published version of the manuscript.

Funding: The work of L.R.Pericchi was partially funded by NIH grants U54CA096300, P20GM103475 and R25MD010399.

Conflicts of Interest: The authors declare no conflict of interest.

Appendix A. Density Function

Suppose $X \sim N(\mu, \sigma^2)$ has a normal distribution and lies within the interval $X \in (a, b)$, $-\infty \leq a < b \leq \infty$. Then X conditional on $a < X < b$ has a truncated normal distribution. Its probability density function, f, for $a \leq X < b$, is given by

$$f(x|\mu, \sigma, a, b) = \frac{\frac{1}{\sigma}\phi(\frac{x-\mu}{\sigma})}{\Phi(\frac{b-\mu}{\sigma}) - \Phi(\frac{a-\mu}{\sigma})}$$

and by $f = 0$ otherwise. Here

$$\phi(\xi) = \frac{1}{\sqrt{2\pi}} \exp(-\frac{1}{2}\xi^2)$$

is the probability density function of the standard normal distribution and $\Phi(\cdot)$ is its cumulative distribution function. If $b = \infty$, then $\Phi(\frac{b-\mu}{\sigma}) = 1$, and similarly, if $a = -\infty$, then $\Phi(\frac{a-\mu}{\sigma}) = 0$. And the cumulative density for the truncated normal distribution is

$$F(x|\mu,\sigma,a,b) = \frac{\Phi(\xi) - \Phi(\alpha)}{Z},$$

where $\xi = \frac{x-\mu}{\sigma}$ and $Z = \Phi(\beta) - \Phi(\alpha)$.

Appendix B. Moments and Entropy

Let $\alpha = \frac{a-\mu}{\sigma}$ and $\beta = \frac{b-\mu}{\sigma}$. For two-sided truncation:

$$\mathbb{E}(X|a < X < b) = \mu + \sigma \frac{\phi(\alpha) - \phi(\beta)}{\Phi(\beta) - \Phi(\alpha)},$$

$$Var(X|a < X < b) = \sigma^2 \left[1 + \frac{\alpha\phi(\alpha) - \beta\phi(\beta)}{\Phi(\beta) - \Phi(\alpha)} - \left(\frac{\phi(\alpha) - \phi(\beta)}{\Phi(\beta) - \Phi(\alpha)} \right)^2 \right].$$

For one sided truncation (upper tail):

$$\mathbb{E}(X|X > a) = \mu + \sigma\lambda(\alpha)$$
$$Var(X|X > a) = \sigma^2[1 - \delta(\alpha)],$$

where $\alpha = \frac{a-\mu}{\sigma}$, $\lambda(\alpha) = \frac{\phi(\alpha)}{1-\Phi(\alpha)}$ and $\delta(\alpha) = \lambda(\alpha)[\lambda(\alpha) - \alpha]$.
For one sided truncation (lower tail):

$$\mathbb{E}(X|X < b) = \mu - \sigma\frac{\phi(\beta)}{\Phi(\beta)}$$

$$Var(X|X < b) = \sigma^2\left[1 - \beta\frac{\phi(\beta)}{\Phi(\beta)} - \left(\frac{\phi(\beta)}{\Phi(\beta)}\right)^2\right].$$

More generally, the moment generating function for truncated normal distribution is

$$e^{\mu t + \sigma^2 t^2/2} \cdot \left[\frac{\Phi(\beta - \sigma t) - \Phi(\alpha - \sigma t)}{\Phi(\beta) - \Phi(\alpha)} \right].$$

For a density $f(x)$ defined over a continuous variable, the *entropy* is given by

$$H[x] = -\int f(x) \log f(x) dx.$$

And the entropy for a truncated normal density is

$$\log(\sqrt{2\pi e}\sigma Z) + \frac{\alpha\phi(\alpha) - \beta\phi(\beta)}{2Z}.$$

References

1. Salmeron, D.; Cano, J.A.; Robert, C.P. Objective Bayesian hypothesis testing in binomial regression models with integral prior distributions. *Stat. Sin.* **2015**, *25*, 1009–1023. [CrossRef]
2. Leon-Novelo, L.; Moreno, E.; Casella, G. Objective Bayes model selection in probit models. *Stat. Med.* **2012**, *31*, 353–365. [CrossRef] [PubMed]
3. Jaakkola, T.S.; Jordan, M.I. Bayesian parameter estimation via variational methods. *Stat. Comput.* **2000**, *10*, 25–37. [CrossRef]

4. Girolami, M.; Rogers, S. Variational Bayesian multinomial probit regression with Gaussian process priors. *Neural Comput.* **2006**, *18*, 1790–1817. [CrossRef]
5. Consonni, G.; Marin, J.M. Mean-field variational approximate Bayesian inference for latent variable models. *Comput. Stat. Data Anal.* **2007**, *52*, 790–798. [CrossRef]
6. Ormerod, J.T.; Wand, M.P. Explaining variational approximations. *Am. Stat.* **2010**, *64*, 140–153. [CrossRef]
7. Grimmer, J. An introduction to Bayesian inference via variational approximations. *Political Anal.* **2010**, *19*, 32–47. [CrossRef]
8. Blei, D.M.; Kucukelbir, A.; McAuliffe, J.D. Variational inference: A review for statisticians. *J. Am. Stat. Assoc.* **2017**, *112*, 859–877. [CrossRef]
9. Pérez, M.E.; Pericchi, L.R.; Ramírez, I.C. The Scaled Beta2 distribution as a robust prior for scales. *Bayesian Anal.* **2017**, *12*, 615–637. [CrossRef]
10. Mulder, J.; Pericchi, L.R. The matrix-F prior for estimating and testing covariance matrices. *Bayesian Anal.* **2018**, *13*, 1193–1214. [CrossRef]
11. Berger, J.O.; Pericchi, L.R. Objective Bayesian Methods for Model Selection: Introduction and Comparison. In *Model Selection*; Institute of Mathematical Statistics: Beachwood, OH, USA, 2001; pp. 135–207.
12. Pericchi, L.R. Model selection and hypothesis testing based on objective probabilities and Bayes factors. *Handb. Stat.* **2005**, *25*, 115–149.
13. Scott, J.G.; Berger, J.O. Bayes and empirical-Bayes multiplicity adjustment in the variable-selection problem. *Ann. Stat.* **2010**, *38*, 2587–2619. [CrossRef]
14. Jeffreys, H. *The Theory of Probability*; OUP: Oxford, UK, 1961.
15. Berger, J.O.; Pericchi, L.R. The intrinsic Bayes factor for model selection and prediction. *J. Am. Stat. Assoc.* **1996**, *91*, 109–122. [CrossRef]
16. Leamer, E.E. *Specification Searches: Ad Hoc Inference with Nonexperimental Data*; Wiley: New York, NY, USA, 1978; Volume 53.
17. Albert, J.H.; Chib, S. Bayesian analysis of binary and polychotomous response data. *J. Am. Stat. Assoc.* **1993**, *88*, 669–679. [CrossRef]
18. Tanner, M.A.; Wong, W.H. The calculation of posterior distributions by data augmentation. *J. Am. Stat. Assoc.* **1987**, *82*, 528–540. [CrossRef]
19. Berger, J.O.; Pericchi, L.R. The intrinsic Bayes factor for linear models. *Bayesian Stat.* **1996**, *5*, 25–44.
20. Casella, G.; Moreno, E. Objective Bayesian variable selection. *J. Am. Stat. Assoc.* **2006**, *101*, 157–167. [CrossRef]
21. Moreno, E.; Bertolino, F.; Racugno, W. An intrinsic limiting procedure for model selection and hypotheses testing. *J. Am. Stat. Assoc.* **1998**, *93*, 1451–1460. [CrossRef]
22. Bishop, C.M. *Pattern Recognition and Machine Learning*; Springer: Berlin/Heidelberg, Germany, 2006.
23. Jordan, M.I.; Ghahramani, Z.; Jaakkola, T.S.; Saul, L.K. An introduction to variational methods for graphical models. *Mach. Learn.* **1999**, *37*, 183–233. [CrossRef]
24. Parisi, G.; Shankar, R. Statistical field theory. *Phys. Today* **1988**, *41*, 110. [CrossRef]
25. Boyd, S.; Vandenberghe, L. *Convex Optimization*; Cambridge University Press: Cambridge, UK, 2004.
26. Berger, J.; Pericchi, L. Training samples in objective Bayesian model selection. *Ann. Stat.* **2004**, *32*, 841–869. [CrossRef]
27. Beal, M.J. *Variational Algorithms for Approximate Bayesian Inference*; University College London: London, UK, 2003.
28. Kruschke, J. *Doing Bayesian Data Analysis: A Tutorial with R, JAGS, and Stan*; Academic Press: Cambridge, MA, USA, 2014.
29. McElreath, R. *Statistical Rethinking: A Bayesian Course with Examples in R and Stan*; Chapman and Hall/CRC: Boca Raton, FL, USA, 2018.
30. Giordano, R.J.; Broderick, T.; Jordan, M.I. Linear response methods for accurate covariance estimates from mean field variational Bayes. In Proceedings of the Advances in Neural Information Processing Systems, Montreal, QC, USA, 7–12 December 2015; pp. 1441–1449.

© 2020 by the authors. Licensee MDPI, Basel, Switzerland. This article is an open access article distributed under the terms and conditions of the Creative Commons Attribution (CC BY) license (http://creativecommons.org/licenses/by/4.0/).

Article

A New Multi-Attribute Emergency Decision-Making Algorithm Based on Intuitionistic Fuzzy Cross-Entropy and Comprehensive Grey Correlation Analysis

Ping Li [1], Ying Ji [1,*], Zhong Wu [1] and Shao-Jian Qu [2]

1. Business School, University of Shanghai for Science and Technology, Shanghai 200000, China; liping185699744@163.com (P.L.); wuzhong@usst.edu.cn (Z.W.)
2. Management Engineering School, University of Nanjing for Information Science & Technology, Nanjing 210000, China; qushaojian@163.com
* Correspondence: jiying@usst.edu.cn

Received: 11 June 2020; Accepted: 9 July 2020; Published: 14 July 2020

Abstract: Intuitionistic fuzzy distance measurement is an effective method to study multi-attribute emergency decision-making (MAEDM) problems. Unfortunately, the traditional intuitionistic fuzzy distance measurement method cannot accurately reflect the difference between membership and non-membership data, where it is easy to cause information confusion. Therefore, from the intuitionistic fuzzy number (IFN), this paper constructs a decision-making model based on intuitionistic fuzzy cross-entropy and a comprehensive grey correlation analysis algorithm. For the MAEDM problems of completely unknown and partially known attribute weights, this method establishes a grey correlation analysis algorithm based on the objective evaluation value and subjective preference value of decision makers (DMs), which makes up for the shortcomings of traditional model information loss and greatly improves the accuracy of MAEDM. Finally, taking the Wenchuan Earthquake on May 12th 2008 as a case study, this paper constructs and solves the ranking problem of shelters. Through the sensitivity comparison analysis, when the grey resolution coefficient increases from 0.4 to 1.0, the ranking result of building shelters remains stable. Compared to the traditional intuitionistic fuzzy distance, this method is shown to be more reliable.

Keywords: multi-attribute emergency decision-making; intuitionistic fuzzy cross-entropy; grey correlation analysis; earthquake shelters; attribute weights

1. Introduction

At present, earthquakes, fires, novel coronavirus infections, and other frequent disasters have caused great loss to human beings. Owing to the uncertainty and fuzziness of such emergency problems, it is difficult for decision makers (DMs) to determine alternatives with real numbers to make quick decisions. The accurate processing of information has become an unavoidable problem in the development of the emergency decision [1–3] field. Under this urgent demand, fuzzy set theory, which can deal well with the uncertainty of decision-making problems, came into being [4]. Fuzzy sets [5,6] use membership as a single scale to reflect the support and opposition of DMs to objective things. However, with the development of decision theory, it is difficult to accurately describe the uncertainty of objective things by fuzzy sets alone. Based on this, Atanassov, a Bulgarian professor, put forward the concept of the intuitionistic fuzzy set (IFS) in the 1980s [7,8]. He used membership degree and non-membership degree to express the support, opposition, and hesitation of decision information. Compared to the fuzzy set, the IFS can describe the natural attributes of objective things more accurately [9–11].

The IFS is a new mathematical tool for dealing with uncertain and complex information efficiently, which is widely used in the field of multi-attribute decision-making (MADM) [12–14]. In recent years, scholars have made great progress in the research of intuitionistic fuzzy multi-attribute decision-making (IFMADM). The similarity measure is one of the most important decision-making methods in IFMADM. Xu et al. [15] systematically analyzed the similarity measurement formula based on geometric distance, set theory, and intuitionistic fuzzy matching degree. In order to improve the measurement accuracy of the similarity of the IFS, Park et al. [16] and Hu et al. [17] used the similarity measurement formula based on intuitionistic fuzzy entropy for the intuitionistic fuzzy number (IFN) and interval IFN, respectively, and optimized the alternatives. The IFS can represent the uncertainty of decision information well, but there are some difficulties in data comparison. Score function and precise function are effective means for data comparison and ranking in IFMADM. Chen et al. [18] were the first experts to study the score function of the IFN. They used the difference between membership and non-membership in the IFN to construct a function to compare the size relationship of the IFN, which is the basis of IFMADM. On the basis of score function, Hong et al. [19] proposed an intuitionistic fuzzy precise function, which greatly improved the efficiency of decision-making. The classical multi-attribute method has a wide range of development and application in the field of intuitionistic fuzzy. Table 1 summarizes some main methods of IFMADM.

Table 1. A brief overview of preprocessing methods in intuitionistic fuzzy multi-attribute decision-making (IFMADM).

Literatures	Methods
Xu [15], Park et al. [16]	Similarity measure
Hu et al. [17]	Similarity measure, Fuzzy entropy
Chen et al. [18]	Score function
Hong et al. [19]	Intuitionistic fuzzy precise function
Wu et al. [20]	AHP, Score judgment matrix
Keshavarzfarda et al. [21]	AHP, DEMATEL
Chatterjee et al. [22], Liao et al. [23]	TOPSIS, VIKOR
Wu et al. [24], Vahdani et al. [25], Yu et al. [26]	ELECTRE, PROMETHEE
Meng et al. [27]	Prospect theory
Luo et al. [28]	Regret theory

Unfortunately, natural disasters, such as fires and floods, often lead to unexpected and disastrous consequences. A large number of emergency decision-making problems have evolved into MADM. Up to now, domestic and foreign scholars have conducted in-depth research in this field. Xu et al. [29] proposed a two-stage method to support the consensus-building process of large-scale MADM and applied it to earthquake shelter selection. Taking a fire and explosion accident as the study, Xu et al. [30] defined a generalized asymmetric language D number and proposed the corresponding MADM fusion algorithm, which verified the effectiveness of the method. Li et al. [31] proposed a risk decision analysis method based on the TODIM (an acronym in Portuguese of interactive and MADM) method to solve the emergency evacuation problem of tourist attractions, in which the attribute value and the probability of state occurrence are in the interval number format. This method solves this kind of emergency decision-making problem well, which shows that it is more effective than the traditional method. Based on an example of ship collision, Xiong et al. [32] used two intelligent algorithms, multi-attribute differential evolution algorithm and non-dominant sorting genetic algorithm, to verify the feasibility and effectiveness of the model. From the prediction model of the triple exponential smoothing method, Wang et al. [33] proposed an MADM additive weighting method, weighted product method, and elimination selection transformation reality method to sort the recycled electric vehicles, which provided an effective solution for managers and researchers in the electric vehicle industry and improved the efficiency of the electric vehicle industry. For the multi-attribute group decision-making problem of community sustainable development emergency response, Wu et al. [34] proposed a method based on subjective imprecise estimation of the reliability of binary language vocabulary, which greatly improved the efficiency of MADM. Karimi et al. [35] introduced the best and worst algorithm to solve the MADM problem in the fuzzy environment and applied this method to the evaluation of hospital maintenance, which proves the satisfactory performance of this method. Based on the above analysis, the MADM method is widely used in the field of emergency decision-making, which can solve the uncertainty well in the case of emergency. Table 2 summarizes some applications of the MADM method in emergency situations.

Table 2. A brief literature list on the applications of multi-attribute decision-making (MADM) methods in emergency situations.

Literatures	Methods	Applications
Xu et al. [29]	Two-stage theory	Earthquake shelter selection
Xu et al. [30]	Generalized asymmetric language	Fire and explosion accident
Li et al. [31]	Risk decision analysis	Electric vehicle industry
Xiong et al. [32]	Evolution and non-dominant sorting genetic algorithm	Ship collision
Wang et al. [33]	Additive weighting	Electric vehicle industry
Wu et al. [34]	Subjective imprecise estimation of binary language	Community development
Karimi et al. [35]	The best and worst algorithm	Hospital maintenance

The above method is effective in solving the multi-attribute emergency decision-making (MAEDM) problem in a fuzzy environment. However, it has some limitations in the following aspects.

(1) In the case of emergency, DMs often have a certain subjective preference for alternatives, which is rarely studied.
(2) The traditional intuitionistic fuzzy distance measurement accuracy is not high. It is easy to have a situation where the IFN cannot be compared, which makes the decision result produce errors.
(3) For MAEDM problems with unknown or partially unknown attribute weights, the research is not deep enough and needs further analysis.
(4) There is no corresponding sensitivity analysis for the ranking results of alternatives, which fails to fully explain the reliability and stability of the evaluation mechanism.

According to the above limitations, the motivation of this paper is summarized as follows:

(1) With the increasing complexity of the global environment, many scholars focus on the field of emergency decision-making. Intuitionistic fuzzy multi-attribute emergency decision-making (IFMAEDM) is the focus of the current research.
(2) It is necessary to propose a distance measurement method based on the IFN, which can get rid of the shortcomings of traditional distance measurement and improve the reliability of decision results.
(3) The research on the uncertainty of attribute weight is the key problem in MAEDM. How to determine the weight is always the core of decision-making.
(4) The evaluation mechanism of the ranking results of alternatives can make the decision results more reliable.

Therefore, based on intuitionistic fuzzy and grey correlation analysis, this paper proposes a method to solve MAEDM by using intuitionistic fuzzy cross-entropy distance. First, the average information entropy of intuitionistic fuzzy is defined, and the measurement method of cross-entropy distance of intuitionistic fuzzy is given. On this basis, considering the unknown and known attribute weights, an optimization model with the subjective preference of the DMs is established and solved. Secondly, the intuitionistic fuzzy decision matrix is obtained according to the objective attribute evaluation of DMs. The intuitionistic fuzzy cross-entropy distance matrix is constructed by combining the objective evaluation value and subjective preference value of alternatives. Then, the attribute weight is determined according to the adjusted intuitionistic fuzzy average information entropy. By using the method of grey correlation analysis, the comprehensive grey relation coefficient of each alternative is obtained, and the order of alternatives is generated. Therefore, a new method is proposed to solve the MAEDM problem by using intuitionistic fuzzy cross-entropy and grey correlation analysis. The important contributions of this paper are mainly reflected in six aspects. (1) The intuitionistic fuzzy cross-entropy distance is defined. (2) A multi-attribute emergency decision with subjective preference is considered. (3) The uncertainty of attribute weight is discussed and solved by intuitionistic fuzzy information entropy. (4) The grey correlation analysis method is applied to MAEDM, which makes full use of decision-making information such as membership, non-membership, and hesitation. (5) According to the grey resolution coefficient, the sensitivity analysis is carried out to verify the reliability and stability of the decision results. (6) Compared to the traditional intuitionistic fuzzy distance, this method is shown to be more stable.

The remainder of this paper is organized as follows. Section 2 defines some basic knowledge of intuitionistic fuzzy theory and introduces the concept of intuitionistic fuzzy cross-entropy distance. In Section 3, a MAEDM model based on intuitionistic fuzzy cross-entropy and comprehensive grey correlation analysis is constructed. In Section 4, taking the ranking of earthquake shelters as an example, the practical application of this method is illustrated by comparing to the traditional intuitionistic fuzzy method. Lastly, Section 5 is the conclusion of the method proposed in this paper and the prospect of future research.

2. Preliminaries

This section first reviews some basic concepts and definitions of intuitionistic fuzzy theory.

As the preference relationship in fuzzy theory is often assigned by the complementary 0.1–0.9 five-scale, we believe that the distribution of the levels between opposition and support is uniform and symmetric. However, in an actual situation, some problems require the use of a non-consistent and asymmetric distribution to evaluate variables, such as the marginal utility decline rate in economics. Therefore, it is very popular to solve this kind of asymmetric problem by fuzzy set theory.

Definition 1 [4]. *If the domain X is a non-empty set, a fuzzy set is defined as:*

$$A = \{< x, \mu_A(x) | x \in X\} \tag{1}$$

which is characterized by a membership function $\mu_A : X \to [0,1]$, where $\mu_A(x)$ denotes the degree of membership of the element x to the set A.

Ordinary fuzzy sets can only represent membership function, which refers to the support degree of an alternative without non-membership degree information. Therefore, Atanassov [7,8] extended the fuzzy set to the IFS. It is shown as follows:

Definition 2 [7]. *If the domain X is a non-empty set, then the intuitionistic fuzzy set A on X can be expressed as:*

$$A = \{< x, \mu_A(x), \nu_A(x) > | x \in X\} \tag{2}$$

where $\mu_A(x)$ and $\nu_A(x)$ are the membership degree and non-membership degree of the element x belonging to A in the domain X, respectively,

$$\mu_A : X \to [0,1], x \in X \to \mu_A(x) \in [0,1]$$
$$\nu_A : X \to [0,1], x \in X \to \nu_A(x) \in [0,1],$$

It satisfies $0 \leq \mu_A(x) + \nu_A(x) \leq 1$; let

$$\pi_A = 1 - \mu_A(x) - \nu_A(x) \tag{3}$$

denote the degree of hesitation or uncertainty that element x in X belongs to IFS A, obviously for any $x \in X$, with the condition $0 \leq \pi_A \leq 1$.

Example 1. *Take an example to illustrate the specific meaning of the IFS. Suppose there is an IFS $A = \{< x, 0.7, 0.2 > | x \in X\}$, which indicates that the membership degree of IFS X is 0.7, the non-membership degree is 0.2, and the hesitation degree is 0.1. If we use this set to represent the voting process, assuming that the number of participants is 10, then 7 people support it, 2 oppose it, and 1 hesitates to remain neutral.*

Definition 3 [36]. *Let $\alpha_A = (\mu_A, \nu_A)$ and $\alpha_B = (\mu_B, \nu_B)$ be the two intuitionistic fuzzy numbers. Then, the normalized Hamming distance between α_A and α_B is defined as follows:*

$$d(\alpha_A, \alpha_B) = \frac{1}{2}(|\mu_A - \mu_B| + |\nu_A - \nu_B|) \tag{4}$$

where $\mu_A \in [0,1]$, $\nu_A \in [0,1]$ and $0 \leq \mu_A + \nu_A \leq 1$; meanwhile, all intuitionistic fuzzy numbers are expressed as θ. Obviously, the fuzzy number $a^+ = (1,0)$ is the maximum value in the fuzzy set, and $a^- = (0,1)$ is the minimum value in the set.

Geometric distance is not suitable for processing fuzzy decision information. According to the traditional distance model, Xu [15] proposed the distance measure formula of the intuitionistic fuzzy set:

Definition 4. *Suppose d is a mapping: $d: (\phi(x))^2 \to [0,1]$. If there are intuitionistic fuzzy sets,*

$$A = \{< x, \mu_A(x), \nu_A(x) > | x \in X\}$$
$$B = \{< x, \mu_B(x), \nu_B(x) > | x \in X\},$$
$$C = \{< x, \mu_C(x), \nu_C(x) > | x \in X\}$$

then the distance measure between the IFSs is

$$d_{Xu} = \left[\frac{1}{2n}\sum_{j=1}^{n}\left(\begin{array}{c}|\mu_A(x_j) - \mu_B(x_j)|^\lambda + |\nu_A(x_j) - \nu_B(x_j)|^\lambda \\ + |\pi_A(x_j) - \pi_B(x_j)|^\lambda\end{array}\right)\right]^{\frac{1}{\lambda}} \tag{5}$$

where $\lambda \geq 1$. When $\lambda = 1$, d_{Xu} degenerates into Hamming distance with IFS:

$$d_H = \frac{1}{2n}\sum_{j=1}^{n}\left(\begin{array}{c}|\mu_A(x_j) - \mu_B(x_j)| + |\nu_A(x_j) - \nu_B(x_j)| \\ + |\pi_A(x_j) - \pi_B(x_j)|\end{array}\right) \tag{6}$$

When $\lambda = 2$, d_{Xu} degenerates into Euclidean distance with IFS:

$$d_E = \left[\frac{1}{2n}\sum_{j=1}^{n}\left(\begin{array}{c}|\mu_A(x_j) - \mu_B(x_j)|^2 + |\nu_A(x_j) - \nu_B(x_j)|^2 \\ + |\pi_A(x_j) - \pi_B(x_j)|^2\end{array}\right)\right]^{\frac{1}{2}} \tag{7}$$

Hamming and Euclidean distance formulas are an extension of intuitionistic fuzzy distance.

Considering the attribute weight vector of $x_j (j = 1, 2, \ldots n)$, $\omega = (\omega_1, \omega_2, \ldots, \omega_n)^T$, satisfies $0 \leq \omega_j \leq 1$ and $\sum_{j=1}^{n} \omega_j = 1$, and the above two distance formulas d_H and d_E can be expressed as:

$$d_{H\omega} = \frac{1}{2n} \sum_{j=1}^{n} \omega_j \left(\begin{array}{c} |\mu_A(x_j) - \mu_B(x_j)| + |\nu_A(x_j) - \nu_B(x_j)| \\ + |\pi_A(x_j) - \pi_B(x_j)| \end{array} \right) \quad (8)$$

$$d_{E\omega} = \left[\frac{1}{2n} \sum_{j=1}^{n} \omega_j \left(\begin{array}{c} |\mu_A(x_j) - \mu_B(x_j)|^2 + |\nu_A(x_j) - \nu_B(x_j)|^2 \\ + |\pi_A(x_j) - \pi_B(x_j)|^2 \end{array} \right) \right]^{\frac{1}{2}} \quad (9)$$

It is not difficult to see from the formula that all intuitionistic fuzzy distances satisfy the following properties:

(1) $0 \leq d(A, B) \leq 1$;
(2) When $A = B$, $d(A, B) = 0$
(3) $d(A, B) = d(B, A)$;
(4) If $A \subseteq B \subseteq C$, $d(A, B) \leq d(A, C)$ and $d(B, C) \leq d(A, C)$.

In order to define the concept of intuitionistic fuzzy cross-entropy, the definition of information entropy is introduced. The average level of residual information after information redundancy eliminated is called information entropy, which is used to measure the uncertainty of information source in the communication process.

Definition 5. *There is a discrete random variable* $X = \{x_1, x_2, \ldots, x_n\}$ *that can be represented as:* $I = \left\{ \begin{array}{c} x_1, x_2, \cdots, x_n \\ p_1, p_2, \cdots, p_n \end{array} \right\}$, *where* $P = (p_1, p_2, \ldots, p_n)$ *is the probability of discrete random variable X satisfying* $0 \leq p_j \leq 1$ *and* $\sum_{i=1}^{n} p_j = 1$; *then, the information entropy of I can be expressed as*

$$I = -\eta \sum_{j=1}^{n} p_j \log_c p_j \quad (10)$$

The constant η means the unit of measurement of information entropy, which is a constant greater than 0, and the base number c of the logarithmic function in the formula can take a non-negative constant. In particular, when $c = 2$, the unit of information entropy is bit. When $c = e$, the unit of information entropy is nat. When $c = 10$, its unit is dit. In general calculation, $\eta = 1$, $c = 2$.

Burillo et al. [37] extended the basic idea of information entropy to the field of intuitionistic fuzzy, and creatively used it to describe the uncertainty of the IFS.

Definition 6. *Let* $X = \{x_1, x_2, \ldots x_n\}$ *be a domain and* $A = \{< x, \mu_A(x), \nu_A(x) > | x \in X\}$ *be an IFS on X. The intuitionistic fuzzy entropy of A can be expressed as:*

$$E_{LH}(A) = \frac{1}{n} \sum_{i=1}^{n} \frac{1 - |\mu_A(x_i) - \nu_A(x_i)| + \pi_A(x_i)}{1 + |\mu_A(x_i) - \nu_A(x_i)| + \pi_A(x_i)} \quad (11)$$

Definition 7. *Another equivalent transformation of intuitionistic fuzzy entropy* E_{LH} *is:*

$$E(A) = \frac{1}{n} \sum_{i=1}^{n} \frac{1 - \max(\mu_A(x_i) - \nu_A(x_i))}{1 - \min(\mu_A(x_i) - \nu_A(x_i))}. \quad (12)$$

Proof. Model (11) and model (12) are equivalent.

$$\begin{aligned}
E_{LH}(A) &= \frac{1}{n}\sum_{i=1}^{n}\frac{1-|\mu_A(x_i)-v_A(x_i)|+\pi_A(x_i)}{1+|\mu_A(x_i)-v_A(x_i)|+\pi_A(x_i)} \\
&= \frac{1}{n}\sum_{i=1}^{n}\frac{1-|\mu_A(x_i)-v_A(x_i)|+1-\mu_A(x_i)-v_A(x_i)}{1+|\mu_A(x_i)-v_A(x_i)|+1-\mu_A(x_i)-v_A(x_i)} \\
&= \frac{1}{n}\sum_{i=1}^{n}\frac{2-|\mu_A(x_i)-v_A(x_i)|-(\mu_A(x_i)+v_A(x_i))}{2+|\mu_A(x_i)-v_A(x_i)|-(\mu_A(x_i)+v_A(x_i))} \\
&= \frac{1}{n}\sum_{i=1}^{n}\frac{1-\frac{1}{2}(|\mu_A(x_i)-v_A(x_i)|+|\mu_A(x_i)+v_A(x_i)|)}{1-\frac{1}{2}(|\mu_A(x_i)+v_A(x_i)|-|\mu_A(x_i)-v_A(x_i)|)} \\
&= \frac{1}{n}\sum_{i=1}^{n}\frac{1-\max(\mu_A(x_i)-v_A(x_i))}{1-\min(\mu_A(x_i)-v_A(x_i))} = E(A)
\end{aligned}$$

□

Definition 7 is more concise in form and simpler in calculation. It eliminates the influence of hesitation and is a better expression of intuitionistic fuzzy entropy.

For the MAEDM problem discussed in this paper, when the attributes are completely unknown, it is necessary to calculate the average information entropy of each attribute. Combining with the intuitionistic fuzzy entropy, the intuitionistic fuzzy cross-entropy distance is defined as:

Definition 8. *Suppose there is a domain* $X = \{x_1, x_2, \ldots, x_n\}$, *where A and B are two IFSs on X,*

$$A = \{< x_j, \mu_A(x_j), v_A(x_j) > |x_j \in X\}$$
$$B = \{< x_j, \mu_B(x_j), v_B(x_j) > |x_j \in X\}$$

then, the intuitionistic fuzzy cross-entropy distance formula of A and B is [38]:

$$\begin{aligned}
CE(A,B) = &\sum_{j=1}^{n}\left\{\frac{1+\mu_A(x_j)-v_A(x_j)}{2}\times \right.\\
&\left. \log_2\frac{1+\mu_A(x_j)-v_A(x_j)}{1/2[1+\mu_A(x_j)-v_A(x_j)+1+\mu_B(x_j)-v_B(x_j)]}\right\} \\
&+\sum_{j=1}^{n}\left\{\frac{1-\mu_A(x_j)+v_A(x_j)}{2}\times \right.\\
&\left. \log_2\frac{1-\mu_A(x_j)+v_A(x_j)}{1/2[1-\mu_A(x_j)+v_A(x_j)+1-\mu_B(x_j)+v_B(x_j)]}\right\}
\end{aligned} \quad (13)$$

As the intuitionistic fuzzy cross-entropy $CE(A, B)$ *does not satisfy the symmetry, considering the problems of emergency decision-making, let*

$$CE^*(A,B) = CE(A,B) + CE(B,A) \quad (14)$$

define the intuitionistic fuzzy cross-entropy distance combined with the characteristics of multi-attribute.

Theorem 1. *Referring to the properties of the intuitionistic fuzzy geometric distance formula, the intuitionistic fuzzy cross-entropy satisfies the following properties:*

(1) $0 \leq CE^*(A,B)$;
(2) If $A = B$, $CE^*(A,B) = 0$;
(3) If $A \subseteq B \subseteq C$, then $CE^*(A,B) \leq CE^*(A,C)$ and $CE^*(B,C) \leq CE^*(A,C)$.

Proof. As

$$CE(A,B) = \sum_{j=1}^{n} \left\{ \frac{1+\mu_A(x_j)-v_A(x_j)}{2} \times \log_2 \frac{1+\mu_A(x_j)-v_A(x_j)}{1/2[1+\mu_A(x_j)-v_A(x_j)+1+\mu_B(x_j)-v_B(x_j)]} \right\}$$
$$+ \sum_{j=1}^{n} \left\{ \frac{1-\mu_A(x_j)+v_A(x_j)}{2} \times \log_2 \frac{1-\mu_A(x_j)+v_A(x_j)}{1/2[1-\mu_A(x_j)+v_A(x_j)+1-\mu_B(x_j)+v_B(x_j)]} \right\},$$

and model (13) has been given, the following exists

$$-CE(A,B) = -\sum_{j=1}^{n}\left\{\frac{1+\mu_A(x_j)-v_A(x_j)}{2} \times \log_2 \frac{1+\mu_A(x_j)-v_A(x_j)}{1/2[1+\mu_A(x_j)-v_A(x_j)+1+\mu_B(x_j)-v_B(x_j)]}\right\}$$
$$+ \sum_{j=1}^{n}\left\{\frac{1-\mu_A(x_j)+v_A(x_j)}{2} \times \log_2 \frac{1-\mu_A(x_j)+v_A(x_j)}{1/2[1-\mu_A(x_j)+v_A(x_j)+1-\mu_B(x_j)+v_B(x_j)]}\right\}$$
$$= \sum_{j=1}^{n}\left\{\frac{1+\mu_A(x_j)-v_A(x_j)}{2} \times \log_2 \frac{1/2[1+\mu_A(x_j)-v_A(x_j)+1+\mu_B(x_j)-v_B(x_j)]}{1+\mu_A(x_j)-v_A(x_j)}\right\}$$
$$+ \sum_{j=1}^{n}\left\{\frac{1-\mu_A(x_j)+v_A(x_j)}{2} \times \log_2 \frac{1/2[1-\mu_A(x_j)+v_A(x_j)+1-\mu_B(x_j)+v_B(x_j)]}{1-\mu_A(x_j)+v_A(x_j)}\right\}$$

As the above logarithmic function is strictly convex, according to the relevant properties,

$$\begin{aligned} f(a_1x_1+a_2x_2+\ldots+a_nx_n) &\leq \\ a_1f(x_1)+a_2f(x_2)+\ldots+a_nf(x_n) \end{aligned}, \tag{15}$$

therefore, we can obtain the following expression,

$$-CE(A,B) \leq \sum_{j=1}^{n} \log_2\left\{\frac{1+\mu_A(x_j)-v_A(x_j)}{2} \times \frac{1/2[1+\mu_A(x_j)-v_A(x_j)+1+\mu_B(x_j)-v_B(x_j)]}{1+\mu_A(x_j)-v_A(x_j)}\right\}$$
$$+ \sum_{j=1}^{n} \log_2\left\{\frac{1-\mu_A(x_j)+v_A(x_j)}{2} \times \frac{1/2[1-\mu_A(x_j)+v_A(x_j)+1-\mu_B(x_j)+v_B(x_j)]}{1-\mu_A(x_j)+v_A(x_j)}\right\}$$
$$\leq \log_2\{[(1+\mu_A(x_j)-v_A(x_j))+(1-\mu_B(x_j)+v_B(x_j))+ (1+\mu_B(x_j)-v_B(x_j))+(1-\mu_A(x_j)+v_A(x_j))]/4\}=0$$

Through the above proof, obviously, $CE(A,B) \geq 0$ and $CE(B,A) \geq 0$, and the same can be obtained. According to model (13) and (14), we can prove that $CE^*(A,B) \geq 0$. □

Proof. When $A = B$, there are the following relationships: $\mu_A(x_j) = \mu_B(x_j), v_A(x_j) = v_B(x_j)$. By substituting it into the model (13), we can obtain the conclusion $CE(A,B) = 0$, $CE(B,A) = 0$. Then, combining model (14), we can prove that $CE^*(A,B) = 0$. □

Proof. According to the understanding of the geometric intuitionistic fuzzy distance formula, it is not difficult to prove that the size of the fuzzy cross-entropy set is positively correlated with the size of distance. Let us assume that with $A \subseteq B \subseteq C$, we have $\mu_A(x_i) \leq \mu_B(x_i) \leq \mu_C(x_i)$ and $v_A(x_i) \leq v_B(x_i) \leq v_C(x_i)$. The following conclusions can be drawn: $\mu_A(x_i) - v_A(x_i) \leq \mu_B(x_i) - v_B(x_i) \leq \mu_C(x_i) - v_C(x_i)$. For the sake of proving convenience, $\mu_A(x_i) - v_A(x_i)$, $\mu_B(x_i) - v_B(x_i)$, and $\mu_C(x_i) - v_C(x_i)$ are recorded as a, b, c, respectively, and satisfy $-1 \leq a \leq b \leq c \leq 1$. Comparing the size relationship between two intuitionistic fuzzy cross-entropies can be done by subtraction. $\Delta CE^* = CE^*(A,C) - CE^*(A,B)$ can be transformed into:

$$\begin{aligned}\Delta CE^* &= \frac{1+a}{2}\log_2\frac{1+a}{1/2[1+a+1+c]} + \frac{1-a}{2}\log_2\frac{1-a}{1/2[1-a+1-c]} \\ &+ \frac{1+c}{2}\log_2\frac{1+c}{1/2[1+c+1+a]} + \frac{1-c}{2}\log_2\frac{1-c}{1/2[1-c+1-a]} \\ &- \frac{1+a}{2}\log_2\frac{1+a}{1/2[1+a+1+b]} - \frac{1-a}{2}\log_2\frac{1-a}{1/2[1-a+1-b]} \\ &- \frac{1+b}{2}\log_2\frac{1+b}{1/2[1+b+1+a]} - \frac{1-b}{2}\log_2\frac{1-b}{1/2[1-b+1-a]}\end{aligned},$$

thus,

$$\begin{aligned}-\Delta CE^* &= \frac{1+a}{2}\log_2\frac{1/2[1+a+1+c]}{1+a} + \frac{1-a}{2}\log_2\frac{1/2[1-a+1-c]}{1-a} \\ &+ \frac{1+c}{2}\log_2\frac{1/2[1+c+1+a]}{1+c} + \frac{1-c}{2}\log_2\frac{1/2[1-c+1-a]}{1-c} \\ &- \frac{1+a}{2}\log_2\frac{1/2[1+a+1+b]}{1+a} - \frac{1-a}{2}\log_2\frac{1/2[1-a+1-b]}{1-a} \\ &- \frac{1+b}{2}\log_2\frac{1/2[1+b+1+a]}{1+b} - \frac{1-b}{2}\log_2\frac{1/2[1-b+1-a]}{1-b}\end{aligned}.$$

As the $-\Delta CE^*$ is a strictly convex function, it has the property (15). It satisfies

$$-\Delta CE^* \leq \log_2\left\{\begin{aligned}&\frac{1+a}{2}\times\frac{1/2(1+a+1+c)}{1+a} + \frac{1-a}{2}\times\frac{1/2(1-a+1-c)}{1-a} \\ &+ \frac{1+c}{2}\times\frac{1/2(1+c+1+a)}{1+c} + \frac{1-c}{2}\times\frac{1/2(1-c+1-a)}{1-c}\end{aligned}\right\} - \log_2\left\{\begin{aligned}&\frac{1+a}{2}\times\frac{1/2(1+a+1+b)}{1+a} - \frac{1-a}{2}\times\frac{1/2(1-a+1-b)}{1-a} \\ &+ \frac{1+b}{2}\times\frac{1/2(1+b+1+a)}{1+b} + \frac{1-b}{2}\times\frac{1/2(1-b+1-a)}{1-b}\end{aligned}\right\} = 0.$$

Obviously, with $-\Delta CE^* \leq 0$, which is $\Delta CE^* \geq 0$, we can easily obtain $CE^*(A,C) \geq CE^*(A,B)$. The same reasoning can be proved, $CE^*(A,C) - CE^*(B,C) \geq 0$; thus, $CE^*(A,C) \geq CE^*(B,C)$. □

It can be seen from property (1) that the fuzzy entropy distance is non-negative. Property (2) means that when two IFSs are completely equal, the minimum intuitionistic fuzzy cross-entropy distance is equal to 0; thus, cross-entropy can be used to measure the difference degree or distance between two IFSs. Property (3) provides a sufficient basis for the comparison of intuitionistic fuzzy cross-entropy distance. Intuitionistic fuzzy cross-entropy extends the meaning of information entropy, which can be used to measure the fuzzy degree and unknown degree between IFSs on the basis of preserving the complete information of the original IFS. The greater the distance between two IFSs, the greater the cross-entropy of the fuzzy numbers. However, the traditional intuitionistic fuzzy distance measurement method cannot accurately reflect the differences between the data.

Based on this, a group of simple data can be used to compare the traditional intuitionistic fuzzy distance and fuzzy cross-entropy distance to show the reliability and stability of cross-entropy used to measure the degree of fuzzy.

Example 2. *Suppose that there are three voting activities with a population of 10. The voting can be represented by three groups of fuzzy numbers: $\alpha_1 = (0.6, 0.3)$, $\alpha_2 = (0.5, 0.4)$, $\alpha_3 = (0.4, 0.2)$. First, we use the traditional Hamming and Euclidean distance model (6) and model (7), respectively, to solve $d_H(\alpha_1, \alpha_3) = d_H(\alpha_2, \alpha_3) = 0.3$ and $d_E(\alpha_1, \alpha_3) = d_E(\alpha_2, \alpha_3) = 0.2646$. Obviously, it can be seen from the calculation results that two traditional distance formulas cannot measure the distance between fuzzy numbers α_1 and α_3, or α_2 and α_3, which is the disadvantage of the classical intuitionistic fuzzy distance measurement method. It is solved by the intuitionistic fuzzy cross-entropy distance method, $CE^*(\alpha_1, \alpha_3) = 0.0037$ and $CE^*(\alpha_2, \alpha_3) = 0.0101$.*

The results show that the distance between α_1 and α_3 is closer than that of the traditional intuitionistic fuzzy distance. Therefore, it is more effective to introduce intuitionistic fuzzy cross-entropy to deal with uncertainty decision information.

3. A Multi-Attribute Emergency Decision Model Based on Intuitionistic Fuzzy Cross-Entropy and Grey Correlation Analysis

This section analyzes the IFMAEDM problem in which DMs have a certain subjective preference for alternatives.

3.1. Problem Description

Taking the Wenchuan earthquake on May 12th 2008 as a study case, the government needs to build a batch of temporary shelters to rescue the victims in the disaster area. Considering the impact of earthquakes, the government has a certain priority (subjective preference) for the construction of regional shelters. After determining the geographical location, disaster risk, rescue facilities, and feasibility, a number of rescues in disaster-affected areas began in an orderly manner. The whole decision-making process aims to find the optimal solution through intuitionistic fuzzy cross-entropy and grey correlation analysis, which determines the area where the shelter is built first. It can be abstractly understood as: The decision-maker (government) gives the IFN representing the attribute value (agree, disagree, neutral) (μ_{ij}, ν_{ij}) from a series of alternatives (disaster-affected areas) $A_i (i = 1, 2, \ldots m)$ according to the objective evaluation attribute (specific factors of disaster situation) $C_j (j = 1, 2, \ldots n)$, which denotes that the decision maker's approval degree is μ_{ij}, objection degree is ν_{ij}, and neutrality degree is $\pi_{ij} = 1 - \mu_{ij} - \nu_{ij}$ for alternative A_i under the condition of attribute C_j. The attribute weight is expressed in ω_j and satisfies $0 \le \omega_j (j = 1, 2, \ldots n) \le 1$ and $\sum_{j=1}^{n} \omega_j = 1$. The IFN meets the following conditions: $0 \le \mu_{ij}, \nu_{ij}, \pi_{ij} \le 1$. Using a fuzzy number to construct multi-attribute intuitionistic fuzzy decision matrix R_{mn}, the expression form is shown in Table 3:

Table 3. Intuitionistic fuzzy decision matrix.

Alternative	C_1	C_2	...	C_n
A_1	(μ_{11}, ν_{11})	(μ_{12}, ν_{12})	...	(μ_{1n}, ν_{1n})
A_2	(μ_{21}, ν_{21})	(μ_{22}, ν_{22})	...	(μ_{2n}, ν_{2n})
...	...	...	...	...
A_m	(μ_{m1}, ν_{m1})	(μ_{m2}, ν_{m2})	...	(μ_{mn}, ν_{mn})

Analyzing the Wenchuan earthquake, DMs have a certain subjective preference for alternatives, which need to consider the severity of the disaster area. The preference value is also IFN $c_i = (\sigma_i, \delta_i)(i = 1, 2, \ldots m)$. The following content uses the method of intuitionistic fuzzy cross-entropy and grey correlation analysis to build the optimal decision model and solve it.

3.2. Steps of Intuitionistic Fuzzy Cross-Entropy and Grey Correlation Analysis Algorithm

For the uncertain MAEDM problem with certain subjective preference, taking the Wenchuan earthquake shelter ranking problem for analysis, the comprehensive algorithm of intuitionistic fuzzy cross-entropy and grey correlation analysis is used to solve it. The specific steps are as follows (see Figure 1 for the flow framework):

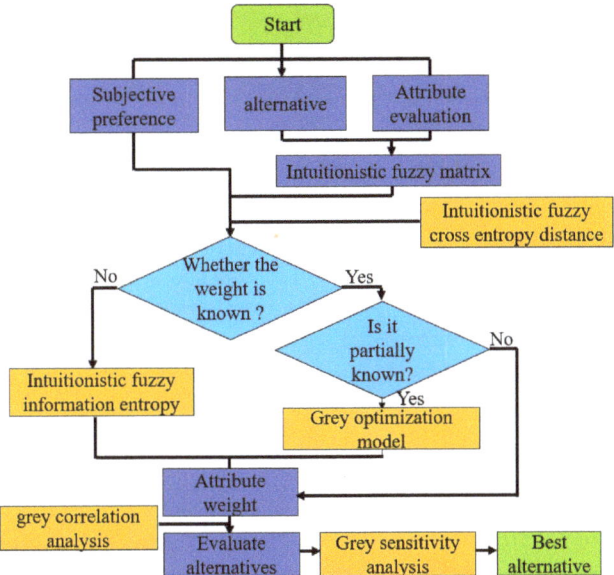

Figure 1. Algorithm framework of intuitionistic fuzzy cross-entropy and grey correlation analysis.

Step 1. According to the data given in the background of the Wenchuan earthquake case, alternative A_i, objective evaluation attribute value C_j, decision maker's subjective preference value c_i, and intuitionistic fuzzy evaluation decision matrix R_{mn} are determined.

Step 2. Using intuitionistic fuzzy cross-entropy distance to solve the grey correlation coefficient between the objective evaluation value of alternatives and the subjective preference value of DMs, the formula is expressed as:

$$\theta_{ij} = \frac{\min\limits_i \min\limits_j CE^*_{ij} + \xi \max\limits_i \max\limits_j CE^*_{ij}}{CE^*_{ij} + \xi \max\limits_i \max\limits_j CE^*_{ij}}, \quad (16)$$

ξ is called the grey resolution coefficient, and the value range is $0 \leq \xi \leq 1$, which is often set as $\xi = 0.5$. It satisfies $0 \leq \theta_{ij}(i = 1, 2, \ldots m; j = 1, 2, \ldots n) \leq 1$. The larger the grey correlation coefficient θ_{ij}, the closer the objective evaluation value and subjective preference value. In model (16), CE^*_{ij} is the intuitionistic fuzzy cross-entropy distance, and the specific formula is as follows:

$$\begin{aligned} CE^*_{ij} &= \frac{1+\mu_{ij}-\nu_{ij}}{2} \times \log_2 \frac{1+\mu_{ij}-\nu_{ij}}{1/2[1+\mu_{ij}-\nu_{ij}+1+\sigma_i-\delta_i]} \\ &+ \frac{1-\mu_{ij}+\nu_{ij}}{2} \times \log_2 \frac{1-\mu_{ij}+\nu_{ij}}{1/2[1-\mu_{ij}+\nu_{ij}+1-\sigma_i+\delta_i]} \\ &+ \frac{1+\sigma_i-\delta_i}{2} \times \log_2 \frac{1+\sigma_i-\delta_i}{1/2[1+\sigma_i-\delta_i+1+\mu_{ij}-\nu_{ij}]} \\ &+ \frac{1-\sigma_i+\delta_i}{2} \times \log_2 \frac{1-\sigma_i+\delta_i}{1/2[1-\sigma_i+\delta_i+1-\mu_{iji}+\nu_{ij}]} \end{aligned}, \quad (17)$$

Step 3. On the basis of the solution method of the grey correlation coefficient given in model (16), the weight of each attribute is calculated to determine the comprehensive correlation coefficient θ_i of each alternative. The following three cases are discussed: The attribute weight is completely unknown, completely known, and the value range is known.

Case 1. Attribute weight is completely unknown. In order to determine the attribute weight, the average information entropy of each attribute must be obtained. On the basis of intuitionistic fuzzy entropy, the calculation method of information entropy is as follows:

$$E(C_j) = -\frac{1}{\ln m} \sum_{i=1}^{m} \frac{CE^*_{ij}}{\sum_{i=1}^{m} CE^*_{ij}} \ln \frac{CE^*_{ij}}{\sum_{i=1}^{m} CE^*_{ij}}, \qquad (18)$$

The natural logarithm is taken to make the entropy value return to 1 and ensure the boundedness of information entropy. By transforming the formula of average information entropy, we can obtain the method of solving attribute weight:

$$\omega_j = \frac{1 - E(C_j)}{\sum_{k=1}^{n} [1 - E(C_k)]} (j = 1, 2, \ldots n), \qquad (19)$$

The weight parameters of each attribute can be determined and substituted,

$$\theta_i = \sum_{j=1}^{n} \theta_{ij} \omega_j (i = 1, 2, \ldots m; j = 1, 2, \ldots n), \qquad (20)$$

In model (20), the comprehensive correlation coefficient of alternatives θ_i can be aggregated.

Case 2. Attribute weights are fully known. Under the condition that the attribute is completely known, the grey correlation coefficient θ_{ij} of each alternative attribute is obtained by using model (16), and the comprehensive correlation degree θ_i of the alternative is obtained by combining model (20).

Case 3. The value range of attribute weight is known. Based on the maximum approach between weights with a known range of values and the subjective decision maker's preference, a linear programming model with attribute weight as a variable is constructed,

$$\max Y(\omega_j) = \sum_{i=1}^{m} \sum_{j=1}^{n} \theta_{ij} \omega_j (j = 1, 2, \cdots n)$$
$$s.t. \begin{cases} \sum_{j=1}^{n} \omega_j = 1, \omega_j \in W \\ 0 \leq \omega_j, (j = 1, 2, \cdots n), (i = 1, 2, \cdots m) \end{cases}, \qquad (21)$$

In this way, the weight parameters of each attribute can be determined.

The weight ω_j of each attribute can be calculated by establishing the optimization model of the maximum comprehensive grey correlation coefficient θ_i:

$$\theta_i = \sum_{j=1}^{n} \theta_{ij} \omega_j (i = 1, 2, \ldots m; j = 1, 2, \ldots n), \qquad (22)$$

The corresponding linear programming model is constructed by programming software Matlab (R2017b) to solve the code, and the attribute weight of each alternative is obtained. Then, the model is substituted into (20) to determine the comprehensive correlation degree θ_i.

Step 4. Based on the comprehensive correlation coefficient obtained under three different attribute weights in Step 3, the alternatives of the earthquake shelter are ranked according to the size relationship. The larger the θ_i, the better the alternative, which is in the front row.

Step 5. The sensitivity analysis is made by setting different values of the grey resolution constant in the correlation coefficient, and the difference of ranking alternatives under different resolution coefficients is compared and analyzed.

4. A Numerical Case Study on the Ranking of Wenchuan Earthquake Shelters

In this section, the traditional intuitionistic fuzzy distance and the intuitionistic fuzzy cross-entropy distance are used to analyze and compare the ranking of earthquake shelters.

4.1. Intuitionistic Fuzzy Cross-Entropy Distance and Grey Correlation Analysis

The stability and reliability of the method of intuitionistic fuzzy cross-entropy and the grey correlation coefficient are analyzed through comparative experiments. Assume that the government carries out shelter assessment and optimization for the five areas with a large disaster impact, and use A, B, C, D, and E to represent them. The government analyzes and evaluates the geographical location C_1, disaster risk C_2, rescue facilities C_3, and feasibility C_4 of the five disaster areas. The decision-maker adopts an IFN to express the objective evaluation value of alternatives under different attributes, and the intuitionistic fuzzy decision matrix $R_{5\times 4}$ is shown in Table 4.

Table 4. Objective evaluation value of each alternative.

Alternative	C_1	C_2	C_3	C_4
A	(0.4, 0.3)	(0.6, 0.3)	(0.5, 0.4)	(0.2, 0.7)
B	(0.5, 0.4)	(0.5, 0.3)	(0.2, 0.7)	(0.7, 0.1)
C	(0.4, 0.3)	(0.3, 0.5)	(0.6, 0.2)	(0.5, 0.2)
D	(0.5, 0.5)	(0.4, 0.5)	(0.4, 0.4)	(0.5, 0.4)
E	(0.6, 0.3)	(0.6, 0.4)	(0.3, 0.6)	(0.6, 0.3)

The decision maker's subjective preference values for alternatives A, B, C, D, and E are also expressed by IFNs: $c_1 = (0.5, 0.4)$, $c_2 = (0.6, 0.3)$, $c_3 = (0.4, 0.3)$, $c_4 = (0.4, 0.5)$, and $c_5 = (0.6, 0.2)$. In order to choose the best alternative to build a shelter in the earthquake disaster area, the government adopts the intuitionistic fuzzy cross-entropy and grey correlation analysis method to make a decision.

Step 1. Determine the values of alternative A, B, C, D, and E; the objective evaluation attribute values C_1, C_2, C_3, C_4; the decision makers' objective evaluation matrix $R_{5\times 4}$; and subjective preference values c_1, c_2, c_3, c_4, c_5.

Step 2. According to model (17), the intuitionistic fuzzy cross-entropy distance between the objective evaluation value and the subjective preference value of each alternative is calculated to form the distance matrix:

$$CE^*_{5\times 4} = \begin{bmatrix} 0.0000 & 0.0151 & 0.0000 & 0.1378 \\ 0.0151 & 0.0038 & 0.2402 & 0.0411 \\ 0.0000 & 0.0327 & 0.0348 & 0.0151 \\ 0.0036 & 0.0000 & 0.0036 & 0.0145 \\ 0.0041 & 0.0159 & 0.1810 & 0.0041 \end{bmatrix}$$

Step 3. Assuming that the grey resolution coefficient is $\xi = 0.5$, the grey correlation coefficient between the decision-maker's subjective preference value and the objective evaluation value is calculated according to model (16). The coefficient matrix is as follows:

$$\theta_{5\times 4} = \begin{bmatrix} 1.0000 & 0.8883 & 1.0000 & 0.4657 \\ 0.8883 & 0.9693 & 0.3333 & 0.7450 \\ 1.0000 & 0.7860 & 0.7753 & 0.8883 \\ 0.9709 & 1.0000 & 0.9709 & 0.8923 \\ 0.9670 & 0.8831 & 0.3989 & 0.9670 \end{bmatrix}$$

Step 4. Calculate the attribute weight ω_j according to the known information provided by the above case. When the attribute weight is known, the model is relatively easy to solve. The following focuses on the analysis of two situations: The attribute weight is completely unknown and the attribute weight range is known.

Case 1. The weight of attributes is completely unknown. According to the idea of intuitionistic fuzzy entropy, the average intuitionistic fuzzy entropy of the attribute is obtained by combining model (18): $E(C_1)= 0.5424$, $E(C_2)= 0.7385$, $E(C_3)= 0.5837$, $E(C_4)= 0.6498$. Then, according to model (19), we obtain the attribute weight $\omega_1 = 0.3080$, $\omega_2 = 0.1761$, $\omega_3 = 0.2802$ and $\omega_4 = 0.2357$. The attribute weight obtained is substituted into model (22), and the comprehensive grey correlation coefficient θ_i of the alternatives under the attribute condition is calculated: $\theta_1= 0.8544$, $\theta_2= 0.7133$, $\theta_3= 0.8730$, $\theta_4= 0.9575$, and $\theta_5= 0.7930$. From the comprehensive grey correlation coefficient θ_i of the alternatives, the result is $\theta_4 > \theta_3 > \theta_1 > \theta_5 > \theta_2$ and $D > C > A > E > B$. Therefore, the alternative D is the best and the government should give priority to building earthquake shelters in the region.

For proving the superiority and stability of the intuitionistic fuzzy cross-entropy and the comprehensive grey correlation analysis algorithm proposed in this paper, different resolution coefficients ξ are set for sensitivity analysis to compare and analyze whether the above alternatives will produce fluctuations. Set $\xi =0.40, 0.50, 0.60, 0.70, 0.80, 0.90, 1.00$. The results of the comprehensive correlation coefficient are shown in Table 5. The ranking results of alternatives did not fluctuate with the change in resolution coefficient.

Table 5. Comprehensive grey correlation coefficient of alternatives under different grey resolution coefficients based on completely unknown attribute weights.

Alternative	$\xi=0.40$	$\xi=0.50$	$\xi=0.60$	$\xi=0.70$	$\xi=0.80$	$\xi=0.90$	$\xi=1.00$
A	0.8372	0.8544	0.8681	0.8793	0.8887	0.8967	0.9037
B	0.6807	0.7133	0.7388	0.7596	0.7769	0.7917	0.8045
C	0.8488	0.8730	0.8906	0.9039	0.9143	0.9226	0.9295
D	0.9479	0.9575	0.9641	0.9689	0.9726	0.9755	0.9779
E	0.7697	0.7930	0.8114	0.8266	0.8393	0.8501	0.8595

In order to verify the reliability and stability of the method proposed in this paper more intuitively, we use Python graphics to carry out simulation experiments on the sequencing and gray resolution coefficient of each alternative, and the specific results are shown in Figure 2 (G is the grey resolution coefficient).

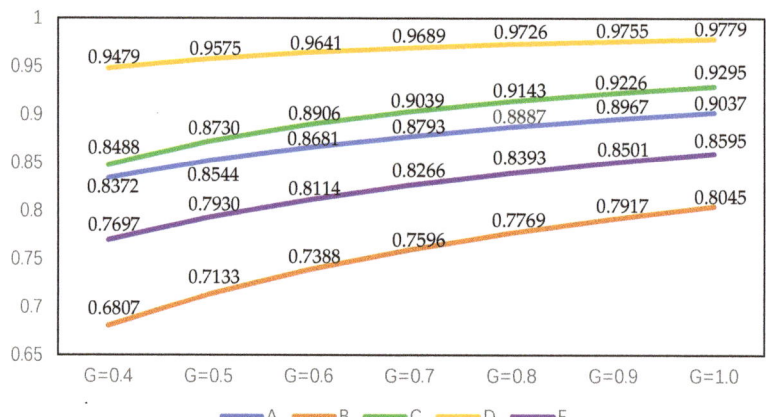

Figure 2. Ranking results of alternatives with different grey resolution coefficients based on completely unknown attribute weights.

It can be seen from Figure 2 that in the seven experiments of sensitivity analysis of grey resolution coefficient, the ranking results of alternatives have not changed, and $D > C > A > E > B$ is always maintained. The simulation experiment shows that D is the best alternative to build a shelter in the earthquake disaster area, and the decision result does not fluctuate, which shows the strong stability.

Case 2. The value range of attribute weight is known: $0.30 \leq \omega_1 \leq 0.32$, $0.17 \leq \omega_2 \leq 0.20$, $0.25 \leq \omega_3 \leq 0.28$, and $0.20 \leq \omega_4 \leq 0.24$. Through the linear programming model (21), the objective function Y to maximize the grey correlation coefficient of alternatives is constructed and solved:

$$\max Y(\omega_j) = \begin{matrix} 4.8262\omega_1 + 4.5267\omega_2 \\ +3.4784\omega_3 + 3.9583\omega_4 \end{matrix}, \text{ s.t.} \begin{cases} 0.30 \leq \omega_1 \leq 0.32 \\ 0.17 \leq \omega_2 \leq 0.20 \\ 0.25 \leq \omega_3 \leq 0.28 \\ 0.20 \leq \omega_4 \leq 0.24 \\ \omega_1 + \omega_2 + \omega_3 + \omega_4 = 1 \\ 0 \leq \omega_j \leq 1, (j = 1, 2, 3, 4) \end{cases} \quad (23)$$

The attribute weight is $\omega_1 = 0.30$, $\omega_2 = 0.18$, $\omega_3 = 0.28$, and $\omega_4 = 0.24$ by MATLAB. Combined with model (22), the comprehensive grey correlation coefficient of each alternative is obtained: $\theta_1 = 0.8517$, $\theta_2 = 0.7131$, $\theta_3 = 0.8718$, $\theta_4 = 0.9573$, $\theta_5 = 0.7928$. According to the comprehensive grey correlation coefficient θ_i of the alternatives, we can obtain $\theta_4 > \theta_3 > \theta_1 > \theta_5 > \theta_2$. Therefore, the order of alternatives is $D > C > A > E > B$, and, thus, alternative D is the best. The government should give priority to building earthquake shelters in area D, which is the same as the decision-making result when the attribute weight is unknown.

In order to further verify the stability and superiority of the algorithm of intuitionistic fuzzy cross-entropy and comprehensive grey correlation analysis when the attribute weight range is known, different resolution coefficients are also set for sensitivity analysis, and the optimal alternative and decision results are compared. Taking ξ =0.40, 0.50, 0.60, 0.70, 0.80, 0.90, and 1.00, and attribute weight and comprehensive grey correlation analysis when the attribute weight range is known, different resolution coefficients are also set for sensitivity analysis, and the optimal alternative and decision results are compared. The attribute weight and comprehensive grey correlation coefficient of each alternative are shown in Tables 6 and 7. From the table data, the change in the grey resolution coefficient does not affect the attribute weight and the decision-making result of the alternative, which is still $D > C > A > E > B$. It is always the best alternative to build the seismic shelter in the D area. In addition, when the weight is completely unknown, the comprehensive grey correlation coefficient of the alternatives is higher than that of the alternatives with known range of attribute weight.

Table 6. Attribute weight values under different grey resolution coefficients.

Alternative	ξ=0.40	ξ=0.50	ξ=0.60	ξ=0.70	ξ=0.80	ξ=0.90	ξ=1.00
ω_1	0.30	0.30	0.30	0.30	0.30	0.30	0.30
ω_2	0.18	0.18	0.18	0.18	0.18	0.18	0.18
ω_3	0.28	0.28	0.28	0.28	0.28	0.28	0.28
ω_4	0.24	0.24	0.24	0.24	0.24	0.24	0.24

Table 7. Comprehensive grey correlation coefficient of alternatives under different grey resolution coefficients based on known range of attribute weight.

Alternative	ξ=0.40	ξ=0.50	ξ=0.60	ξ=0.70	ξ=0.80	ξ=0.90	ξ=1.00
A	0.8341	0.8517	0.8656	0.8771	0.8867	0.8948	0.9019
B	0.6805	0.7131	0.7387	0.7595	0.7768	0.7916	0.8044
C	0.8473	0.8718	0.8895	0.9029	0.9134	0.9218	0.9288
D	0.9476	0.9573	0.9639	0.9688	0.9725	0.9754	0.9778
E	0.7695	0.7928	0.8113	0.8264	0.8392	0.8500	0.8594

More importantly, when the grey resolution coefficient fluctuates from 0.4 to 1.0, whether the weight is known or unknown, the change range of the comprehensive grey correlation coefficient of alternative D is the smallest, which is 0.0300 and 0.0302, respectively (see Table 8). Alternative B is always the worst, and its fluctuation is also the largest, which is 0.1438 and 0.1239, respectively. Based on this, the stability of the proposed method is proved.

Table 8. Change degree of comprehensive grey correlation coefficient of alternatives under fluctuation of grey resolution coefficient.

Alternative	Δθ (Unknown Weight)	Δθ (Weight Range Known)
A	0.0665	0.0678
B	0.1238	0.1239
C	0.0807	0.0815
D	0.0300	0.0302
E	0.0898	0.0899

From Table 7, Python simulation results are shown in Figure 3. Compared to Figure 2, the comprehensive grey correlation coefficient decreases but does not change the overall trend of each alternative, and the decision results remain unchanged. Whether the attribute weights are known or not, the optimal alternative and ranking results are the same, which shows the superiority and stability of the method.

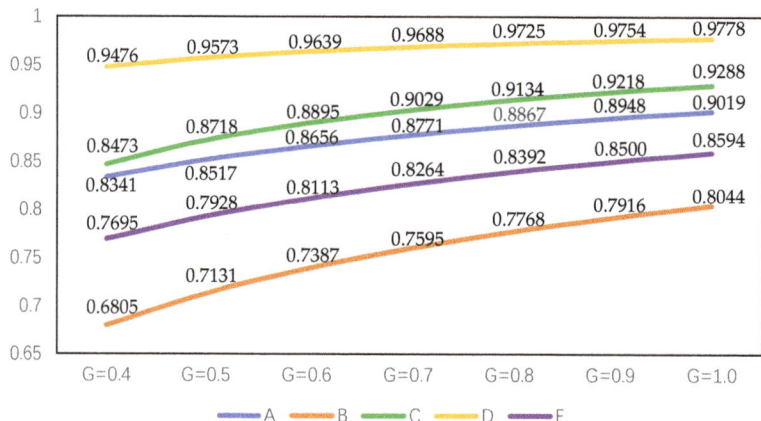

Figure 3. Ranking results of alternatives with different grey resolution coefficients based on known attribute weight range.

Through the above comparative analysis, the intuitionistic fuzzy entropy and grey correlation analysis method has achieved good results in solving the MAEDM problems. In this way, the ranking results have strong stability and environmental adaptability.

4.2. Traditional Intuitionistic Fuzzy Distance and Grey Correlation Analysis

Based on the data given by the above problem of ranking earthquake shelters, the traditional intuitionistic fuzzy distance and grey correlation degree are used to analyze and give the ranking results.

The traditional intuitionistic fuzzy distance model (4) has been given; thus, the corresponding grey correlation coefficient ε_{ij} is

$$\varepsilon_{ij} = \frac{\min_i \min_j d(r_{ij}, c_i) + \xi \max_i \max_j d(r_{ij}, c_i)}{d(r_{ij}, c_i) + \xi \max_i \max_j d(r_{ij}, c_i)} \quad (24)$$

where r_{ij} denotes the objective evaluation value, c_i denotes the subjective preference information, and grey resolution coefficient $\xi = 0.50$.

Step 1. Calculating the grey correlation coefficient of each alternative between the objective evaluation value and subjective preference information.

$$\varepsilon_{5\times 4} = \begin{bmatrix} 0.6667 & 0.6667 & 1.0000 & 0.4000 \\ 0.6667 & 0.8000 & 0.3333 & 0.5714 \\ 1.0000 & 0.5714 & 0.5714 & 0.6667 \\ 0.8000 & 1.0000 & 0.8000 & 0.6667 \\ 0.8000 & 0.6667 & 0.8000 & 0.8000 \end{bmatrix}$$

Step 2. Determining the attribute weight. Due to the fact that the range of attribute weight values is known, utilize model (21) to establish the following single-objective programming model:

$$\max Z(\omega_j) = \begin{array}{c} 3.9334\omega_1 + 3.7048\omega_2 \\ +3.5047\omega_3 + 3.1048\omega_4 \end{array} \quad s.\,t. \begin{cases} 0.30 \leq \omega_1 \leq 0.32 \\ 0.17 \leq \omega_2 \leq 0.20 \\ 0.25 \leq \omega_3 \leq 0.28 \\ 0.20 \leq \omega_4 \leq 0.24 \\ \omega_1 + \omega_2 + \omega_3 + \omega_4 = 1 \\ 0 \leq \omega_j \leq 1, (j = 1,2,3,4) \end{cases} \quad (25)$$

Solving this model, attribute weight can be obtained: $\omega_1 = 0.30$, $\omega_2 = 0.18$, $\omega_3 = 0.28$, and $\omega_4 = 0.24$.

Step 3. On the basis of model (20), the comprehensive grey correlation coefficient is calculated: $\varepsilon_1 = 0.6960$, $\varepsilon_2 = 0.5745$, $\varepsilon_3 = 0.7229$, $\varepsilon_4 = 0.8040$, $\varepsilon_5 = 0.7760$.

Step 4. Determining the alternatives ranking. Rank the alternatives according to the size of the comprehensive grey correlation coefficient ε_j. Thus, $D > E > C > A > B$ is the ranking result.

4.3. Comparative Analysis

Based on the ranking problem of earthquake shelters, this paper makes a comparative analysis from two aspects:

(1). The attribute weight is completely unknown and the attribute weight range is known

For a more intuitive comparison, it is further explored based on Figures 2 and 3. Regardless of whether the attribute weight is known or unknown, the ranking results of alternatives maintain high stability. The best alternative is always D, and the worst is always B. The comprehensive grey correlation coefficient of the alternative is positively correlated with the grey resolution coefficient, which indicates that the larger the resolution coefficient, the greater the correlation coefficient of the corresponding alternative.

Moreover, in the case of unknown weight, the comprehensive grey correlation coefficient of each alternative is always better than that of the known weight range, which also indirectly proves the fact that attribute weights are uncertain in most fields of decision problems (see Figures 4 and 5). In addition, the results obtained by using a reasonable method to determine the attribute weights are more practical.

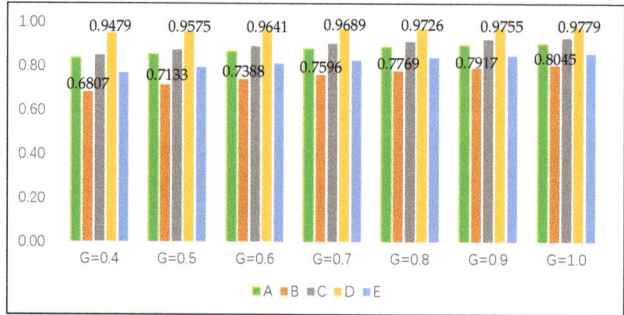

Figure 4. The alternatives with different grey resolution coefficients based on completely unknown attribute weights.

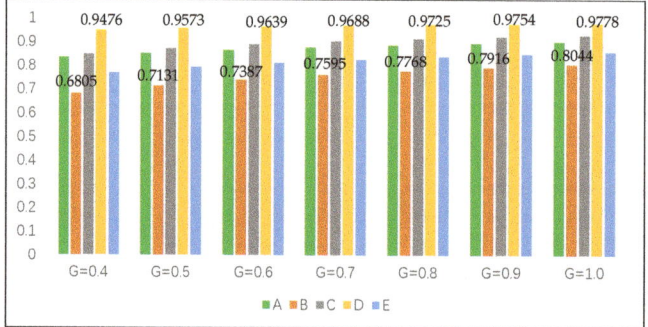

Figure 5. The alternatives with different grey resolution coefficients based on known attribute weight range.

Meanwhile, based on the data in Table 8, we can further analyze the volatility of the comprehensive grey correlation coefficient in two cases. From Figure 6 (deviation 1 represents unknown weights and deviation 2 represents known weights range), the deviation curves of the comprehensive grey correlation coefficient in the two kinds of weights situation almost coincide. However, when the weight is unknown, the fluctuation amplitude of the comprehensive grey correlation coefficient is still less than that of the known attribute weight range.

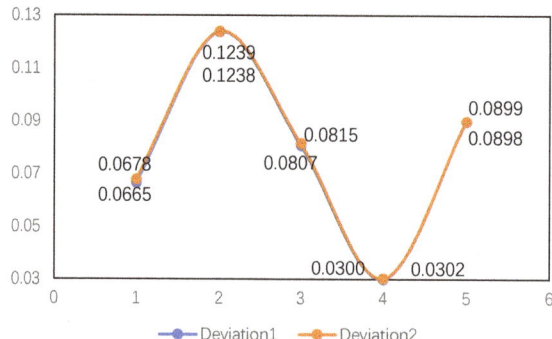

Figure 6. Deviation of comprehensive grey correlation coefficient in two cases.

Through the comparative analysis, we can see that the ranking result with unknown weight is more reasonable and more consistent with the uncertainty of the decision environment in MAEDM problems.

(2). The traditional intuitionistic fuzzy distance with the intuitionistic fuzzy cross-entropy distance

Through the above solution, the ranking results of the intuitionistic fuzzy cross-entropy method is $D > C > A > E > B$. Under the sufficient sensitivity analysis, the results maintain a high stability. However, by using the traditional intuitionistic fuzzy distance method, the result of ranking becomes $D > E > C > A > B$. Although the ranking result has little change, the best alternative is still D and the worst one is B (see Table 9). This also fully proves that the method based on intuitionistic fuzzy cross-entropy and grey correlation analysis proposed in this paper has strong stability.

Table 9. Ranking results under different methods.

Methods	Ranking Results
The traditional intuitionistic fuzzy distance	$D > E > C > A > B$
The intuitionistic fuzzy cross-entropy distance (unknown weight)	$D > C > A > E > B$
The intuitionistic fuzzy cross-entropy distance (weight range known)	$D > C > A > E > B$

According to the above two groups of comparative analysis, it can be concluded from many aspects that D is the best alternative. For the decision maker to make rescue measures, it is the most reasonable decision to give priority to the establishment of earthquake shelters in the D area.

5. Conclusions

This paper presents a new MAEDM method based on intuitionistic fuzzy cross-entropy and comprehensive grey correlation analysis. The main contributions are as follows: (1) Overcome the limitations of the traditional intuitionistic fuzzy geometric distance algorithm, and introduce the intuitionistic fuzzy cross-entropy distance measurement method, which can not only retain the integrity of decision information, but also directly reflect the differences between intuitionistic fuzzy data. (2) This paper focuses on the weight problem in MAEDM, and analyzes and compares the known and unknown attribute weights, which greatly improves the reliability and stability of decision-making results. (3) By using the method of grey correlation analysis, the fitting degree between the objective evaluation value and the subjective preference value of the decision maker can be fully considered. On this basis, a sensitivity analysis is made for the grey resolution coefficient to make the ranking result more reasonable. (4) The intuitionistic fuzzy cross-entropy and grey correlation analysis algorithm are introduced into the emergency decision-making problems such as the location ranking of shelters in earthquake disaster areas, which greatly reduces the risk of decision-making. (5) By comparing the traditional intuitionistic fuzzy distance to the intuitionistic fuzzy cross-entropy, the validity of the proposed method is verified.

Unfortunately, the method proposed in this paper is applicable to the emergency decision-making problems with certain subjective preference. For the emergency problems with which the decision maker has no obvious preference, the method needs to be further studied. In addition, considering more attribute indicators to rank alternatives may obtain more convincing results.

These aspects will become the research hotspot in the future: (1) In the MAEDM, the attribute weight problem will become a research focus. Considering the time factor, it may be an interesting topic to develop the weight into a dynamic field in the future. (2) The decision maker's preference relation and attribute weight often have great uncertainty. It is an effective method to discuss the multi-attribute emergency decision by using a more reliable robust optimization [39–41].

Author Contributions: Conceptualization, P.L.; Data curation, P.L.; Formal analysis, P.L.; Funding acquisition, P.L.; Investigation, P.L.; Methodology, Y.J. and P.L.; Project administration, S.-J.Q.; Resources, Y.J. and S.-J.Q.; Software, P.L.; Supervision, Y.J., Z.W. and S.-J.Q.; Validation, P.L., Y.J. and Z.W.; Visualization, P.L.; Writing—original draft, P.L.; Writing—review & editing, Y.J. and Z.W. All authors have read and agreed to the published version of the manuscript.

Funding: The work is supported by the National Social Science Foundation of China (No. 17BGL083).

Acknowledgments: Thanks is given to my tutor for the guidance of this paper, which greatly improved the quality of the article. Thank you for providing this academic platform for me to submit my manuscript. We are very grateful to the editors and referees for their careful reading and constructive suggestions on the manuscript.

Conflicts of Interest: This article has never been published in any journal or institution, and there will be no conflict of interest.

References

1. Liang, Y.; Tu, Y.; Ju, Y.; Shen, W. A multi-granularity proportional hesitant fuzzy linguistic TODIM method and its application to emergency decision making. *Int. J. Disaster Risk Reduct.* **2019**, *36*, 101081. [CrossRef]
2. Gao, J.; Xu, Z.; Liang, Z.; Liao, H. Expected consistency-based emergency decision making with incomplete probabilistic linguistic preference relations. *Knowl. Based Syst.* **2019**, *176*, 15–28. [CrossRef]
3. Nassereddine, M.; Azar, A.; Rajabzadeh, A.; Afsar, A. Decision making application in collaborative emergency response: A new PROMETHEE preference function. *Int. J. Disaster Risk Reduct.* **2019**, *38*, 101221. [CrossRef]
4. Zadeh, L.A. Fuzzy sets. *Inf. Control* **1965**, *83*, 338–353. [CrossRef]
5. Hu, D.; Jiang, T.; Yu, X.C. Construction of non-convex fuzzy sets and its application. *Neurocomputing* **2020**, *393*, 175–186. [CrossRef]
6. Garg, H.; Chen, S.M. Multiattribute group decision making based on neutrality aggregation operators of q-rung orthopair fuzzy sets. *Inf. Sci.* **2020**, *517*, 427–447. [CrossRef]
7. Atanassov, K.T. Intuitionistic fuzzy sets. *Fuzzy Sets Syst.* **1986**, *20*, 87–96. [CrossRef]
8. Atanassov, K.T.; Gargov, G. Interval-valued intuitionistic fuzzy sets. *Fuzzy Sets Syst.* **1989**, *31*, 343–349. [CrossRef]
9. Krawczak, M.; Szkatuła, G. On matching of intuitionistic fuzzy sets. *Inf. Sci.* **2020**, *517*, 254–274. [CrossRef]
10. Ngan, R.T.; Son, L.H.; Ali, M.; Tamir, D.E.; Rishe, N.D.; Kandel, A. Representing complex intuitionistic fuzzy set by quaternion numbers and applications to decision making. *Appl. Soft Comput. J.* **2020**, *87*, 105961. [CrossRef]
11. Arora, J.; Tushir, M. An Enhanced Spatial Intuitionistic Fuzzy C-means Clustering for Image Segmentation. *Procedia Comput. Sci.* **2020**, *167*, 646–655. [CrossRef]
12. Wang, F.; Wan, S.P. Possibility degree and divergence degree based method for interval-valued intuitionistic fuzzy multi-attribute group decision making. *Exp. Syst. Appl.* **2020**, *141*, 112929. [CrossRef]
13. Liu, P.S.; Diao, H.Y.; Zou, L.; Deng, A.S. Uncertain multi-attribute group decision making based on linguistic-valued intuitionistic fuzzy preference relations. *Inf. Sci.* **2020**, *508*, 293–308. [CrossRef]
14. Gao, Y.; Li, D.S.; Zhong, H. A novel target threat assessment method based on three-way decisions under intuitionistic fuzzy multi-attribute decision making environment. *Eng. Appl. Artif. Intell.* **2020**, *87*, 103276. [CrossRef]
15. Xu, Z.S. Some similarity measures of intuitionistic fuzzy sets and their applications to multiple attribute decision making. *Fuzzy Optim. Decis. Mak.* **2007**, *6*, 109–121. [CrossRef]
16. Park, J.H.; Hwang, J.H.; Park, W.J. Similarity measure on intuitionistic fuzzy sets. *J. Cent. South Univ.* **2013**, *20*, 2233–2238. [CrossRef]
17. Hu, K.; Li, J. The entropy and similarity measure of interval valued intuitionistic fuzzy sets and their relationship. *Int. J. Fuzzy Syst.* **2013**, *15*, 279–288.
18. Chen, S.M.; Tan, J.M. Handling multicriteria fuzzy decision-making problems based on vague set theory. *Fuzzy Sets Syst.* **1994**, *67*, 163–172. [CrossRef]
19. Hong, D.H.; Choi, C.H. Multicriteria fuzzy decision-making problems based on vague set theory. *Fuzzy Sets Syst.* **2000**, *114*, 103–113. [CrossRef]
20. Wu, J.; Huang, H.B.; Cao, Q.W. Research on AHP with interval-valued intuitionistic fuzzy sets and its application in multi-criteria decision making problems. *Appl. Math. Model.* **2013**, *37*, 9898–9906. [CrossRef]
21. Keshavarzfarda, R.; Makui, A. An IF-DEMATEL-AHP based on triangular intuitionistic fuzzy numbers. *Decis. Sci. Lett.* **2015**, *4*, 237–246. [CrossRef]
22. Chatterjee, K.; Kar, B.M.; Kar, S. Strategic decisions using intuitionistic fuzzy VIKOR method for information system outsourcing. In Proceedings of the 2013 International Symposium on Computational and Business Intelligence, New Delhi, India, 24–26 August 2013; pp. 123–126.

23. Liao, H.C.; Xu, Z.S. VIKOR-based method for hesitant fuzzy multi-criteria decision making. *Fuzzy Optim. Decis. Mak.* **2013**, *12*, 373–392. [CrossRef]
24. Wu, M.C.; Chen, T.Y. The ELECTRE multicriteria analysis approach based on Atanassov's intuitionistic fuzzy sets. *Exp. Syst. Appl.* **2011**, *38*, 12318–12327. [CrossRef]
25. Vahdani, B.; Mousavi, S.M.; Tavakkoli, M.R. A new de-sign of the elimination and choice translating reality method for multi-criteria group decision-making in an intuitionistic fuzzy environment. *Appl. Math. Model.* **2013**, *37*, 1781–1799. [CrossRef]
26. Yu, Z.; Xu, Z.; Ma, Y. Prioritized multi-criteria decision making based on the idea of PROMETHEE. *Procedia Comput. Sci.* **2013**, *17*, 449–456. [CrossRef]
27. Meng, F. An approach to Antanassov's interval-valued intuitionistic fuzzy multi-attribute decision making based on prospect theory. *Int. J. Comput. Intell. Syst.* **2015**, *8*, 591–605. [CrossRef]
28. Luo, Y.; Wei, G. Multiple attribute decision making with intuitionistic fuzzy information and uncertain attribute weights using minimization of regret. In Proceedings of the 2009 4th IEEE Conference on Industrial Electronics and Applications, Xi'an, China, 25–27 May 2009; pp. 3720–3723.
29. Xu, Y.J.; Wen, X.W.; Zhang, W.C. A two-stage consensus method for large-scale multi-attribute group decision making with an application to earthquake shelter selection. *Comput. Ind. Eng.* **2018**, *116*, 113–129. [CrossRef]
30. Xu, X.H.; Wang, L.L.; Chen, X.H.; Liu, B.S. Large group emergency decision-making method with linguistic risk appetites based on criteria mining. *Knowl. Based Syst.* **2019**, *182*, 104849. [CrossRef]
31. Li, M.Y.; Cao, P.P. Extended TODIM method for multi-attribute risk decision making problems in emergency response. *Comput. Ind. Eng.* **2019**, *135*, 1286–1293. [CrossRef]
32. Xiong, W.T.; Van Gelder, P.H.A.J.M.; Yang, K.W. A decision support method for design and operationalization of search and rescue in maritime emergency. *Ocean Eng.* **2020**, *207*, 107399. [CrossRef]
33. Wang, Z.G.; Hao, H.; Gao, F.; Zhang, Q.; Zhang, J.; Zhou, Y.J. Multi-attribute decision making on reverse logistics based on DEA-TOPSIS: A study of the Shanghai End-of-life vehicles industry. *J. Clean. Prod.* **2019**, *214*, 730–737. [CrossRef]
34. Wu, Q.; Wu, P.; Zhou, L.G.; Chen, H.Y.; Guan, X.J. Some new Hamacher aggregation operators under single-valued neutrophic 2-tuple linguistic environment and their applications to multi-attribute group decision making. *Comput. Ind. Eng.* **2018**, *116*, 144–162. [CrossRef]
35. Karimi, H.; Sadeghi-Dastaki, M.; Javan, M. A fully fuzzy best–worst multi attribute decision making method with tri-angular fuzzy number: A case study of maintenance assessment in the hospitals. *Appl. Soft Comput. J.* **2020**, *86*, 105882. [CrossRef]
36. Xu, Z.S. Models for multiple attribute decision-making with intuitionistic fuzzy information, International Journal of Un-certainty. *Fuzziness Knowl. Based Syst.* **2007**, *15*, 285–297. [CrossRef]
37. Burillo, P.; Bustince, H. Entropy on intuitionistic fuzzy sets and on interval-valued fuzzy sets. *Fuzzy Sets Syst.* **1996**, *78*, 305–316. [CrossRef]
38. Ye, J. Fuzzy cross entropy of interval-valued intuitionistic fuzzy sets and its optimal decision-making method based on the weigths of alternatives. *Exp. Syst. Appl.* **2011**, *38*, 6179–6183. [CrossRef]
39. Ji, Y.M.; Qi, M.L. A robust optimization approach for decontamination planning of emergency planning zone: Facility location and assignment plan. *Socio Econ. Plan. Sci.* **2020**, *70*, 100740. [CrossRef]
40. Dey, A.; Zaman, K. A robust optimization approach for solving two-person games under interval uncertainty. *Comput. Oper. Res.* **2020**, *119*, 104937. [CrossRef]
41. Ji, Y.; Qu, S.J.; Wu, Z.; Liu, Z.M. A Fuzzy-Robust Weighted Approach for Multicriteria Bilevel Games. *IEEE Trans. Ind. Inf.* **2020**, *16*, 5369–5376. [CrossRef]

© 2020 by the authors. Licensee MDPI, Basel, Switzerland. This article is an open access article distributed under the terms and conditions of the Creative Commons Attribution (CC BY) license (http://creativecommons.org/licenses/by/4.0/).

Article

Cointegration and Unit Root Tests: A Fully Bayesian Approach

Marcio A. Diniz [1,*], Carlos A. B. Pereira [2] and Julio M. Stern [3]

[1] Statistics Department, Universidade Federal de S. Carlos, Rod. Washington Luis, km 235, S. Carlos 13565-905, Brazil
[2] Statistics Department, Universidade de S. Paulo, São Paulo 01000, Brazil; cpereira@ime.usp.br
[3] Applied Mathematics Department, Universidade de S. Paulo, São Paulo 01000, Brazil; jstern@ime.usp.br
* Correspondence: marcio.alves.diniz@gmail.com

Received: 3 August 2020; Accepted: 27 August 2020; Published: 31 August 2020

Abstract: To perform statistical inference for time series, one should be able to assess if they present deterministic or stochastic trends. For univariate analysis, one way to detect stochastic trends is to test if the series has unit roots, and for multivariate studies it is often relevant to search for stationary linear relationships between the series, or if they cointegrate. The main goal of this article is to briefly review the shortcomings of unit root and cointegration tests proposed by the Bayesian approach of statistical inference and to show how they can be overcome by the Full Bayesian Significance Test (FBST), a procedure designed to test sharp or precise hypothesis. We will compare its performance with the most used frequentist alternatives, namely, the Augmented Dickey–Fuller for unit roots and the maximum eigenvalue test for cointegration.

Keywords: time series; Bayesian inference; hypothesis testing; unit root; cointegration

---

Several times series present deterministic or stochastic trends, which imply that the effects of these trends on the level of the series are permanent. Consequently, the mean and variance of the series will not be constant and will not revert to a long-term value. This feature reflects the fact that the stochastic processes generating these series are not (weakly) stationary, imposing problems to perform inductive inference using the most traditional estimators or predictors. This is so because the usual properties of these procedures will not be valid under such conditions.

Therefore, when modeling non-stationary time series, one should be able to properly detrend the used series, either by directly modeling the trend by deterministic functions, or by transforming the series to remove stochastic trends. To determine which strategy is the suitable solution, several statistical tests were developed since the 1970s by the frequentist school of statistical inference.

The Augmented Dickey–Fuller (ADF) test is one of the most popular tests used to assess if a time series has a stochastic trend or, for series described by auto-regressive models, if they have a unit root. When one is searching for long term relationships between multiple series under analysis, it is crucial to know if there are stationary linear combinations of these series, i.e., if the series are cointegrated. Cointegration tests were developed, also by the frequentist school, in the late 1980s [1] and early 1990s [2]. Only in the late 1980s did the Bayesian approach to test the presence of unit roots start to be developed.

Both unit root and cointegration tests may be considered tests on precise or sharp hypotheses, i.e., those in which the dimension of the parameter space under the tested hypothesis is smaller than the dimension of the unrestricted parameter space. Testing sharp hypotheses poses major difficulties for either the frequentist or Bayesian paradigms, such as the need to eliminate nuisance parameters.

The main goal of this article is to briefly review the shortcomings of the tests proposed by the Bayesian school and how they can be overcome by the Full Bayesian Significance Test (FBST). More specifically, we will compare its performance with the most used frequentist alternatives, the ADF for unit roots, and the maximum eigenvalue test for cointegration. Since this is a review article, it is important to remark that the results presented here were published elsewhere by the same authors, see [3,4].

To accomplish this objective, we will define the FBST in the next section, also showing how it can be implemented in a general context. The following section discusses the problems of testing the existence of unit roots in univariate time series and how the Bayesian tests approach the problem. Section 4 then shows how the FBST is applied to test if a time series has unit roots and illustrates this with applications on a real data set. In the sequel, we discuss the Bayesian alternatives to cointegration tests and then apply the FBST to test for cointegration using real data sets. We conclude with some remarks and possible extensions for future work.

1. FBST

The Full Bayesian Significance Test was proposed in [5] mainly to deal with sharp hypotheses. The procedure has several properties, see [6,7], most interestingly the fact that it is only based on posterior densities, thus avoiding the necessity of complications such as the elimination of nuisance parameters or the adoption of priors with positive probabilities attached to sets of zero Lebesgue measure.

We shall consider general statistical models in which the parameter space is denoted by $\Theta \subseteq \mathbb{R}^m$, $m \in \mathbb{N}$. A sharp hypothesis H assumes that θ, the parameter vector of the chosen statistical model, belongs to a sub-manifold Θ_H of smaller dimensions than Θ. This implies, for continuous parameter spaces, that the subset Θ_H has null Lebesgue measure whenever H is sharp. The sample space, the set of all possible values of the observable random variables (or vectors), is here denoted by $\mathcal{X}$.

Following the Bayesian paradigm, let $h(\cdot)$ be a probability prior density over Θ, $\mathbf{x} \in \mathcal{X}$, the observed sample (scalar or vector), and $L(\cdot \mid \mathbf{x})$ the likelihood derived from data $\mathbf{x}$. To evaluate the Bayesian evidence based on the FBST, the sole relevant entity is the posterior probability density for θ given $\mathbf{x}$,

$$g(\theta \mid \mathbf{x}) \propto h(\theta) \cdot L(\theta \mid \mathbf{x}).$$

It is important to highlight that the procedure may be used when the parameter space is discrete. However, when the posterior probability distribution over Θ is absolutely continuous, the FBST appears as a more suitable alternative to significance hypothesis testing. For notational simplicity, we will denote Θ_H by H in the sequel.

Let $r(\theta)$ be a reference density on Θ such that the function $s(\theta) = g(\theta \mid \mathbf{x})/r(\theta)$ is a *relative surprise*, (see [8], pp. 145–146) function. The reference density is important because it guarantees that the FBST is invariant to reparametrizations, even when $r(\theta)$ is improper, see [6,9]. Thus, when considering $r(\theta)$ proportional to a constant, the surprise function will be, in practical terms, equivalent to the posterior distribution. For the applications considered in this article, we will use the improper uniform density as reference density on Θ. The authors of [10] remark that it is possible to generalize the procedure using other reference densities such as neutral, invariant, maximum-entropy or non-informative priors, if they are available and desirable.

Definition 1 (Tangent set). *Considering a sharp hypothesis $H : \theta \in \Theta_H$, the tangential set of the hypothesis given the sample is given by*

$$\mathbb{T}_\mathbf{x} = \{\theta \in \Theta : s(\theta) > s^*\}. \tag{1}$$

where $s^* = \sup_{\theta \in H} s(\theta)$.

Notice that the tangent set $\mathbb{T}_\mathbf{x}$ is the highest relative surprise set, that is, the set of points of the parameter space with higher relative surprise than any point in H, being *tangential* to H in this sense. This approach takes into consideration the statistical model in which the hypothesis is defined, using several components of the model to define an evidential measure favoring the hypothesis.

Definition 2 (Evidence). *The Bayesian evidence value against H, $\overline{ev}$, is defined as*

$$\overline{ev} = P(\theta \in \mathbb{T}_\mathbf{x} \mid \mathbf{x}) = \int_{\mathbb{T}_\mathbf{x}} dG_\mathbf{x}(\theta), \tag{2}$$

where $G_\mathbf{x}(\theta)$ denotes the posterior distribution function of θ and the above integral is of the Riemann–Stieltjes type.

Definition 2 sets $\overline{ev}$ as the posterior probability of the tangent set that is interpreted as an evidence value against H. Hence, the evidence value supporting H is the complement of $\overline{ev}$, namely, $ev = 1 - \overline{ev}$. Notwithstanding, ev is not evidence against $A : \theta \notin \Theta_H$, the alternative hypothesis (which is not sharp anyway). Equivalently, $\overline{ev}$ is not evidence in favor of A, although it is against H.

Definition 3 (Test). *The FBST is the procedure that rejects H whenever $ev = 1 - \overline{ev}$ is smaller than a critical level, ev_c.*

Thus, we are left with the problem of deciding the critical level ev_c for each particular application. We briefly discuss this and other practical issues in the following subsection.

1.1. Practical Implementation: Critical Values and Numerical Computation

Since ev (also called e-value) is a statistic, it has a sampling distribution derived from the adopted statistical model and in principle this distribution could be used to find a threshold value. If the likelihood and the posterior distribution satisfy certain regularity conditions. See [11], p. 436. [12] proved that, asymptotically, there is a relationship between ev and the p-values obtained from the frequentist likelihood ratio procedure used to test the same hypotheses. This fact provides a way to find, at least asymptotically, a critical value to ev to reject the hypothesis being tested.

In a recent review [7], the authors discuss different ways to provide a threshold for ev. Among these alternatives, we highlight the standardized e-value, which follows, asymptotically, the uniform distribution on $(0,1)$. See also [13] for more on the standardized version of ev.

One could also try to define the FBST as a Bayes test derived from a particular loss function and the respective minimization of the posterior expected loss. Following this strategy, [10] showed that there are loss functions which result in ev as a Bayes estimator of $\phi = \mathbb{I}_H(\theta)$, where $\mathbb{I}_A(x)$ denotes the indicator function, being equal to one if $x \in A$ and zero otherwise, $x \notin A$. Hence, the FBST is in fact a Bayes procedure in the formal sense as defined by Wald in [14].

Table 1. Pseudocode to implement the FBST.

General algorithm: compute ev supporting hypothesis $H : \theta \in \Theta_H$
1. Specify the statistical model (likelihood) and prior distribution on Θ.
2. Specify the reference density, $r(\theta)$, and derive the relative surprise function, $s(\theta)$.
3. Find s^*, the maximum value of $s(\theta)$ under the constraint $\theta \in H$.
4. Integrate the posterior distribution on the tangent set—Equation (2)—to find $\overline{ev}$.
5. Find $ev = 1 - \overline{ev}$.

To compute the evidence value supporting H defined in the last section, we need to follow the steps showed in Table 1. Appendix A provides detailed information about the computational resources and codes used to implement the FBST in the examples presented in this work. After defining the statistical model and prior, it is simple to find the surprise function, $s(\theta)$. In step 3, one should find the point of the parameter space in H that maximizes $s(\theta)$, that is, to solve a problem of constrained numerical maximization. In several applications, this step does not present a closed form solution, requiring the use of numerical optimizers.

Step 4 involves the integration of the posterior distribution on a subset of Θ, the tangent set $\mathbb{T}_\mathbf{x}$ that can be highly complex. Once more, since in many cases it is fairly difficult to find an explicit expression for $\mathbb{T}_x$, one may use various numerical techniques to compute the integral. If it is possible to generate random samples from the posterior distribution, Monte Carlo integration provides an estimate of ev, as we will show in this work. Another alternative is to use approximation techniques, such as those proposed in [15], based on a Laplace approximation. We discuss how to implement such approximations for unit root and cointegration tests in [3,4].

2. Bayesian Unit Root Tests

Before presenting the Bayesian procedures used to test the presence of unit roots, let us fix notation. We will denote by y_t the t-th value of a univariate time series observed in $t = 1, \ldots, T + p$ dates, where T and p are positive integers. The usual approach is to assume that the series under analysis is described by an auto-regressive process with p lags, $AR(p)$, meaning that the data generating process is fully described by a stochastic difference equation of order p, possibly with an intercept or drift and a deterministic linear trend, i.e.,

$$y_t = \mu + \delta \cdot t + \phi_1 y_{t-1} + \ldots + \phi_p y_{t-p} + \varepsilon_t \tag{3}$$

with ε_t i.i.d. $N(0, \sigma^2)$ for $t = 1, \ldots, T + p$. Using the lag or backshift operator B, we denote y_{t-k} by $B^k y_t$, allowing us to rewrite (3) as

$$(1 - \phi_1 B - \ldots \phi_p B^p) y_t = \mu + \delta \cdot t + \varepsilon_t \tag{4}$$

where $\phi(B) = (1 - \phi_1 B - \ldots \phi_p B^p)$ is the autoregressive polynomial. The difference Equation (3) will be stable, implying that the process generating $\{y_t\}_{t=1}^{T+p}$ is (weakly) stationary, whenever the roots of the characteristic polynomial $\phi(z)$, $z \in \mathbb{C}$, lie outside the unit circle, since there may be complex roots. The set of polynomial operators, such as lag polynomials like $\phi(B)$, induces an algebra that is isomorphic to the algebra of polynomials in real or complex variables, see [16].

If some of the roots lie exactly on the unit circle, it is said that the process has unit roots. In order to test such a hypothesis statistically, (3) is rewritten as

$$\Delta y_t = \mu + \delta \cdot t + \Gamma_0 \, y_{t-1} + \Gamma_1 \Delta y_{t-1} + \ldots + \Gamma_{p-1} \Delta y_{t-p+1} + \varepsilon_t \tag{5}$$

where $\Delta y_t = y_t - y_{t-1}$, $\Gamma_0 = \phi_1 + \ldots + \phi_p - 1$ and $\Gamma_i = -\sum_{j=i+1}^{p} \phi_j$, for $i = 1, \ldots, p-1$. If the generating process has only one unit root, one root of the complex polynomial $\phi(z)$,

$$1 - \phi_1 z - \phi_2 z^2 \ldots \phi_p z^p,$$

is equal to one, meaning that

$$1 - \phi_1 - \phi_2 - \ldots - \phi_p = 0$$

i.e., $\phi(1) = 0$, and all the other roots are on or outside the unit circle. In this case, $\Gamma_0 = 0$, the hypothesis that will be tested when modeling (5). Even though tests based on these assumptions verify if the process has a single unit root, there are generalizations based on the same principles that test the existence of multiple unit roots, see [17].

The search for Bayesian unit root tests began in the late 1980s. As far as we know, [18,19] were the first works to propose a Bayesian approach for unit root tests. The frequentist critics of these articles received a proper answer in [20,21], generating a fruitful debate that produced a long list of papers in the literature of Bayesian time series. A good summary of the debate and the Bayesian papers that resulted from it is presented in [22]. We will present here only the most relevant strategies proposed by the Bayesian school to test for unit roots.

Let $\theta = (\rho, \psi)$ be the parameters vector, in which $\rho = \sum_{i=1}^{p} \phi_i$ and $\psi = (\mu, \delta, \Gamma_1, \ldots, \Gamma_{p-1})$. Assuming σ^2 fixed, the prior density for θ can be factorized as

$$h(\theta) = h_0(\rho) \cdot h_1(\psi \mid \rho).$$

The marginal likelihood for ρ, denoted by L_m, is:

$$L_m(\rho \mid \mathbf{y}) \propto \int_\Psi L(\theta \mid \mathbf{y}) \cdot h_1(\psi \mid \rho) \, d\psi.$$

where $\mathbf{y} = \{y_t\}_{t=1}^{T+p}$ is the observations vector, $L(\theta|\mathbf{y})$ the full likelihood, and Ψ the support of the random vector ψ. This marginal likelihood, associated with a prior for ρ, is the main ingredient used by standard Bayesian procedures to test the existence of unit roots. Even though the procedure varies among authors according to some specific aspects, mentioned below, basically all of them use Bayes factors and posterior probabilities.

One important issue is the specification of the null hypothesis: some authors, starting from [23], consider $H_0 : \rho = 1$ against $H_1 : \rho < 1$. Starting from [24], this is the way the frequentist school addresses the problem, but following this approach no explosive value for ρ is considered. The decision theoretic Bayesian approach solved the problem using the posterior probabilities ratio or Bayes factor:

$$B_{01} = \frac{L_m(\rho = 1 \mid \mathbf{y})}{\int_0^1 L_m(\rho \mid \mathbf{y}) \cdot h_0(\rho) \, d\rho}.$$

Advocates of this solution argue that one of the advantages of this approach is that the null and the alternative hypotheses are given equal weight. However, the expression above is not defined if $h_0(\rho)$ is not a proper density since the denominator of the Bayes factor is equal to the predictive density, defined just if $h_0(\rho)$ is a proper density. There are also problems if $L_m(\rho = 1|\mathbf{y})$ is zero or infinite.

The problem is approached by [20,25] by testing $H_0 : \rho \geq 1$ against $H_1 : \rho < 1$, considering explicitly explosive values for ρ. The main advantage of this strategy is the possibility to compute posterior probabilities like

$$P(\rho > 1 \mid \mathbf{y}) = \int_1^\infty g_m(\rho \mid \mathbf{y}) \, d\rho$$

defined even for improper priors on ρ, where g_m is the marginal posterior for ρ.

In [26], the authors do not choose ρ as the parameter of interest, examining instead the largest absolute value of the roots of the characteristic polynomial and then verifying if it is smaller or larger than one. Usually, this value is slightly smaller than ρ, but the authors argue that this small difference may be important. When this approach is used, unit roots are found less frequently. For an AR(3) model with a constant and deterministic trend, [26] derives the posterior density for the dominant root for the 14 series used in [27] and concluded the following: for eleven of the series, the dominant root was smaller than one, that is to say, the series were trend-stationary. These results were based on flat priors for the autoregressive parameters and the deterministic trend coefficient.

Another controversy is about the prior over ρ: [20] argues that the difference between the results given by the frequentist and Bayesian inferences is due to the fact that the flat prior proposed in [18] overweights the stationary region of ρ. Hence, he derived a Jeffreys prior for the AR(1) model: this prior quickly diverges as ρ increases and becomes larger than one. The obtained posterior led to the same results of [27], which will be discussed in detail in the following section. The critics of the approach adopted by Phillips in [20] judged the Jeffreys prior as unrealistic, from a subjective point of view. See the comments on Phillips's paper on the *Journal of Applied Econometrics*, volume 6, number 4, 1991. The subsequent papers of the same number support the Bayesian approach. This is a nonsensical objection if one considers that the Jeffreys prior is crucial to ensure an invariant inferential procedure, and invariance is a highly desirable property, for either objective or subjective reasons. See [28] for more on invariance in physics and statistical models.

A final controversial point concerns the modeling of initial observations. If the likelihood explicitly models the initial observed values (it is an *exact* likelihood), the process is implicitly considered stationary. In fact, when it is known that the process is stationary, and it is believed that the data generating process is working for a long period, it is reasonable to assume that the parameters of the model determine the marginal distribution of the initial observations. In the simplest AR(1) model, this would imply that $y_1 \sim N(0, \sigma^2/(1-\rho^2))$. In this scenario, to perform the inference conditional on the first observation would discard relevant information. On the other hand, there is no marginal distribution defined for y_1 if the generating process is not stationary. Then, it is valid to use a likelihood conditional on initial observations. For the models presented here, we always work with the conditional likelihood. As argued in [18], inferences for stationary models are little affected by using conditional likelihoods, especially for large samples. He compares these inferences with the ones based on exact likelihoods under explicit modeling for initial observations.

3. Implementing the FBST for Unit Root Testing

We will now describe how to use the FBST to test for the presence of unit roots referring to the general model (5). It is also possible to include $q \in \mathbb{N}$ moving average terms in (3) to model the process, a case that will not be covered in this article but that, in principle, shall not imply major problems for the FBST.

$$\Delta y_t = \mu + \delta \cdot t + \Gamma_0 \, y_{t-1} + \Gamma_1 \Delta y_{t-1} + \ldots + \Gamma_{p-1} \Delta y_{t-p+1} + \varepsilon_t, \tag{5}$$

where $\varepsilon_t \overset{i.i.d.}{\sim} N(0, \sigma^2)$ for $t = 1, \ldots, T+p$, recalling also that the hypothesis being tested is $\Gamma_0 = 0$. We slightly change the notation of the last section now using ψ to denote the vector $(\mu, \delta, \Gamma_0, \ldots, \Gamma_{p-1})$ and setting $\theta = (\psi, \sigma)$.

Recalling the steps to implement the FBST displayed in Table 1, we have just specified the statistical model. The likelihood, conditional on the first p observations, derived from the Gaussian model is

$$L(\theta \mid \mathbf{y}) = (2\pi)^{-T/2} \sigma^{-T} \exp\left\{ -\frac{1}{2\sigma^2} \cdot \sum_{t=p+1}^{T+p} \varepsilon_t^2 \right\}, \tag{6}$$

in which $\varepsilon_t = \Delta y_t - \mu - \delta \cdot t - \Gamma_0 y_{t-1} - \Gamma_1 \Delta y_{t-1} - \ldots - \Gamma_{p-1} \Delta y_{t-p+1}$. To complete step 1 of Table 1, we need a prior distribution for θ. For all the series modeled in this article, we will use the following non informative prior:

$$h(\theta) = h(\psi, \sigma) \propto 1/\sigma. \tag{7}$$

We are aware of the problems caused by improper priors applied to this problem when one uses alternative approaches, like those mentioned by [22]. However, one of our goals is to show how the FBST can be implemented even for a potentially problematic prior like this one. To write the posterior, we use the following notation:

$$\Delta Y = \begin{bmatrix} \Delta y_{p+1} \\ \Delta y_{p+2} \\ \vdots \\ \Delta y_{T+p} \end{bmatrix}, \; X = \begin{bmatrix} 1 & p+1 & y_p & \Delta y_p & \cdots & \Delta y_2 \\ 1 & p+2 & y_{p+1} & \Delta y_{p+1} & \cdots & \Delta y_3 \\ \vdots & \vdots & \vdots & \vdots & & \vdots \\ 1 & T+p & y_{T+p-1} & \Delta y_{T+p-1} & \cdots & \Delta y_{T+1} \end{bmatrix}, \; \psi = \begin{bmatrix} \mu \\ \delta \\ \Gamma_0 \\ \vdots \\ \Gamma_{p-1} \end{bmatrix},$$

being ΔY of dimension $T \times 1$, X of dimension $T \times (p+2)$ and ψ, $(p+2) \times 1$. Thanks to this notation, we can write, using primes to denote transposed matrices:

$$\sum_{t=p+1}^{T+p} \varepsilon_t^2 = (\Delta Y - X\psi)'(\Delta Y - X\psi) = (\Delta Y - \widehat{\Delta Y})'(\Delta Y - \widehat{\Delta Y}) + (\psi - \widehat{\psi})' X'X (\psi - \widehat{\psi}),$$

where $\widehat{\psi} = (X'X)^{-1}X' \cdot \Delta Y$ is the ordinary least squares (OLS) estimator of ψ and $\widehat{\Delta Y} = X\widehat{\psi}$ its prediction for ΔY. Thus, the full posterior is

$$g(\theta \mid \mathbf{y}) \propto \sigma^{-(T+1)} \exp\left\{-\frac{1}{2\sigma^2}[(\Delta Y - \widehat{\Delta Y})'(\Delta Y - \widehat{\Delta Y}) + (\psi - \widehat{\psi})'X'X(\psi - \widehat{\psi})]\right\}, \quad (8)$$

a Normal-Inverse Gamma density.

Step 2 demands a reference density in order to define the relative surprise function. Since we will use the improper density $r(\theta) \propto 1$, the surprise function will be equivalent to the posterior distribution in our applications. Given this, to find s^* (Step 3), we need to find the maximum value of the posterior under the hypothesis being tested, in our case, $\Gamma_0 = 0$.

This maximization step is fairly simple to implement given the modeling choices made here: Gaussian likelihood, non informative prior and reference density proportional to a constant. The restricted (assuming H) posterior distribution is

$$g_r(\theta_r \mid \mathbf{y}) \propto \sigma^{-(T+1)} \exp\left\{-\frac{1}{2\sigma^2}[(\Delta Y - \widehat{\Delta Y}_r)'(\Delta Y - \widehat{\Delta Y}_r) + (\psi_r - \widehat{\psi}_r)'X_r'X_r(\psi_r - \widehat{\psi}_r)]\right\}, \quad (9)$$

in which $\theta_r = (\psi_r, \sigma)$, ψ_r being vector ψ without Γ_0,

$$X_r = \begin{bmatrix} 1 & p+1 & \Delta y_p & \cdots & \Delta y_2 \\ 1 & p+2 & \Delta y_{p+1} & \cdots & \Delta y_3 \\ \vdots & \vdots & \vdots & \vdots & \vdots \\ 1 & T+p & \Delta y_{T+p-1} & \cdots & \Delta y_{T+1} \end{bmatrix}, \widehat{\psi}_r = (X_r'X_r)^{-1}X_r' \cdot \Delta Y, \text{ and } \widehat{\Delta Y}_r = X_r\widehat{\psi}_r,$$

that is, X_r is simply matrix X above without its third column, since under $H: \Gamma_0 = 0$ and the coefficient of the third column of X is Γ_0—see Equation (5)—$\widehat{\psi}_r$ is a least squares estimator of ψ_r and $\widehat{\Delta Y}_r$ denotes the predicted values for ΔY given by the restricted model. From (9), it is easy to show that the maximum a posteriori (MAP) estimator of θ_r is given by $(\widehat{\psi}_r, \widehat{\sigma}_r)$, with

$$\widehat{\sigma}_r = \sqrt{\frac{(\Delta Y - \widehat{\Delta Y}_r)'(\Delta Y - \widehat{\Delta Y}_r)}{T+1}}.$$

Plugging the values of $\widehat{\psi}_r$ and $\widehat{\sigma}_r$ into (9), we find s^*, as requested in Step 3. Step 4 will also be easy to implement thanks to structure of the models assumed in this section. Since the full posterior, (8), is a Normal-Inverse Gamma density, a simple Gibbs sampler allows us to obtain a random sample from such distribution, suggesting a Monte Carlo approach to compute $\overline{ev}$. From (8), the conditional posteriors of ψ and σ are, respectively,

$$g_\psi(\psi \mid \sigma, \mathbf{y}) \propto N(\widehat{\psi}, \sigma^2(X'X)^{-1}) \quad (10)$$

$$g_{\sigma^2}(\sigma^2 \mid \psi, \mathbf{y}) \propto IG\left(\frac{T+1}{2}, H\right) \quad (11)$$

in which $H = 0.5[(\Delta Y - \widehat{\Delta Y})'(\Delta Y - \widehat{\Delta Y}) + (\psi - \widehat{\psi})'X'X(\psi - \widehat{\psi})]$, IG denotes the Inverse-Gamma distribution and $\widehat{\psi}$ is the OLS estimator of ψ, as above. Appendix B brings the parametrization and the probability density function of the Inverse-Gamma distribution. With a sizable random sample from the full posterior, we estimate $\overline{ev}$ as the proportion of sampled vectors that generate a value for the posterior greater than s^*, found in Step 3. Hence, in Step 5, we only compute one minus the estimate of $\overline{ev}$ found in Step 4. The whole procedure is summarized in Table 2. For the implementations in this article we sampled 51,000 vectors from (8) and discarded the first 1,000 as a burn-in sample.

Table 2. Pseudocode to implement the FBST to unit root tests.

General algorithm: compute ev supporting hypothesis $H : \Gamma_0 = 0$ in model (5)
1. Statistical model: Gaussian; prior: $h(\theta) \propto 1/\sigma$.
2. Reference density: $r(\theta) \propto 1$; relative surprise function: $g(\theta \mid \mathbf{y})$.
3. Find s^*: (9) evaluated at $\hat{\psi}_r$ and $\hat{\sigma}_r$.
4. Gibbs sampler (from Equations (10) and (11)) to obtain N random samples of parameter vectors from (8). Evaluate the posterior at the sampled vectors and estimate $\overline{ev}$ as the proportion of N in which the evaluated values are larger than s^*.
5. Find $ev = 1 - \overline{ev}$.

Results

We implemented the FBST as described above to test the presence of unit roots in 14 U.S. macroeconomic time series, all with annual frequency, first mentioned in [27]. We used the extended series, analyzed in [23]. Appendix A brings more information on the data set and the computational resources and codes used to obtain the results displayed in Table 3 below.

Table 3 reports the names of the tested series, the number of available observations or sample size, the adopted value for p—as denoted in Equation (8)—if a linear (deterministic) trend was included in the model or not, the ADF test statistic and its respective p-value. We have used the computer package described in [29] to find the ADF p-values, available in the R library urca. The last two columns bring the posterior probability of non-stationarity, $\Gamma_0 \geq 0$, and the FBST e-values for the specified models. In order to obtain comparable results, we have adopted the models chosen by [22] for all the series. All the models considered the intercept or constant term, μ in (8).

The results show that the non-stationary posterior probabilities are quite distant from the ADF p-values. These results were highlighted in [18,19]. Considering the simplest AR(1) model, they argued that, once frequentist inference is based on the distribution of $\hat{\rho}|\rho = 1$, the non-stationary posterior probabilities provide counterintuitive conclusions since the referred distribution is skewed. Their main argument is that Bayesian inference uses a distribution (marginal posterior of ρ) that is not skewed.

As mentioned before, ref. [20] claims that the difference in results between frequentist and Bayesian approaches is due to the flat prior that puts much weight on the stationary region. He proposed the use of Jeffreys priors, which restored the conclusions drawn by the frequentist test. Phillips argued that the flat prior was, actually, informative when used in time series models like those for unit root tests. Using simulations, he shows that " *[the use of a] flat prior has a tendency to bias the posterior towards stationarity. ... even when [the estimate] is close to unity, there may still be a non negligible downward bias in the [flat] posterior probabilities*". Notwithstanding, the e-values reported in the last column are quite close to the ADF p-values even using the flat prior criticized by Phillips.

Table 3. Unit root tests for the extended Nelson and Plosser data set.

Series	Sample Size	p	Trend	ADF	p-Value	$P(\Gamma_0 \geq 0\|y)$	e-Value
Real GNP	80	2	yes	-3.52	0.044	0.0005	0.040
Nominal GNP	80	2	yes	-2.06	0.559	0.0238	0.523
Real GNP per capita	80	2	yes	-3.59	0.037	0.0004	0.034
Industrial prod.	129	2	yes	-3.62	0.032	0.0003	0.028
Employment	99	2	yes	-3.47	0.048	0.0004	0.043
Unemployment rate	99	4	no	-4.04	0.019	0.0001	0.020
GNP deflator	100	2	yes	-1.62	0.778	0.0584	0.762
Consumer prices	129	4	yes	-1.22	0.902	0.1154	0.983
Nominal wages	89	2	yes	-2.40	0.377	0.0106	0.341
Real wages	89	2	yes	-1.71	0.739	0.0475	0.715
Money stock	100	2	yes	-2.91	0.164	0.0029	0.147
Velocity	119	2	yes	-1.62	0.779	0.0620	0.777
Bond yield	89	4	no	-1.35	0.602	0.0962	0.936
Stock prices	118	2	yes	-2.44	0.357	0.0103	0.349

4. Bayesian Cointegration Tests

Before starting our brief review of the most relevant Bayesian cointegration tests, we fix notation and present the definitions to which we will refer in the sequel.

All the tests mentioned here are based on the following multivariate framework. Let $Y_t = [y_{1t} \ldots y_{nt}]'$ be a vector with $n \in \mathbb{N}$ time series, all of them assumed to be integrated of order $d \in \mathbb{N}$, i.e., have d unit roots. The series are said to be cointegrated if there is a nontrivial linear combination of them that has $b \in \mathbb{N}$ unit roots, $b < d$. We will assume that, as in most applications, $d = 1$ and $b = 0$, meaning that, if the time series in Y_t is cointegrated, there is a linear combination $a'Y_t$ that is stationary, where $a \in \mathbb{R}^n$ is the cointegrating vector. Since the linear combination $a'Y_t$ is often motivated by problems found in economics, it is called a long-run equilibrium relationship. The explanation is that non-stationary time series that are related by a long-run relationship cannot drift too far from the equilibrium because economic forces will act to restore the relationship.

Notice also that: (i) the cointegrating vector is not uniquely determined since, for any scalar s, $(s \cdot a)$ is a cointegrating vector; and (ii) if Y_t has more than two series, it is possible that there is more than one cointegrating vector generating a stationary linear combination.

It is assumed that the data generating process of Y_t is described by the following vector autoregression with $p \in \mathbb{N}$ lags, denoted VAR(p), and given by:

$$Y_t = c + \Phi_0 D_t + \Phi_1 Y_{t-1} + \ldots + \Phi_p Y_{t-p} + E_t, \quad (12)$$

in which c is a $(n \times 1)$ vector of constants, D_t a vector $(n \times 1)$ with some deterministic variable, such as deterministic trends or seasonal dummies, Φ_i are $(n \times n)$ coefficients matrices and E_t is a $(n \times 1)$ stochastic vector with multivariate normal distribution with null expected value and covariance matrix Ω, denoted $N_n(0, \Omega)$. This dynamic model is assumed valid for $t = 1, \ldots, T + p$, the available span of observations of Y_t. As in the univariate case, one may include moving average terms in (12), i.e., lags for E_t, but this, in principle, would not cause major problems in the Bayesian framework. Model (12) can be rewritten using the lag or backshift operator as

$$(I_n - \Phi_1 B - \ldots - \Phi_p B^p)Y_t = c + \Phi_0 D_t + E_t, \quad (13)$$

where $\Phi(B) = I_n - \Phi_1 B - \ldots - \Phi_p B^p$ is the (multivariate) autoregressive polynomial and I_n denotes the n-dimensional identity matrix. The associate characteristic polynomial in this context will be the determinant of $\Phi(z)$, $z \in \mathbb{C}$. If all the roots of the characteristic polynomial lie outside the unit circle, it is possible to show that Y_t has a stationary representation—see [30]—such as Equation (12). In order to determine if this is the case, model (12) is rewritten as an (vectorial) error correction model (VECM):

$$\Delta Y_t = c + \Phi_0 D_t + \Gamma_1 \Delta Y_{t-1} + \ldots + \Gamma_{p-1} \Delta Y_{t-p+1} + \Pi Y_{t-1} + E_t, \quad (14)$$

where $\Delta Y_t = [\Delta y_{1t} \ldots \Delta y_{nt}]'$, $\Gamma_i = -(\Phi_{i+1} + \ldots \Phi_p)$ for $i = 1, 2, \ldots, p-1$ and $\Pi = -\Phi(1) = -(I_n - \Phi_1 - \ldots - \Phi_p)$. It is possible to show that, when all the roots of $\det(\Phi(z))$ are outside the unit circle, matrix Π in (14) has full rank, i.e., all the n eigenvalues of Π are n non null. If the rank of Π is null, this matrix cannot be distinguished from a null matrix, implying that the series in Y_t has at least one unit root and a valid representation is a VAR of order $p - 1$, i.e., model (14) without the term ΠY_{t-1}. It is possible that the series in Y_t has two unit roots each, implying that the correct VECM must be written with $\Delta^2 Y_t$ as a dependent variable.

Finally, if the $(n \times n)$ matrix Π has rank r, $0 < r < n$, it has $n - r$ non null eigenvalues, implying that the series in Y_t has at least one unit root and its valid representation is given by the VECM in Equation (14). In this case, $\Pi = \alpha \beta'$, where α and β are matrices $(n \times r)$ of rank r. Matrix β denotes the one with the cointegrating vectors and matrix α is called the loading matrix, since it contains the weights of the equilibrium relationships. The tests developed in [2] focus on the rank of matrix Π.

The pioneer Bayesian works to study VAR models and reduced rank regressions are [31–33]. However, the main concern of these papers is to estimate the model parameters and their (marginal) posterior distributions. The usual approach is to assume a given rank for the long run matrix Π, and proceed with all the computations conditional on the given rank. The Bayesian initiatives to test the rank of the referred matrix are recent, the main reference for Bayesian inference on VECM's being [34].

To justify inferential procedures based on prespecified ranks of matrix Π, [22] argued that an empirical cointegration analysis should be based on economic theory, which proposes models obeying equilibrium relationships. According to this view, cointegration research should be "confirmatory" rather than "exploratory". Even though the advocated conditional inference is of simple implementation and very useful for small samples, [22] recognized that tests for the rank of matrix Π should be developed. To our knowledge, few initiatives with this purpose were developed up to now.

One common approach to test sharp hypotheses in the Bayesian framework is by means of Bayes factors. Testing the rank of matrix Π by Bayes factors implies several computational complications and requires the use of proper priors, as shown in [35]. Following an informal approach, [33] obtained the posterior distribution of the ordered eigenvalues of the "squared" long run matrix, $\Pi' \cdot \Pi$, obtained from a VAR model without assuming the existence of cointegration relations. As the long run matrix has a reduced rank, it has some null eigenvalues, and this should be revealed by the fact that the smallest eigenvalues should have a lot of probability mass accumulated on values close to zero. The computations can be made straightforwardly, simulating values for the long run matrix from its (marginal) posterior distribution, which is a matrix t-Student distribution under the non informative prior (16), also considered in the sequel.

Another common procedure is to estimate the rank of Π as the value r that maximizes the (marginal) posterior distribution of the rank. Conditioned on such an estimate, one proceeds to derive the full posterior and eventually estimate the cointegration space, i.e., the linear space spanned by β.

A different approach was proposed by [36], who used the Posterior Information Criterion (PIC), developed in [37], as a criterion to choose the mode of the posterior distribution of the rank of Π. However, as highlighted in [34], one of the advantages of the Bayesian approach is the possibility to incorporate the uncertainty about the parameters in the analysis, represented by the posterior distribution of the rank and, whatever the tool the scientist uses to infer the value of r, it is derived from this posterior distribution.

The authors of [38] nested the reduced rank models in an unrestricted VAR and used Metropolis–Hastings sampling with the Savage–Dickey density ratio—see [39]—to estimate the Bayes Factors of all the models with incomplete rank up to the model with full rank. The Bayes Factor derivation requires the estimation of an error correction factor for the incomplete rank. This factor, however, is not defined for improper priors due to a problem known as *Bartlett paradox*, which arises whenever one compares models of different dimensions. The difficulty is relevant in the present case because, after deriving the rank posterior density, one may consider that models of different dimensions are being compared. The paradox is stated informally as: improper priors should be avoided when one computes Bayes Factors (except for parameters common to both models) as they depend on arbitrary constants (that are integrals).

More recently, [40] developed an efficient procedure to obtain the posterior distribution of the rank using a uniform proper prior over the cointegration space linearly normalized. The author derived solutions for the posterior probabilities for the null rank and for the full rank of Π. The posterior probabilities of each intermediate rank are derived from the posterior samples of the matrices that compose the long run matrix (α and β), properly normalized, under each rank and using the method proposed by [41].

5. Implementing the FBST as a Cointegration Test

This section describes how to implement the FBST to test for cointegration. We will proceed in the same spirit of Section 3, i.e., describing the steps given in Table 1 to implement the test for cointegration.

Let us begin recalling the VECM given by Equation (14):

$$\Delta \mathbf{Y}_t = \mathbf{c} + \mathbf{\Phi}_0 \mathbf{D}_t + \mathbf{\Gamma}_1 \Delta \mathbf{Y}_{t-1} + \ldots + \mathbf{\Gamma}_{p-1} \Delta \mathbf{Y}_{t-p+1} + \mathbf{\Pi} \mathbf{Y}_{t-1} + \mathbf{E}_t, \tag{14}$$

$t = 1, \ldots, T + p$, in which $\mathbf{E}_t \overset{i.i.d.}{\sim} N_n(\mathbf{0}, \Sigma)$ with $\mathbf{0}$ a null vector of dimension $n \times 1$ and Ω a symmetric positive definite real matrix. Notice that these assumptions already specify the statistical model (Gaussian) and its implied likelihood. Before giving it explicitly, let us rewrite Equation (14) using matrix notation:

$$\Delta \mathbf{Y} = \mathbf{Z} \cdot \eta + \mathbf{E} \tag{15}$$

where $\Delta \mathbf{Y} = \begin{bmatrix} \Delta \mathbf{Y}'_{p+1} \\ \Delta \mathbf{Y}'_{p+2} \\ \vdots \\ \Delta \mathbf{Y}'_{T+p} \end{bmatrix}$, $\mathbf{Z} = \begin{bmatrix} 1 & \mathbf{D}'_{p+1} & \Delta \mathbf{Y}'_p & \cdots & \Delta \mathbf{Y}'_2 & \mathbf{Y}'_p \\ 1 & \mathbf{D}'_{p+2} & \Delta \mathbf{Y}'_{p+1} & \cdots & \Delta \mathbf{Y}'_3 & \mathbf{Y}'_{p+1} \\ \vdots & \vdots & \vdots & & \vdots & \vdots \\ 1 & \mathbf{D}'_{T+p} & \Delta \mathbf{Y}'_{T-1} & \cdots & \Delta \mathbf{Y}'_{T+p-1} & \mathbf{Y}'_{T+p-1} \end{bmatrix}$, $\eta = \begin{bmatrix} \mathbf{c}' \\ \mathbf{\Phi}_0 \\ \mathbf{\Gamma}_1 \\ \vdots \\ \mathbf{\Gamma}_{p-1} \\ \mathbf{\Pi} \end{bmatrix}$

and the error vector is given by $\mathbf{E} \sim MN_{T \times n}(0, I_T, \Omega)$, denoting the matrix normal distribution. See Appendix B for more information on this distribution. Now the parameter vector is given by $\Theta = (\eta, \Omega)$.

Notice that $\Delta \mathbf{Y}$ is formed by piling up T transposed vectors $\Delta \mathbf{Y}_t$, thus resulting in a matrix with T lines and n columns (n is the number of time series in vector $\mathbf{Y}_t$), those being also dimensions of matrix $\mathbf{E}$. Matriz $\mathbf{Z}$ is constructed likewise—always piling up the transposed vectors—resulting in a matrix with T lines and $pn + n + 1$ columns. Finally, matrix η has the matrices of coefficients, all piled up properly, resulting in a matrix with $pn + n + 1$ lines and n columns.

Given the assumptions above, $\Delta \mathbf{Y} \sim MN_{T \times n}(\mathbf{Z} \cdot \eta, I_T, \Omega)$, implying that the likelihood is

$$L(\Theta \mid \mathbf{y}) \propto |\Omega|^{-T/2} \exp\left\{ -\frac{1}{2} \cdot \mathrm{tr}[\Omega^{-1}(\Delta \mathbf{Y} - \mathbf{Z} \cdot \eta)'(\Delta \mathbf{Y} - \mathbf{Z} \cdot \eta)] \right\}$$

where $\mathbf{y}$ denotes the set of observed values of vectors $\mathbf{Y}_t$ for $t = 1, \ldots, T + p$. As in Section 3, we will consider an improper prior for Θ, given by

$$h(\Theta) = h(\eta, \Omega) \propto |\Omega|^{-(n+1)/2}, \tag{16}$$

and our reference density, $r(\Theta)$, will be proportional to a constant, leading to a surprise function equivalent to the (full) posterior distribution. These choices correspond to steps 1 and 2 of Table 1. These modeling choices imply the following posterior density:

$$\begin{aligned} g(\Theta \mid \mathbf{y}) &\propto |\Omega|^{-(T+n+1)/2} \exp\left\{ -\tfrac{1}{2} \cdot \mathrm{tr}[\Omega^{-1}(\Delta \mathbf{Y} - \mathbf{Z} \cdot \eta)'(\Delta \mathbf{Y} - \mathbf{Z} \cdot \eta)] \right\} \\ &= |\Omega|^{-(T+n+1)/2} \exp\left\{ -\tfrac{1}{2} \cdot \mathrm{tr}\{\Omega^{-1}[\mathbf{S} + (\eta - \hat{\eta})' \cdot \mathbf{Z}'\mathbf{Z} \cdot (\eta - \hat{\eta})]\} \right\} \end{aligned} \tag{17}$$

where $\hat{\eta} = (\mathbf{Z}'\mathbf{Z})^{-1}\mathbf{Z}'\Delta \mathbf{Y}$ and $\mathbf{S} = \Delta \mathbf{Y}'\Delta \mathbf{Y} - \Delta \mathbf{Y}'\mathbf{Z}(\mathbf{Z}'\mathbf{Z})^{-1}\mathbf{Z}'\Delta \mathbf{Y}$.

To implement Step 3 of Table 1, we need to find the maximum a posteriori of (17) under the constraint $\Theta \subset \Theta_H$, i.e., we need to maximize the posterior in Θ_H. Since we are testing the rank of matrix Π, as discussed in the beginning of Section 4, it is necessary to maximize the posterior assuming the rank of Π is r, $0 \leq r \leq n$. Thanks to the modeling choices made here—Gaussian likelihood and Equation (16) as prior—our posterior is almost identical to a Gaussian likelihood, allowing us to find this maximum using a strategy similar to that proposed by [2], who derived the maximum of the (Gaussian) likelihood function assuming a reduced rank for Π. We will summarize Johansen's algorithm, providing in Appendix C a heuristic argument of why it indeed provides the maximum value of the posterior under the assumed hypotheses.

We begin estimating a VAR($p-1$) model for $\Delta \mathbf{Y}_t$ with all the explanatory variables shown in (14) except for $\mathbf{Y}_{t-1}$. Using the matrix notation established above, this corresponds to estimate

$$\Delta \mathbf{Y} = \mathbf{Z}_1 \cdot \eta_1 + \mathbf{U},$$

where $\mathbf{Z}_1 = \begin{bmatrix} 1 & \mathbf{D}'_{p+1} & \Delta \mathbf{Y}'_p & \cdots & \Delta \mathbf{Y}'_2 \\ 1 & \mathbf{D}'_{p+2} & \Delta \mathbf{Y}'_{p+1} & \cdots & \Delta \mathbf{Y}'_3 \\ \vdots & \vdots & \vdots & & \vdots \\ 1 & \mathbf{D}'_{T+p} & \Delta \mathbf{Y}'_{T-1} & \cdots & \Delta \mathbf{Y}'_{T+p-1} \end{bmatrix}$ and $\eta_1 = \begin{bmatrix} \mathbf{e}' \\ \tau_0 \\ v_1 \\ \vdots \\ v_{p-1} \end{bmatrix}$ showing that $\mathbf{Z}_1$ is obtained

from matrix $\mathbf{Z}$ extracting its last n columns, exactly those corresponding to $\mathbf{Y}_{t-1}$.

We also estimate a second set of auxiliary equations, regressing $\mathbf{Y}_{t-1}$ on a vector of constants and $\mathbf{D}_t, \Delta \mathbf{Y}_{t-1}, \ldots, \Delta \mathbf{Y}_{t-p+1}$. By piling up all the (transposed) vectors $\mathbf{Y}'_{t-1}$ for $t = p+1, \ldots, T+p$, we have a $(T \times n)$ matrix, denoted by $\mathbf{Y}_{-1}$. As above, these equations can be represented by

$$\mathbf{Y}_{-1} = \mathbf{Z}_1 \cdot \eta_2 + \mathbf{V},$$

where $\mathbf{Y}_{-1} = \begin{bmatrix} \mathbf{Y}'_p \\ \mathbf{Y}'_{p+1} \\ \vdots \\ \mathbf{Y}'_{T+p-1} \end{bmatrix}$ and $\eta_2 = \begin{bmatrix} \mathbf{m}' \\ \nu_0 \\ \zeta_1 \\ \vdots \\ \zeta_{p-1} \end{bmatrix}$.

Considering the OLS estimates of these sets of equations and their respective estimated residuals, we may write

$$\widehat{\Delta \mathbf{Y}} = \mathbf{Z}_1 \cdot \widehat{\eta}_1 + \widehat{\mathbf{U}} \tag{18}$$

$$\widehat{\mathbf{Y}}_{-1} = \mathbf{Z}_1 \cdot \widehat{\eta}_2 + \widehat{\mathbf{V}} \tag{19}$$

where $\widehat{\eta}_1 = (\mathbf{Z}'_1 \mathbf{Z}_1)^{-1} \mathbf{Z}'_1 \cdot \Delta \mathbf{Y}$, $\widehat{\eta}_2 = (\mathbf{Z}'_1 \mathbf{Z}_1)^{-1} \mathbf{Z}'_1 \cdot \mathbf{Y}_{-1}$, $\widehat{\mathbf{U}}$ and $\widehat{\mathbf{V}}$ are the respective matrices of estimated residuals. Thanks to the Frisch–Waugh–Lovell theorem—see [42] theorem 3.3 or [43] Section 2.4—it is possible to show that the estimated residuals of these auxiliary regressions are related by Π in the following regressions:

$$\widehat{\mathbf{U}} = \Pi \widehat{\mathbf{V}} + \widehat{\mathbf{W}}. \tag{20}$$

One can prove that the OLS estimates of Π obtained from (15) and from (20) are numerically identical, as the estimated residuals $\widehat{\mathbf{E}}$ and $\widehat{\mathbf{W}}$.

The second stage of Johansen's algorithm requires the computation of the following sample covariance matrices of the OLS residuals obtained above:

$$\widehat{\Sigma}_{VV} = \frac{1}{T} \cdot \widehat{\mathbf{V}}' \widehat{\mathbf{V}} \qquad \widehat{\Sigma}_{UU} = \frac{1}{T} \cdot \widehat{\mathbf{U}}' \widehat{\mathbf{U}}$$

$$\widehat{\Sigma}_{UV} = \frac{1}{T} \cdot \widehat{\mathbf{U}}' \widehat{\mathbf{V}} \qquad \widehat{\Sigma}_{VU} = \widehat{\Sigma}'_{UV}$$

and, from these, we find the n eigenvalues of matrix

$$\widehat{\Sigma}_{VV}^{-1} \cdot \widehat{\Sigma}_{VU} \cdot \widehat{\Sigma}_{UU}^{-1} \cdot \widehat{\Sigma}_{UV},$$

ordering them decreasingly $\widehat{\lambda}_1 > \widehat{\lambda}_2 > \ldots > \widehat{\lambda}_n$. The maximum value attained by the log posterior subject to the constraint that there are r ($0 \leq r \leq n$) cointegration relationships is

$$\ell^* = K - \frac{(T+n+1)}{2} \cdot \log |\widehat{\Sigma}_{UU}| - \frac{T+n+1}{2} \cdot \sum_{i=1}^{r} \log(1 - \widehat{\lambda}_i), \tag{21}$$

where K is a constant that depends only on T, n and $\mathbf{y}$ by means of the marginal distribution of the data set, $\mathbf{y}$. Since ℓ^* represents the maximum of the log-posterior, to obtain s^*, one should take $s^* = \exp(\ell^*)$, completing step 3 of Table 1.

As in Section 3, we compute $\overline{ev}$ in step 4 by means of a Monte Carlo algorithm. It is easy to factor the full posterior (17) as a product of a (matrix) normal and an Inverse-Wishart, suggesting a Gibbs sampler to generate random samples from the full posterior. See Appendix B for more on the Inverse-Wishart distribution. Thus, the conditional posteriors for η and Ω are, respectively,

$$g_\eta(\eta \mid \Omega, \mathbf{y}) \propto MN_{n \times k}(\widehat{\eta}, (\mathbf{Z}'\mathbf{Z})^{-1}, \Omega) \tag{22}$$

$$g_\Omega(\Omega \mid \eta, \mathbf{y}) \propto IW(\Omega | \mathbf{S} + (\eta - \widehat{\eta})' \cdot \mathbf{Z}'\mathbf{Z} \cdot (\eta - \widehat{\eta}), T) \tag{23}$$

where $\mathbf{S} = \Delta\mathbf{Y}'\Delta\mathbf{Y} - \Delta\mathbf{Y}'\mathbf{Z}(\mathbf{Z}'\mathbf{Z})^{-1}\mathbf{Z}'\Delta\mathbf{Y}$, IW denotes the Inverse-Wishart, $k = pn + n + 1$ is the number of lines of η, and $\widehat{\eta}$ its OLS estimator, as above. From a Gibbs sampler set with these conditionals, we obtain a random sample from the full posterior to estimate $\overline{ev}$ as the proportion of sampled vectors that generate a value for the posterior greater than s^*. Finally, we obtain $ev = 1 - \overline{ev}$ in the final step (5). The whole implementation for cointegration tests, following the assumptions made in this section, is summarized in Table 4. See Appendix A for more information on the computational resources needed to implement the steps given by Table 4.

Table 4. Pseudocode to implement the FBST to cointegration tests.

General algorithm: compute ev supporting hypothesis $H : \text{rank}(\Pi) = r$ ($0 \leq r \leq n$) in model (14)
1. Statistical model: Gaussian; prior: $h(\Theta) \propto |\Omega|^{-(n+1)/2}$.
2. Reference density: $r(\Theta) \propto 1$; relative surprise function: $g(\Theta \mid \mathbf{y})$.
3. Find s^*: Johansen's algorithm; obtain ℓ^* from Equation (21) with $s^* = \exp(\ell^*)$.
4. Gibbs sampler (from Equations (22) and (23)) to obtain N random samples of parameter vectors from (17). Evaluate the posterior at the sampled vectors and estimate $\overline{ev}$ as the proportion of N for which the evaluated values are larger than s^*.
5. Find $ev = 1 - \overline{ev}$.

Before presenting the results of the procedure applied to real data sets, it is important to remark one feature of the FBST applied to cointegration tests. The estimated eigenvalues of matrix Π, $\widehat{\lambda}_i$, correspond to the squared canonical correlations between $\Delta\mathbf{Y}_t$ and $\mathbf{Y}_{-1}$ corrected for the variable in $\mathbf{Z}_1$ and therefore lie between 0 and 1. Therefore, (21) shows that $\ell_0^* \leq \ell_1^* \leq \ldots \ell_n^*$, where ℓ_r^* denotes the maximum of the posterior (14) assuming Π has rank $0 \leq r \leq n$. Therefore, one may say that the hypotheses $\text{rank}(\Pi) = r$ are nested, in the sense that the respective e-values obtained by the FBST for these hypotheses are always non-decreasing $ev(0) \leq ev(1) \leq \ldots \leq ev(n)$.

This nested formulation is also present in the frequentist procedure proposed by [2], based on the likelihood ratio statistics for successive ranks of Π. Thus, the FBST should be used, like the maximum eigenvalue test, in a sequential procedure to test for the number of cointegrating relationships. We will show how this should be done in presenting the applied results in the sequel.

Results

Now we present, by means of four examples, the application of FBST as a cointegration test. In all the examples, we have adopted a Gaussian likelihood and the improper prior (16). The Gibbs sampler was implemented as described above, providing 51,000 random vectors from the posterior (17). The first 1000 samples were discarded as a burn-in sample, the remaining 50,000 being used to estimate the integral (2). The tables show the e-value computed from the FBST and the maximum eigenvalue test statistics with their respective *p*-values.

Example 1. *We analyzed four electroencephalography (EEG) signals from a subject that has previously presented epileptic seizures. The original study, [44], had the aim of detecting seizures based on multiple hours of recordings for each individual and the cointegration analysis of the mentioned signals was presented by [45]. In fact, the cointegration hypothesis is tested using the phase processes estimated from the original signals. This is done by passing the signal into the Hilbert transform and then "unwrapping" the resulting transform. Sections 2 and 5 of [45,46] provide more details on the Hilbert transform and unwrapping.*

The labels of the modeled series refer to the electrodes on the scalp. As seen in Figures 1 and 2, the series are called FP1-F7, FP1-F3, FP2-F4, and FP2-F8, where FP refers to the frontal lobes and F refers to a row of electrodes placed behind these. Even numbered electrodes are on the right side and odd numbered electrodes are on the left side. The electrodes for these four signals mirror each other on the left and right sides of the scalp. The recordings of the studied subject, an 11-year-old female, identified a seizure in the interval (measured in seconds) [2956, 2996]. Therefore, like [45], we analyze the period of 41 seconds prior to the seizure—interval [2956, 2996]—and the subsequent 41 seconds—interval [2996, 3036]—the seizure period. In the sequel, we will refer to these as *prior to seizure* and *during seizure*, respectively. Since the sample frequency has 256 measurements per second, there are a total of 10,496 measurements for each of the four signals. Ref. [45] used 40 seconds for each period, obtaining slightly different results.

Figures 1 and 2 display the estimated phases based on the original signals. The model proposed by [45] is a VAR(1), resulting in a VECM given by

$$\Delta \mathbf{Y}_t = \mathbf{c} + \Pi \mathbf{Y}_{t-1} + \mathbf{E}_t. \tag{24}$$

Tables 5 and 6 present the results that essentially lead to the same conclusions obtained by [45], even though they have based their findings on the trace test. See Table 8 of [45].

The comparison between *p*-values and the FBST e-values must be made carefully, the main reason being the fact that *p*-values are not measures supporting the null hypothesis, while e-values provide exactly such a kind of support. That being said, a possible way to compare them is by checking the decision their use recommend regarding the hypothesis being tested, i.e., to reject or not the null hypothesis.

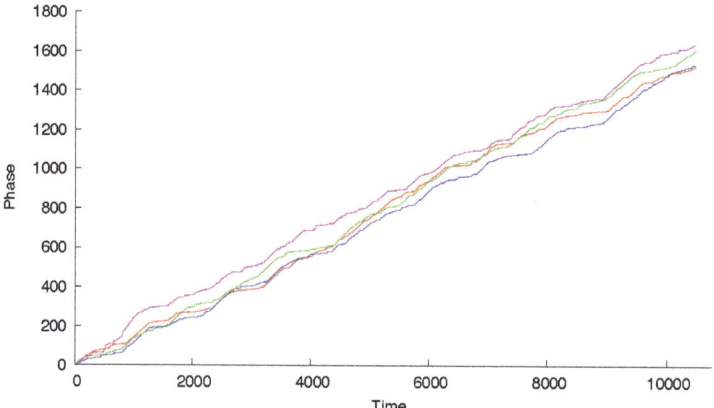

Figure 1. Estimated phase processes prior to a seizure.

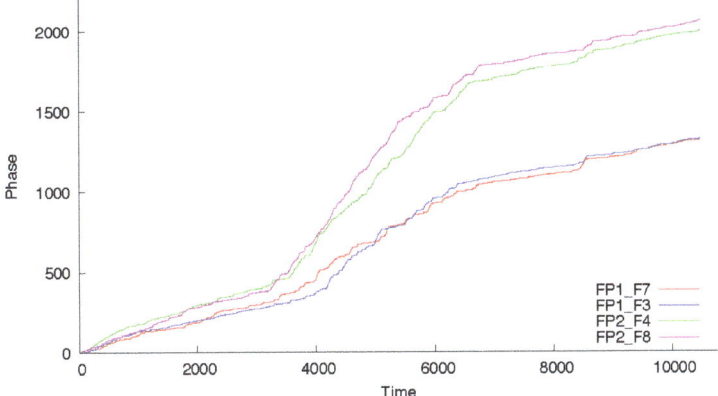

Figure 2. Estimated phase processes during a seizure.

Table 5. FBST and max. eig. test: prior to seizure.

H_0	FBST	Max.	p-Value
$r = 0$	$\simeq 0$	60.966	$\simeq 0$
$r = 1$	0.0691	30.727	0.0010
$r = 2$	0.9990	11.458	0.1337
$r = 3$	$\simeq 1$	0.0812	0.7757

Table 6. FBST and max. eig. test: during seizure.

H_0	FBST	Max.	p-Value
$r = 0$	$\simeq 0$	1120.5	$\simeq 0$
$r = 1$	0.1144	31.563	0.0007
$r = 2$	0.9999	6.5015	0.5574
$r = 3$	$\simeq 1$	1.4383	0.2304

Frequentist tests often adopt a significance level approach: given an observed p-value, the hypothesis is rejected if the p-value is smaller or equal to the mentioned level, usually 0.1, 0.05, or 0.01. Since the cointegration ranks generate nested likelihoods, the hypotheses are tested sequentially, starting with null rank, $r = 0$. For Table 5, adopting a 0.01 significance level, the maximum eigenvalue test would reject $r = 0$ and $r = 1$, and would not reject $r = 2$. The same conclusions follow for Table 6. Thus, the recommended action is to work, for estimation purposes for instance, assuming two cointegration relationships.

The question on which threshold value to adopt for the FBST was already mentioned on Section 1.1, but it is worthwhile to underline it once more. We highly recommend a principled approach deriving the cut-off value from a loss function, which is specific for the problem at hand and the purposes of the analysis. A naive but simpler approach would be to reject the hypothesis if the e-value is smaller than 0.05 or 0.01, emulating the frequentist strategy. Even not recommending this path, since p-values are not supporting measures for the hypothesis being tested while e-values are, the researcher may numerically compare p-values and e-values in a specific scenario. If the researcher derived the p-values from a generalized likelihood ratio test, it is possible to asymptotically compare them. The relationship is: $ev = 1 - F_m[F_{m-h}^{-1}(1 - \mathbf{p})]$, where m is the dimension of the full parameter space, h the dimension of the parameter space under the null hypothesis, F_m the chi-square distribution function with m degrees of freedom and $\mathbf{p}$ the corresponding p-value. See [9,12] for the proof of the asymptotic relationship between e-values and p-values.

Since the maximum eigenvalue test is derived as a likelihood ratio test, this comparison may be done for the results of all the examples presented here, and more appropriately to this example, given its sample size of 10,496 observations. Regarding Tables 5 and 6, one could be in doubt regarding whether to reject or not the hypothesis $r = 1$ since the e-values are larger than 0.01. However, for this model and hypothesis, the e-value corresponding to 0.01 is 0.436. Therefore, in both tables, one could reject the hypothesis and proceed to the next rank that has plenty of evidence in its favor. In conclusion, the practical decisions of both tests (FBST and maximum eigenvalue) would be the same: to not reject $r = 2$.

Example 2 ([47]). *Compare three methods for modeling empirical seasonal temperature forecasts over South America. One of these methods is based on a (possible) long-term cointegration relationship between the temperatures of the quarter March–April–May (MAM) of each year and the temperature of the previous months of November–December–January (NDJ). When there is such a relationship, the authors used the NDJ temperatures (of the previous year) as a predictor for the following MAM season.*

The original data set has monthly temperatures for each coordinate (latitude and longitude) of the covered area. The mentioned series of temperatures (MAM and NDJ) are computed as seasonal averages from this monthly data set by averaging over consecutive three months. Since we have data available from January 1949 to May 2020, the time series of monthly and seasonal average surface temperatures of length 72 for each grid point.

The authors of [47] consider $\mathbf{Y}_t$ as a two-dimensional vector, its first component being the seasonal (average) MAM temperature of year t and the second component the seasonal NDJ temperature of the *previous* year. They consider a VAR(2) without deterministic terms to model the series, resulting in a VECM

$$\Delta \mathbf{Y}_t = \Gamma_1 \Delta \mathbf{Y}_{t-1} + \Pi \mathbf{Y}_{t-1} + \mathbf{E}_t. \tag{25}$$

We have chosen five grid points corresponding to major Brazilian cities to test the cointegration hypothesis of the mentioned seasonal series. The coordinates chosen were the closest ones from: 23.5505° S, 46.6333° W for São Paulo; 22.9068° S, 43.1729° W for Rio de Janeiro; 19.9167° S, 43.9345° W for Belo Horizonte; 15.8267° S, 47.9218° W for Brasília and 12.9777° S, 38.5016° W for Salvador. Figures 3 and 4 show the seasonal temperatures for São Paulo and Brasília, respectively, indicating that the cointegration hypothesis is plausible for both cities.

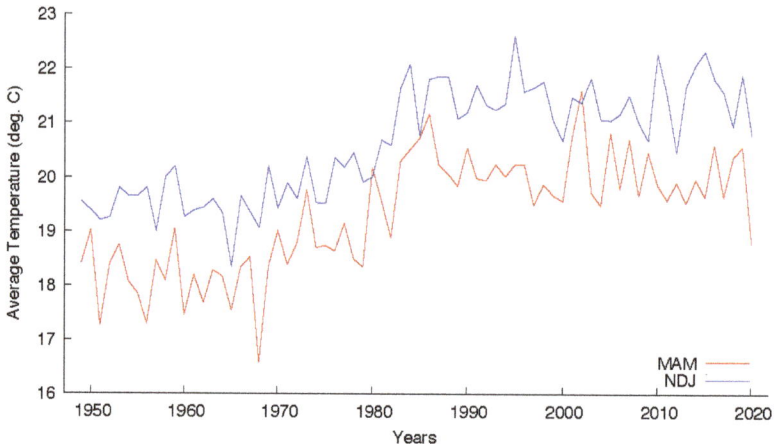

Figure 3. Seasonal (MAM and NDJ) temperatures for São Paulo from 1949 to 2020.

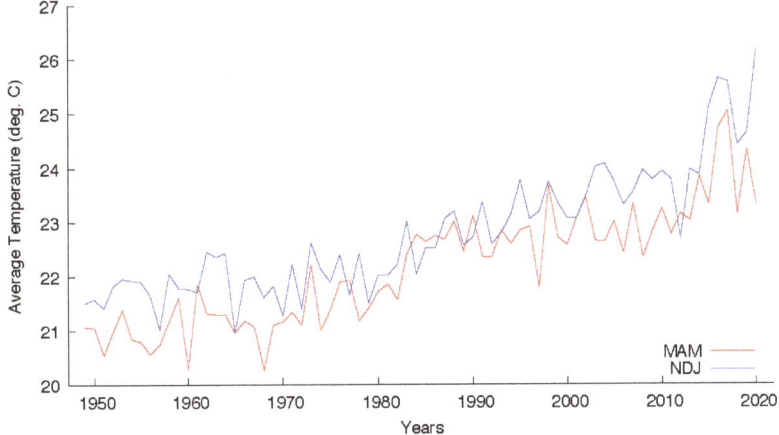

Figure 4. Seasonal (MAM and NDJ) temperatures for Brasília from 1949 to 2020.

Table 7. FBST and maximum eigenvalue test applied to temperature data (MAM and NDJ series) of the mentioned Brazilian cities.

Cities	$H_0 : r = 0$			$H_0 : r = 1$		
	FBST	Max.	p-Value	FBST	Max.	p-Value
São Paulo	0.0012	33.302	$\simeq 0$	$\simeq 1$	0.0893	0.8205
Rio de Janeiro	0.0273	23.294	0.0004	$\simeq 1$	2.43e-5	0.9986
Belo Horizonte	0.0173	24.621	0.0001	$\simeq 1$	0.0963	0.8126
Brasília	0.1129	18.008	0.0045	0.9999	1.3321	0.2892
Salvador	0.0172	24.431	0.0001	$\simeq 1$	0.2450	0.6838

The results are shown in Table 7. Assuming a significance level of 0.01, the maximum eigenvalue test reject the null rank and do not reject $r = 1$ for all the five cities. If we adopt the asymptotic relationship between p-values and e-values for the model under analysis, we obtain an e-value of 0.276 corresponding to a 0.01 p-value for $r = 0$. Therefore, the FBST would also reject the null rank for all the cities. The hypothesis $r = 1$ is not rejected since all the e-values are close to 1, once more agreeing with the maximum eigenvalue test.

One remark about Brasília seems in order. The city was built to be the federal capital, being officially inaugurated on 21 April 1960. The construction began circa 1957 and before that the site had no human occupation. The process of moving all the administration from Rio de Janeiro, the former capital, was slow and only the 1980 census detected a population over 1 million inhabitants. The present population is almost 3.2 million inhabitants living in the Federal District that includes Brasília and minor surrounding cities. Figure 4 indicates that the seasonal temperatures began to rise exactly after 1980.

Example 3. *we applied the FBST to the Finish data set used in their seminal work [2].*

The authors used the logarithms of the series of $M1$ monetary aggregate, inflation rate, real income, and the primary interest rate set by the Bank of Finland to model the money demand, which, in theory, follows a long-term relationship. The sample has 106 quarterly observations of the mentioned variables, starting at the second trimester of 1958 and finishing in the third trimester of 1984. The chosen model was a VAR(2) with unrestricted constant, meaning that the series in $\mathbf{Y}_t$ have one unit root with drift vector **c** and the cointegrating relations may have a non-zero mean. For more information about how to specify deterministic terms in a VAR see [48], chapter 6. Seasonal dummies for the first three quarters of the year were also considered in the model chosen by [2]. Writing the model in the error correction form, we have:

$$\Delta \mathbf{Y}_t = \mathbf{c} + \Phi_{0,1}\mathbf{D}_{1t} + \Phi_{0,2}\mathbf{D}_{2t} + \Phi_{0,3}\mathbf{D}_{3t} + \Gamma_1 \Delta \mathbf{Y}_{t-1} + \Pi \mathbf{Y}_{t-1} + \mathbf{E}_t. \tag{26}$$

where $\Pi = \Phi_1 + \Phi_2 - I_3$, $\Gamma_1 = -\Phi_2$, $\mathbf{c}$ is a vector with constants and $\mathbf{D}_{it}$ denote the seasonal dummies for trimester $i = 1, 2, 3$. The results are displayed in Table 8.

Table 8. FBST and maximum eigenvalue test applied to Finish data of Johansen and Juselius (1990).

H_0	FBST	Max.	p-Value
$r = 0$	0.132	38.489	0.0007
$r = 1$	0.994	26.642	0.0060
$r = 2$	$\simeq 1$	7.8924	0.3983

In [2], the authors concluded that there is, at least, two cointegration vectors, a conclusion that follows if one adopts a 0.01 significance level, for instance. Using the asymptotic relationship between p-values and e-values for Equation (26), we obtain, for $r = 0$, an e-value of 0.998, and, for $r = 1$, an e-value of 0.999, corresponding to a 0.01 p-value. These apparently discrepant values for the e-values are due to the high dimensions of the unrestricted ($m = 58$) and under H_0 ($h = 42$ for $r = 0$ and $h = 43$ for $r = 1$) parameter spaces. Therefore, under this criterion, the FBST also rejects the null rank and $r = 1$ (since $0.132 < 0.998$ and $0.994 < 0.999$, respectively) and does not reject $r = 2$, recommending the same action as the maximum eigenvalue test.

Example 4. *As a final example, we apply the FBST to a US data set discussed in [49]. The observations have annual periodicity and went from 1900 to 1985. We tested for cointegration between real national income, M1 monetary aggregate deflated by the GDP deflator and the commercial papers return rate. The chosen model was a VAR(1) with unrestricted constant. The series were used in natural logarithms and the results follow below:*

Table 9. FBST and maximum eigenvalue test applied to US data of Lucas (2000).

H_0	FBST	Max.	p-Value
$r = 0$	0.042	25.334	0.0101
$r = 1$	0.996	4.2507	0.8271

Table 9 shows that the maximum eigenvalue test rejects $r = 0$ and does not reject $r = 1$ at a 0.05 significance level. Once more adopting the asymptotic relationship between p-values and e-values for the chosen model, we obtain, for $r = 0$, an e-value of 0.247 corresponding to a 0.01 p-value. Thus, under this criterion, the FBST also rejects the null rank and does not reject $r = 1$.

6. Conclusions

In the past few decades, the econometric literature introduced statistical tests to identify unit roots and cointegration relationships in time series. The Bayesian approach applied to these topics advanced considerably after the 1990s, developing interesting alternatives, mostly for unit root testing. The (parametric) frequentist tests mentioned here may not be suitable since these procedures rely on the distribution of the test statistic—usually assuming the hypothesis being tested is true—which depend on a particular a statistical model, usually Gaussian. When the distributions of such statistics cannot be obtained, the procedure is saved by asymptotic results. If the researcher considers different statistical models and the available sample is small, the results of the tests may be quite misleading.

The present work reviewed a simple and powerful Bayesian procedure that can be applied to both purposes: unit root and cointegration testing. We have also shown that the FBST works considerably well even when one uses improper priors, a choice that may preclude the derivation of Bayes Factors, a standard Bayesian procedure in hypotheses testing.

A long series of articles provided in [7] and the references therein, has showed the versatility and properties of FBST, such as: a. the e-value derivation and computation are straightforward from its general definition; b. it uses absolutely no artificial restrictions like a distinct probability measure on the hypothesis set, induced by some specific parametrization; c. it is in strict compliance with the likelihood principle; d. it can conduct the test with any prior distribution; e. it does not need closed conjectures concerning error distributions, even for small samples; f. it is an exact procedure, since it does rely on asymptotic assumptions; and g. it is invariant with respect to the null hypothesis parametrization and with respect to the parameter space parametrization. See [9], p. 253 for this property.

To proceed with this research agenda, it would be interesting to perform more simulation studies with the FBST applied to unit root testing for a larger group of parametric and semi-parametric models (likelihoods). Another possibility is to include moving average terms in the data generating processes and work with Gaussian and non-Gaussian ARMA models. Notice that, given the points made above, these extensions would not impose major problems to the FBST as they would to the frequentist procedures. Regarding cointegration, the same extensions may be studied in future works, although the adoption of statistical models outside the Gaussian family would require further efforts to numerically implement the FBST. We shall also investigate the effect of the prior choice in the estimates of cointegration relations, especially for small samples.

Author Contributions: M.A.D. was responsible for conceptualization, computational implementation of the methods, formal analysis, investigation, and visualization. C.A.B.P.and J.M.S. were responsible for conceptualization, methodology, formal analysis, supervision, and funding acquisition. All the authors were responsible for writing, reviewing, and editing the original text. All authors have read and agreed to the published version of the manuscript.

Funding: This work was also partially funded by CNPq—the Brazilian National Council of Technological and Scientific Development (grants PQ 307648-2018-4, 302767-2017-7, 301892-2015-6, and 308776-2014-3); and FAPESP—the State of São Paulo Research Foundation (grants CEPID Shell-RCGI201450279-4; CEPIDCeMEAI 2013-07375-0). The authors are extremely grateful for the support received from their colleagues, collaborators, users, and critics in the construction works of their research projects.

Acknowledgments: The authors would like to thank J. Østergaard and C. A. Coelho for kindly providing us access to the data sets used in [45,47], respectively. We also would like to thank the support provided by UFSCar—Federal University of São Carlos, USP—University of São Paulo, and UFMS—Federal University of Mato Grosso do Sul.

Conflicts of Interest: The funders had no role in the design of the study; in the collection, analyses, or interpretation of data; in the writing of the manuscript, or in the decision to publish the results.

Appendix A. Computational Resources

The FBST was implemented in all the examples using codes written by the authors in Matlab/Octave programming language. The results displayed in Tables 3 and 5–9 were obtained using GNU Octave version 4.4.1. The only package required to run the routines was the statistics package (version 1.4.1), necessary to simulate vectors of random variables from the distributions mentioned in the text. The codes are briefly described at https://www.ime.usp.br/~jstern/software/, where they can be freely downloaded.

The original data sets used in the examples presented in this work can be obtained from the following sources:

1. Table 3: fourteen U.S. economic time series used by [23]. Available at the R library urca, where it is named "npext".
2. Example 1: the original series used in [44,45] are available at https://physionet.org/content/chbmit/1.0.0/. The data for the subject analyzed in Example 1 is from file chb01_03.edf, found inside folder chb01. To obtain Tables 5 and 6, the data were transformed as described in Example 1.

3. Example 2: the original data set used in [47] is available at https://climexp.knmi.nl/NCEPData/ghcn_cams_05.nc, provided by the Global Historical Climatology Network (GHCN)/Climate Anomaly Monitoring System (CAMS). The data set studied here is the 2 m temperature analysis (0.5 × 0.5) data, a high resolution (0.5 × 0.5 degrees in latitude and longitude) global land surface temperature data set covering the period 1949 to near present, in our case May 2020.
4. Example 3: the original data set with four macroeconomic series used by [2] to estimate the money demand of Finland is available in the R library urca with the name "finland".
5. Example 4: the original data used in [49] can be downloaded from https://www.ime.usp.br/~jstern/software/.

Appendix B. Non-Standard Distributions Used in This Article

Appendix B.1. Inverse-Gamma

The probability density function of the Inverse-Gamma distribution is given

$$f_0(x \mid a, b) = \frac{b^a}{\Gamma(a)} \cdot \left(\frac{1}{x}\right)^{a+1} \exp\left(-\frac{b}{x}\right)$$

for $x > 0$ and zero, otherwise. The parameters a and b are both positive real numbers and Γ is the gamma function.

Appendix B.2. Matrix Normal

The probability density function of the random matrix $\mathbf{X}$ with dimensions $p \times q$ that follows the matrix normal distribution $MN_{p \times q}(\mathbf{M}, \mathbf{U}, \mathbf{V})$ has the form:

$$f_1(\mathbf{X} \mid \mathbf{M}, \mathbf{U}, \mathbf{V}) = \frac{\exp\left(-\frac{1}{2} \operatorname{tr}\left[\mathbf{V}^{-1}(\mathbf{X} - \mathbf{M})' \mathbf{U}^{-1}(\mathbf{X} - \mathbf{M})\right]\right)}{(2\pi)^{pq/2} |\mathbf{V}|^{p/2} |\mathbf{U}|^{q/2}}$$

where $\mathbf{M} \in \mathbb{R}^{p \times q}$, $\mathbf{U} \in \mathbb{R}^{p \times p}$ and $\mathbf{V} \in \mathbb{R}^{q \times q}$, being $\mathbf{U}$ and $\mathbf{V}$ symmetric positive semidefinite matrices. The matrix normal distribution can be characterized by the multivariate normal distribution as follows: $\mathbf{X} \sim MN_{p \times q}(\mathbf{M}, \mathbf{U}, \mathbf{V})$ if and only if $\operatorname{vec}(\mathbf{X}) \sim N_{pq}(\operatorname{vec}(\mathbf{M}), \mathbf{V} \otimes \mathbf{U})$, where $\otimes$ denotes the Kronecker product and vec the vectorization of $\mathbf{M}$.

Appendix B.3. Inverse-Wishart

The probability density function of the Inverse-Wishart distribution is

$$f_2(\mathbf{x} \mid \Lambda, \nu) = \frac{|\Lambda|^{\nu/2}}{2^{\nu p/2} \Gamma_p(\frac{\nu}{2})} |\mathbf{x}|^{-(\nu+p+1)/2} \exp\left[-\frac{1}{2} \operatorname{tr}(\Lambda \mathbf{x}^{-1})\right]$$

where $\mathbf{x}$ and Λ are $p \times p$ positive-definite matrices, and Γ_p is the multivariate gamma function. Notice that we may also write the same density with $\operatorname{tr}(\mathbf{x}^{-1} \Lambda)$ inside the exponential function, as would be convenient in our implementation of the Gibbs sampler in Section 5.

Appendix C. Heuristic Proof of Johansen's Procedure

The goal of this appendix is to provide a brief heuristic explanation of the procedure, discussed in Section 5 that finds the maximum of posterior (17) subject to the hypothesis that matrix Π has reduced rank r, $0 \leq r \leq n$. The procedure is based on the algorithm proposed in [2,50] to maximize a Gaussian likelihood under the same assumption (reduced rank of matrix Π). The formal proof of Johansen's algorithm can be found in [51], chapter 20. As mentioned in Section 5, Johansen's algorithm can be applied to the posterior (17) since this distribution is very close to a (multivariate) Gaussian likelihood.

The first step of the algorithm involves "concentrating" the posterior, i.e., to assume Ω and Π are given and maximize the posterior with respect to all the other parameters in Θ. Hence, let γ denote the matrix η except for matrix Π, i.e., $\gamma' = \begin{bmatrix} \mathbf{c} & \Phi_0' & \Gamma_1' & \cdots & \Gamma_{p-1}' \end{bmatrix}$. The concentrated log-posterior, denoted by $\mathcal{M}$, is found by replacing γ with $\hat{\gamma}(\Pi)$ in (17):

$$\mathcal{M}(\Pi, \Omega \mid \mathbf{y}) = \ln[g(\hat{\gamma}(\Pi); \Pi, \Omega \mid \mathbf{y})] = C + \frac{(T+n+1)}{2} \ln |\Omega^{-1}| - \left\{ -\frac{1}{2} \cdot \text{tr}[\Omega^{-1}(\hat{\mathbf{U}} - \Pi \hat{\mathbf{V}})'(\hat{\mathbf{U}} - \Pi \hat{\mathbf{V}})] \right\} \quad (A1)$$

where C is a constant that depends on T, n and $\mathbf{y}$. The strategy behind concentrating the posterior is that, if we can find the values $\hat{\Omega}$ and $\hat{\Pi}$ that maximize $\mathcal{M}$, then these same values, along with $\hat{\gamma}(\hat{\Pi})$, will maximize (17) under the constraint $\text{rank}(\Pi) = r$. Carrying the concentration on one step further, we can find the value of Ω that maximizes (A1) assuming Π known, giving

$$\hat{\Omega}(\Pi) = \frac{1}{T+n+1} \cdot (\hat{\mathbf{U}} - \Pi \hat{\mathbf{V}})'(\hat{\mathbf{U}} - \Pi \hat{\mathbf{V}}).$$

To evaluate the concentrated log-posterior at $\hat{\Omega}(\Pi)$, notice that

$$\text{tr}\left[\hat{\Omega}(\Pi)^{-1}(\hat{\mathbf{U}} - \Pi \hat{\mathbf{V}})'(\hat{\mathbf{U}} - \Pi \hat{\mathbf{V}})\right] = \text{tr}[(T+n+1)I_n] = n(T+n+1)$$

and, therefore, denoting by $\mathcal{M}^*$ this new concentrated log-posterior, we have

$$\mathcal{M}^*(\Pi \mid \mathbf{y}) = C + \frac{(T+n+1)n}{2} - \frac{(T+n+1)}{2} \ln \left| \frac{1}{T+n+1} (\hat{\mathbf{U}} - \Pi \hat{\mathbf{V}})'(\hat{\mathbf{U}} - \Pi \hat{\mathbf{V}}) \right| \quad (A2)$$

$$= C + \frac{(T+n+1)n}{2} - \frac{(T+n+1)}{2} \ln \left| \frac{T}{T+n+1} \cdot \frac{1}{T} (\hat{\mathbf{U}} - \Pi \hat{\mathbf{V}})'(\hat{\mathbf{U}} - \Pi \hat{\mathbf{V}}) \right| \quad (A3)$$

$$= C + \frac{(T+n+1)n}{2} - \frac{(T+n+1)}{2} \ln \left[\left(\frac{T}{T+n+1} \right)^n \cdot \left| \frac{1}{T} (\hat{\mathbf{U}} - \Pi \hat{\mathbf{V}})'(\hat{\mathbf{U}} - \Pi \hat{\mathbf{V}}) \right| \right] \quad (A4)$$

$$= K - \frac{(T+n+1)}{2} \cdot \ln \left| \frac{1}{T} (\hat{\mathbf{U}} - \Pi \hat{\mathbf{V}})'(\hat{\mathbf{U}} - \Pi \hat{\mathbf{V}}) \right| \quad (A5)$$

where K is a new constant depending only on T, n and $\mathbf{y}$. Equation (A5) represents the maximum value one can achieve for the log-posterior for any given matrix Π. Thus, maximizing the posterior comes down to choosing Π so as to minimize the determinant

$$\left| \frac{1}{T} (\hat{\mathbf{U}} - \Pi \hat{\mathbf{V}})'(\hat{\mathbf{U}} - \Pi \hat{\mathbf{V}}) \right|$$

subject to the constraint $\text{rank}(\Pi) = r$. The solution of this problem demands the analysis of the sample covariance matrices of the OLS residuals $\hat{\mathbf{U}}$ and $\hat{\mathbf{V}}$ and here we only present the final expression for the maximum value achieved for the log-posterior, denoted ℓ^* in Section 5:

$$\ell^* = K - \frac{(T+n+1)}{2} \cdot \ln |\hat{\Sigma}_{UU}| - \frac{T+n+1}{2} \cdot \sum_{i=1}^{r} \ln(1 - \hat{\lambda}_i). \quad (A6)$$

Chapter 20 of [51] provides the formal derivation of (A6).

References

1. Engle, R.F.; Granger, C.W.J. Co-Integration and Error Correction: Representation, Estimation, and Testing. *Econometrica* **1987**, *55*, 251–276. [CrossRef]
2. Johansen, S.; Juselius, K. Maximum likelihood estimation and inference on cointegration—With application to the demand for money. *Oxf. Bull. Econ. Stat.* **1990**, *52*, 169–210. [CrossRef]
3. Diniz, M.A.; Pereira, C.A.B.; Stern, J.M. Unit Roots: Bayesian Significance Test. *Commun. Stats. Theory Methods* **2011**, *40*, 4200–4213. [CrossRef]

4. Diniz, M.A.; Pereira, C.A.B.; Stern, J.M. Cointegration: Bayesian Significance Test. *Commun. Stats. Theory Methods* **2012**, *41*, 3562–3574. [CrossRef]
5. Pereira, C.A.B.; Stern, J.M. Evidence and credibility: Full Bayesian Significance Test for precise hypotheses. *Entropy* **1999**, *1*, 69–80.
6. Pereira, C.A.B.; Stern, J.M.; Wechsler, S. Can a Significance Test Be Genuinely Bayesian. *Bayesian Anal.* **2008**, *1*, 79–100. [CrossRef]
7. Stern, J.M.; Pereira, C.A.B. The e-value: A Fully Bayesian Significance Measure for Precise Statistical Hypotheses and its Research Program. *São Paulo J. Math. Sci.* **2020**. Available online: https://link.springer.com/article/10.1007%2Fs40863-020-00171-7 (accessed on 20 August 2020).
8. Good, I.J. *Good Thinking: The Foundations of Probability and Its Applications*; University of Minnesota Press: Minneapolis, MN, USA, 1983.
9. Stern, J.M. Cognitive Constructivism and the Epistemic Significance of Sharp Statistical Hypotheses in Natural Sciences. *arXiv* **2010**, arXiv:1006.5471. Available online: https://arxiv.org/abs/1006.5471 (accessed on 20 August 2020).
10. Madruga, M.R.; Esteves, L.G.; Wechsler, S. On the Bayesianity of Pereira-Stern tests. *Test* **2001**, *10*, 291–299. [CrossRef]
11. Schervish, M. *Theory of Statistics*; Springer: New York, NY, USA, 1995.
12. Diniz, M.A.; Pereira, C.A.B.; Polpo, A.; Stern, J.M.; Wechsler, S. Relationship between Bayesian and frequentist significance indices. *Int. J. Uncertain. Quantif.* **2012**, *2*, 161–172. [CrossRef]
13. Borges, W.; Stern, J.M. The rules of logic composition for the Bayesian epistemic E-values. *Log. J. IGPL* **2007**, *15*, 401–420. [CrossRef]
14. Wald, A. *Statistical Decision Functions*; John Wiley and Sons: New York, NY, USA, 1950.
15. Tierney, L.; Kadane, J.B. Accurate approximation for posterior moments and marginal densities. *J. Am. Stat. Assoc.* **1986**, *81*, 82–86. [CrossRef]
16. Dhrymes, P.J. *Mathematics for Econometrics*; Springer: New York, NY, USA, 1978.
17. Dickey, D.A.; Pantula, S.G. Determining the Ordering of Differencing in Autoregressive Processes. *J. Bus. Econ. Stat.* **1987**, *5*, 455–461.
18. Sims, C.A. Bayesian skepticism on unit root econometrics. *J. Econ. Dyn. Control* **1988**, *12*, 463–474. [CrossRef]
19. Sims, C.A.; Uhlig, H. Understanding unit rooters: A helicopter tour. *Econometrica* **1991**, *59*, 1591–1600. [CrossRef]
20. Phillips, P.C. To criticize the critics: An objective Bayesian analysis of stochastic trends. *J. Appl. Econ.* **1991**, *6*, 333–364. [CrossRef]
21. Phillips, P.C. Bayesian routes and unit roots: De rebus prioribus semper est disputandum. *J. Appl. Econ.* **1991**, *6*, 435–474. [CrossRef]
22. Bauwens, L.; Lubrano, M.; Richard, J.-F. *Bayesian Inference in Dynamic Econometric Models*; Oxford University Press: Oxford, UK, 1999.
23. Schotman, P.C.; van Dijk, H. K. On Bayesian routes to unit roots. *J. Appl. Econ.* **1991**, *49*, 387–401. [CrossRef]
24. Dickey, D.A.; Fuller, W.A. Distribution of the estimators for autoregressive time series with a unit root. *J. Am. Stat. Assoc.* **1979**, *74*, 427–431.
25. Lubrano, M. Testing for unit roots in a Bayesian framework. *J. Econ.* **1995**, *69*, 81–109. [CrossRef]
26. DeJong, D.; Whiteman, C.H. Reconsidering Trends and random walks in macroeconomic time series. *J. Econ.* **1991**, *28*, 221–254. [CrossRef]
27. Nelson, C.; Plosser, C. Trends and random walks in macroeconomic time series: Some evidence and implications. *J. Monet. Econ.* **1982**, *10*, 139–162. [CrossRef]
28. Stern, J.M. Symmetry, Invariance and Ontology in Physics and Statistics. *Symmetry* **2011**, *3*, 611–635. [CrossRef]
29. MacKinnon, J.G. Approximate asymptotic distribution functions for unit-root and cointegration tests. *J. Bus. Econ. Stat.* **1994**, *12*, 167–176.
30. Johansen, S. *Likelihood-based Inference in Cointegrated Vector Autoregressive Models*; Oxford University Press: Oxford, UK, 1996.
31. DeJong, D. Co-integration and trend-stationary in macroeconomic time series. *J. Econ.* **1992**, *52*, 347–370. [CrossRef]
32. Geweke, J. Bayesian reduced rank regression in econometrics. *J. Econ.* **1996**, *75*, 121–146. [CrossRef]

33. Bauwens, L.; Lubrano, M. *Advances in Econometrics*; JAI Press: Greenwich, CT, USA, 1996.
34. Koop, G.; Strachan, R.; van Dijk, H.K.; Villani, M. *The Palgrave Handbook of Theoretical Econometrics*; Palgrave McMillan: London, UK, 2006.
35. Kleibergen, F.; Paap, R. Priors, posterior odds and Lagrange multiplier statistics in Bayesian analyses of cointegration. *Econ. Inst. Res. Pap.* **1996**. Available online: https://repub.eur.nl/pub/1398/ (accessed on 20 August 2020).
36. Chao, J.; Phillips, P.C. Model selection in partially nonstationary vector autoregressive processes with reduced rank structure. *J. Econ.* **1999**, *91*, 227–271. [CrossRef]
37. Phillips, P.C. Econometric model determination. *Econometrica* **1996**, *59*, 283–306. [CrossRef]
38. Kleibergen, F.; Paap, R. Priors, posterior odds and bayes factors for a Bayesian analysis of cointegration. *J. Econ.* **2002**, *111*, 223–249. [CrossRef]
39. Verdinelli, I.; Wasserman, L. Computing Bayes factors using a generalization of the Savage-Dickey density ratio. *J. Am. Stat. Assoc.* **1995**, *90*, 614–618. [CrossRef]
40. Villani, M. Bayesian reference analysis of cointegration. *Econ. Theory* **2005**, *21*, 326–357. [CrossRef]
41. Chib, S.; Greenberg, E. Understanding the Metropolis-Hastings algorithm. *Am. Stat.* **1995**, *49*, 327–335.
42. Greene, W.H. *Econometric Analysis*; Prentice Hall: Bergen County, NJ, USA, 2008.
43. Davidson, R.; MacKinnon, J.G. *Econometric Theory and Methods*; Oxford University Press: Oxford, UK, 2004.
44. Shoeb, A.H. *Application of Machine Learning to Epileptic Seizure Onset Detection and Treatment*; MIT Press: Cambridge, MA, USA, 2009.
45. Østergaard, J.; Rahbeck, A.; Ditlevsen, S. Oscillating systems with cointegrated phase processes. *J. Math. Biol.* **2017**, *75*, 845–883. [CrossRef] [PubMed]
46. Freeman, W.J. Hilbert transform for brain waves. *Scholarpedia* **2007**, *2*, 1338. [CrossRef]
47. Turasie, A.A.; Coelho, C.A.S. Cointegration modeling for empirical South American seasonal temperature forecasts. *Int. J. Climatol.* **2016**, *36*, 4523–4533. [CrossRef]
48. Lütkepohl, H. *New Introduction to Multiple Time Series Analysis*; Springer: Berlin, Germany, 2005.
49. Lucas, R. Inflation and welfare. *Econometrica* **2000**, *68*, 247–274. [CrossRef]
50. Johansen, S. Statistical analysis of cointegration vectors. *J. Econ. Dyn. Control* **1988**, *12*, 231–254. [CrossRef]
51. Hamilton, J.D. *Time Series Analysis*; Princeton University Press: Princeton, NJ, USA, 1994.

© 2020 by the authors. Licensee MDPI, Basel, Switzerland. This article is an open access article distributed under the terms and conditions of the Creative Commons Attribution (CC BY) license (http://creativecommons.org/licenses/by/4.0/).

Article

A Novel Perspective of the Kalman Filter from the Rényi Entropy

Yarong Luo [1], Chi Guo [1,2,*], Shengyong You [1] and Jingnan Liu [1,2]

1. Global Navigation Satellite System Research Center, Wuhan University, Wuhan 430079, China; yarongluo@whu.edu.cn (Y.L.); shengyongyou@whu.edu.cn (S.Y.); jnliu@whu.edu.cn (J.L.)
2. Artificial Intelligence Institute, Wuhan University, Wuhan 430079, China
* Correspondence: guochi@whu.edu.cn

Received: 21 July 2020; Accepted: 31 August 2020; Published: 3 September 2020

Abstract: Rényi entropy as a generalization of the Shannon entropy allows for different averaging of probabilities of a control parameter α. This paper gives a new perspective of the Kalman filter from the Rényi entropy. Firstly, the Rényi entropy is employed to measure the uncertainty of the multivariate Gaussian probability density function. Then, we calculate the temporal derivative of the Rényi entropy of the Kalman filter's mean square error matrix, which will be minimized to obtain the Kalman filter's gain. Moreover, the continuous Kalman filter approaches a steady state when the temporal derivative of the Rényi entropy is equal to zero, which means that the Rényi entropy will keep stable. As the temporal derivative of the Rényi entropy is independent of parameter α and is the same as the temporal derivative of the Shannon entropy, the result is the same as for Shannon entropy. Finally, an example of an experiment of falling body tracking by radar using an unscented Kalman filter (UKF) in noisy conditions and a loosely coupled navigation experiment are performed to demonstrate the effectiveness of the conclusion.

Keywords: Rényi entropy; discrete Kalman filter; continuous Kalman filter; algebraic Riccati equation; nonlinear differential Riccati equation

1. Introduction

In the late 1940s, Shannon introduced a logarithmic measure of information [1] and a theory that included information entropy (the literature shows that it is related to Boltzmann entropy in statistical mechanics). The more stochastic and unpredictable a variable is, the larger its entropy is. As a measure of information, entropy has been used in various fields, such as information theory, signal processing, information-theoretic learning [2,3], etc. As a generalization of the Shannon entropy, Rényi entropy, named after Alfréd Rényi [4], allows for different averaging of probabilities through a control parameter α, and is usually used to quantify the diversity, uncertainty, or randomness of random variables. Liang [5] presented the evolutionary entropy equations and the uncertainty estimation for Shannon entropy and relative entropy, which is also called Kullback–Leibler divergence [6], within the framework of dynamical systems. However, higher-order Rényi entropy has some better properties than Shannon entropy by setting the control parameter α in most cases.

The Kalman filter [7] and its variants have been widely used in navigation, control, tracking, etc. Many works focus on combining different entropy and entropy-like quantities with the original Kalman filter to improve the performance. When the state space equation is nonlinear, Rényi entropy can be used to measure the nonlinearity [8,9]. Shannon entropy was used to estimate the weight of each particle from the weights of different measurement models for the fusion algorithm in [10]. Quadratic Rényi entropy [11] of innovation has been used as a minimum entropy criterion under a nonlinear and non-Gaussian circumstance [12] in unscented Kalman filter (UKF) [13] and finite mixtures [14]. A generalized density evolution equation [15] and polynomial-based non-linear compensation [16] were used to improve the minimum entropy filtering [17]. Relative entropy has been used to measure the similarity between the probabilistic density functions during the recursive processes of the nonlinear filter [18,19]. As for the nonlinear measurement equation with additive Gaussian noise, relative entropy can be deduced to measure the nonlinearity of the measurement [20], and can also be used to measure the approximation error of the i-th measurement element in the partitioned update Kalman filter [21]. When the state variables and the measurement variables do not belong to strict Gaussian distribution, such as in the seamless indoor/outdoor multi-source fusion positioning problem [22], the estimation error can be measured by the relative entropy. Relative entropy can also be used to calculate the number of particles in the unscented particle filter for mobile robot self-localization [23] and to calculate the sample window size in the cubature Kalman filter (CKF) [24] for attitude estimation [25]. Moreover, it has been verified that the original Kalman filter can be derived by maximizing the relative entropy [26]. Meanwhile, the robust maximum correntropy criterion has been adopted as the optimal criterion to derive the maximum correntropy Kalman filter [27,28]. However, there has been no work on the direct connections between the Rényi entropy and the Kalman filter theory until now.

In this paper, we propose a new perspective of the Kalman filter from the Rényi entropy for the first time, which bridges the gap between the Kalman filter and the Rényi entropy. We calculate the temporal derivative of the Rényi entropy for the Kalman filter mean square error matrix, which provides the optimal recursive solution mathematically and will be minimized to obtain the Kalman filter gain. Moreover, from the physical point of view, the continuous Kalman filter approaches a steady state when the temporal derivative of the Rényi entropy is equal to zero, which also means that the Rényi entropy will keep stable. A numerical experiment of falling body tracking in noisy conditions with radar using the UKF and a practical experiment of loosely-coupled integration are provided to demonstrate the effectiveness of the above conclusion.

The structure of this paper is as follows. In Section II, the definitions and properties of Shannon entropy and Rényi entropy are presented. In Section III, the Kalman filter is derived from the perspective of minimizing the temporal derivative of Rényi entropy, and the connection between the Rényi entropy and the algebraic Riccati equation is explained. In Section IV, experimental results and analysis are given by the simulation of the UKF and the real integrated navigation data. We finally conclude this paper and provide an outlook for future work in Section V.

2. The Connection between the Kalman Filter and the Temporal Derivative of the Rényi Entropy

2.1. Rényi Entropy

To calculate the Rényi entropy of the continuous probability density function (PDF), it is necessary to extend the definition of the Rényi entropy to the continuous form. The Rényi entropy of order α for a continuous random variable with a multivariate Gaussian PDF $p(x)$ is defined [4] and calculated [9] as:

$$H_R^\alpha(x) = \frac{1}{1-\alpha} \ln \int_\mathcal{S} p^\alpha(x) dx = \frac{N}{2} \ln(2\pi \alpha^{\frac{1}{\alpha-1}}) + \frac{1}{2} \ln(\det \Sigma), \quad (1)$$

where $\alpha > 0, \alpha \neq 1$, and α is a parameter providing a family of entropy functions. N is the dimension of the random variable x. $\mathcal{S}$ is the support. Σ is the covariance matrix of $p(x)$.

It is straightforward to show that the temporal derivative of the Rényi entropy is given by [9]:

$$\dot{H}_R^{(\alpha)}(x) = \frac{1}{2} Tr\{\Sigma^{-1} \dot{\Sigma}\}, \quad (2)$$

where $\dot{\Sigma}$ is the temporal derivative of the covariance matrix and $Tr(\cdot)$ is the trace operator.

It is easy to get the Shannon entropy for the multivariate Gaussian PDF by taking the limitation of Equation (1) as α approaches 1. This entropy is given as $H(x) = \frac{N}{2} \ln(2\pi e) + \frac{1}{2} \ln(\det \Sigma)$, and the temporal derivative of the Shannon entropy is given as $\dot{H}(x) = \frac{1}{2} Tr\{\Sigma^{-1} \dot{\Sigma}\}$. It is obvious the temporal of the Shannon entropy is the same as the temporal of the Rényi entropy. Therefore, we will see later that the conclusion can also be derived from the temporal derivative of the Shannon entropy. However, the Rényi entropy for the multivariate Gaussian PDF instead of the temporal derivative of the Rényi entropy will be used by adjusting the free parameter α for different uncertainty measurements in most cases, as the filtering problem has to account for the nonlinearity and the non-Gaussian noise; we adopt the Rényi entropy as the measurement for uncertainty.

2.2. Kalman Filter

Given the continuous-time linear system [29]:

$$\dot{X}(t) = F(t)X(t) + G(t)w(t) \tag{3}$$

$$Z(t) = H(t)X(t) + v(t), \tag{4}$$

where $X(t)$ is the state vector; $F(t)$ is the state transition matrix; $G(t)$ is the system noise driving matrix; $Z(t)$ is the measurement vector; $H(t)$ is the measurement matrix; and $w(t)$ and $v(t)$ are independent white Gaussian noise with zero mean value; their covariance matrices are $Q(t)$ and $R(t)$, respectively:

$$\mathbb{E}[w(t)] = 0, \mathbb{E}[w(t)w^T(\tau)] = Q(t)\delta(t-\tau) \tag{5}$$

$$\mathbb{E}[v(t)] = 0, \mathbb{E}[v(t)v^T(\tau)] = R(t)\delta(t-\tau) \tag{6}$$

$$\mathbb{E}[w(t)v^T(\tau)] = 0, \tag{7}$$

where $\delta(t)$ is the Dirac impulse function, $Q(t)$ is a symmetric non-negative definite matrix, and $R(t)$ is a symmetric positive matrix.

The continuous Kalman filter can be deduced by taking the limit of the discrete Kalman filter. The discrete-time state-space model is arranged as follows [29]:

$$X_k = \Phi_{k|k-1} X_{k-1} + \Gamma_{k|k-1} W_{k-1} \tag{8}$$

$$Z_k = H_k X_k + V_k \tag{9}$$

where X_k is an n-dimensional state vector; Z_k is an m-dimensional measurement vector; $\Phi_{k|k-1}$, $\Gamma_{k|k-1}$, and H_k are the known system structure parameters, which are called the $n \times n$ dimensional one-step state update matrix, the $n \times l$ dimensional system noise distribution matrix, and the $m \times n$ dimensional measurement matrix, respectively; W_{k-1} is the l-dimensional system noise vector, and V_k is the m-dimensional measurement noise vector. Both of them are Gaussian noise vector sequences with zero mean value, and are independent of each other:

$$\mathbb{E}[W_k] = 0, \mathbb{E}[W_k W_j^T] = Q_k \delta_{kj} \tag{10}$$

$$\mathbb{E}[V_k] = 0, \mathbb{E}[V_k V_j^T] = R_k \delta_{kj} \tag{11}$$

$$\mathbb{E}[W_k V_j^T] = 0. \tag{12}$$

The above equation is the basic assumption for the noise requirement in the Kalman filtering state space model, where Q_k is a symmetric non-negative definite matrix, and R_k is a symmetric positive definite matrix. δ_{kj} is the Kronecker δ function.

The covariance parameters Q_k and R_k play roles similar to those of Q and R in the continuous filter, but they do not have the same numerical values. Next, the relationship between the corresponding continuous and discrete filter parameters will be derived.

To achieve the transformation from the continuous form to the discrete form, the relations between Q and R and the corresponding Q_k and R_k for a small step size T_s are needed. According to the linear system theory, the relation between Q and Q_k from Equation (3) to Equation (8) is as follows:

$$\Phi_{k|k-1} = \Phi(t_k, t_{k-1}) \approx e^{\int_{t_{k-1}}^{t_k} F(\tau)d\tau} \tag{13}$$

$$\Gamma_{k|k-1}W_{k-1} = \int_{t_{k-1}}^{t_k} \Phi(t_k, \tau)G(\tau)w(t)d\tau. \tag{14}$$

Denote the discrete-time interval as $T_s = t_k - t_{k-1}$, when $F(t)$ does not change too dramatically within the shorter integral interval $[t_{k-1}, t_k]$. Take the Taylor expansion of $e^{F(t_{k-1})T_s}$ with respect to $F(t_{k-1})T_s$ and set $F(t_{k-1})T_s << I$, so the higher-order terms are negligible and the one-step transition matrix, Equation (13), can be approximated as:

$$\Phi_{k|k-1} \approx e^{F(t_{k-1})T_s} = I + F(t_{k-1})T_s + F^2(t_{k-1})\frac{T_s^2}{2!} + F^3(t_{k-1})\frac{T_s^3}{3!} + \cdots \approx I + F(t_{k-1})T_s. \tag{15}$$

Equation (14) shows that $\Gamma_{k|k-1}W_{k-1}$ is the linear transform of the Gaussian white noise $w(\tau)$; the result remains the normal distribution random vector. Therefore, the first- and second-order statistical characteristics can be used to describe and be equivalent to $\Gamma_{k|k-1}W_{k-1}$. Referring to Equation (5), the mean of $\Gamma_{k|k-1}W_{k-1}$ is given as follows:

$$\mathbb{E}[\Gamma_{k|k-1}W_{k-1}] = \mathbb{E}[\int_{t_{k-1}}^{t_k} \Phi(t_k, \tau)G(\tau)w(\tau)d\tau] = \int_{t_{k-1}}^{t_k} \Phi(t_k, \tau)G(\tau)\mathbb{E}[w(\tau)]d\tau = 0. \tag{16}$$

For the second-order statistical characteristics, when $k \neq j$, the time parameter between the noise $w(\tau_k)$ and $w(\tau_j)$ is independent, so $\Gamma_{k|k-1}W_{k-1}$ and $\Gamma_{j|j-1}W_{j-1}$ are uncorrelated:

$$\mathbb{E}[(\Gamma_{k|k-1}W_{k-1})(\Gamma_{j|j-1}W_{j-1})^T] = 0 \quad (k \neq j). \tag{17}$$

When $k = j$, thus

$$\begin{aligned}\mathbb{E}[(\Gamma_{k|k-1}W_{k-1})(\Gamma_{k|k-1}W_{k-1})^T] &= \mathbb{E}\left\{[\int_{t_{k-1}}^{t_k} \Phi(t_k, \tau)G(\tau)w(\tau)d\tau][\int_{t_{k-1}}^{t_k} \Phi(t_k, s)G(s)w(s)ds]^T\right\} \\ &= \mathbb{E}\left\{\int_{t_{k-1}}^{t_k} \Phi(t_k, \tau)G(\tau)w(\tau)\int_{t_{k-1}}^{t_k} w^T(s)G^T(s)\Phi^T(t_k, s)dsd\tau\right\} \\ &= \int_{t_{k-1}}^{t_k} \Phi(t_k, \tau)G(\tau)\int_{t_{k-1}}^{t_k} \mathbb{E}[w(\tau)w^T(s)]G^T(s)\Phi^T(t_k, s)dsd\tau.\end{aligned} \tag{18}$$

Substituting Equation (5) into the above equation:

$$\begin{aligned}\mathbb{E}[(\Gamma_{k|k-1}W_{k-1})(\Gamma_{k|k-1}W_{k-1})^T] &= \int_{t_{k-1}}^{t_k} \Phi(t_k, \tau)G(\tau)\int_{t_{k-1}}^{t_k} Q(t)\delta(\tau - s)G^T(s)\Phi^T(t_k, s)dsd\tau \\ &= \int_{t_{k-1}}^{t_k} \Phi(t_k, \tau)G(\tau)Q(\tau)G^T(\tau)\Phi^T(t_k, \tau)d\tau.\end{aligned} \tag{19}$$

When the noise control matrix $G(\tau)$ changes slowly during the time interval $[t_{k-1}, t_k]$, Equation (19) becomes:

$$
\begin{aligned}
&\mathbb{E}[(\Gamma_{k|k-1}W_{k-1})(\Gamma_{k|k-1}W_{k-1})^T] \\
&\approx \int_{t_{k-1}}^{t_k} [I+F(t_{k-1})(t_k-\tau)]G(t_{k-1})Q(\tau)G^T(t_{k-1})[I+F(t_{k-1})(t_k-\tau)]^T d\tau \\
&= [I+F(t_{k-1})T_s] \cdot [G(t_{k-1})Q(t_{k-1})G^T(t_{k-1})T_s] \cdot [I+F(t_{k-1})T_s]^T \\
&\quad + \frac{1}{12}F(t_{k-1})G(t_{k-1})Q(t_{k-1})G^T(t_{k-1})F(t_{k-1})^T T_s^3 \\
&\approx \{[I+F(t_{k-1})T_s]G(t_{k-1})\} \cdot [Q(t_{k-1})T_s] \cdot \{[I+F(t_{k-1})T_s]G(t_{k-1})\}^T.
\end{aligned}
\tag{20}
$$

When $F(t_{k-1})T_s \ll I$ is satisfied, the above equation can be further approximated:

$$\mathbb{E}[(\Gamma_{k|k-1}W_{k-1})(\Gamma_{k|k-1}W_{k-1})^T] \approx G(t_{k-1}) \cdot [Q(t_{k-1})T_s] \cdot G^T(t_{k-1}). \tag{21}$$

Comparing the result with Equation (10):

$$\Gamma_{k|k-1} \approx [I+F(t_{k-1})T_s]G(t_{k-1}) \approx G(t_{k-1}) \tag{22}$$

$$\mathbb{E}[W_k W_j^T] = Q_k \delta_{kj} = [Q(t_k)T_s]\delta_{kj}. \tag{23}$$

Notice that [29]:

$$Q_k = Q(t_k)T_s. \tag{24}$$

The derivation of the equation relating to R_k and R is more subtle. In the continuous model, $v(t)$ is white, so simple sampling of $Z(t)$ leads to measurement noise with infinite variance. Hence, in the sampling process, we have to imagine averaging the continuous measurement over the T_s interval to get an equivalent discrete sample. This is justified because x is not the Gaussian white noise and can be approximately constant within the interval.

$$Z_k = \frac{1}{T_s}\int_{t_{k-1}}^{t_k} Z(t)dt = \frac{1}{T_s}\int_{t_{k-1}}^{t_k} [H(t)x(t)+v(t)]dt = H(t_k)x_k + \frac{1}{T_s}\int_{t_{k-1}}^{t_k} v(t)dt. \tag{25}$$

Then, the discrete noise matrix and the continuous noise matrix are equivalent:

$$V_k = \frac{1}{T_s}\int_{t_{k-1}}^{t_k} v(t)dt. \tag{26}$$

From Equation (12), we have:

$$
\begin{aligned}
\mathbb{E}[V_k V_j^T] &= R_k \delta_{kj} = \frac{1}{T_s^2}\int_{t_{k-1}}^{t_k}\int_{t_{j-1}}^{t_j} \mathbb{E}[v(\tau)v(s)]d\tau ds \\
&= \frac{1}{T_s^2}\int_{t_{k-1}}^{t_k}\int_{t_{j-1}}^{t_j} R(\tau)\delta(s-\tau)d\tau ds = \frac{1}{T_s^2}\int_{t_{k-1}}^{t_k} R(\tau)\delta_{kj}d\tau \approx \frac{R(t_k)}{T_s}\delta_{kj}.
\end{aligned}
\tag{27}
$$

Comparing it with Equation (6), we have [29]:

$$R_k = \frac{R(t_k)}{T_s}. \tag{28}$$

2.3. Derivation of the Kalman Filter

Assuming that the optimal state estimation at t_{k-1} is $\hat{X}_{k-1}$, the state estimation error is $\tilde{X}_{k-1}$, and the state estimation covariance matrix is Σ_{k-1}:

$$\tilde{X}_{k-1} = X_{k-1} - \hat{X}_{k-1} \tag{29}$$

and

$$\Sigma_{k-1} = \mathbb{E}[\tilde{X}_{k-1}\tilde{X}_{k-1}^T] = \mathbb{E}[(X_{k-1} - \hat{X}_{k-1})(X_{k-1} - \hat{X}_{k-1})^T]. \tag{30}$$

If we take the expectation operator of both sides of Equation (8), we obtain the state one-step prediction and the state one-step estimation error:

$$X_{k|k-1}^- = \mathbb{E}[X_k] = \mathbb{E}[\Phi_{k|k-1}X_{k-1} + \Gamma_{k|k-1}W_{k-1}] = \Phi_{k|k-1}\mathbb{E}[X_{k-1}] = \Phi_{k|k-1}\hat{X}_{k-1}, \tag{31}$$

$$\tilde{X}_{k|k-1} = X_k - X_{k|k-1}^-. \tag{32}$$

Substituting Equations (8) and (31) into Equation (32) leads to:

$$\begin{aligned}\tilde{X}_{k|k-1} &= (\Phi_{k|k-1}X_{k-1} + \Gamma_{k|k-1}W_{k-1}) - \Phi_{k|k-1}\hat{X}_{k-1} \\ &= \Phi_{k|k-1}(X_{k-1} - \hat{X}_{k-1}) + \Gamma_{k|k-1}W_{k-1} = \Phi_{k|k-1}\tilde{X}_{k-1} + \Gamma_{k|k-1}W_{k-1}.\end{aligned} \tag{33}$$

Since $\tilde{X}_{k-1}$ is uncorrelated with W_{k-1}, we therefore obtain the covariance of the state one-step estimation error $\tilde{X}_{k|k-1}$ as follows:

$$\begin{aligned}\Sigma_{k|k-1} &= \mathbb{E}[\tilde{X}_{k|k-1}\tilde{X}_{k|k-1}^T] = \mathbb{E}[(\Phi_{k|k-1}\tilde{X}_{k-1} + \Gamma_{k|k-1}W_{k-1})(\Phi_{k|k-1}\tilde{X}_{k-1} + \Gamma_{k|k-1}W_{k-1})^T] \\ &= \Phi_{k|k-1}\mathbb{E}[\tilde{X}_{k-1}\tilde{X}_{k-1}^T]\Phi_{k|k-1}^T + \Gamma_{k|k-1}\mathbb{E}[W_{k-1}W_{k-1}^T]\Gamma_{k|k-1}^T \\ &= \Phi_{k|k-1}\Sigma_{k-1}\Phi_{k|k-1}^T + \Gamma_{k|k-1}Q_{k-1}\Gamma_{k|k-1}^T.\end{aligned} \tag{34}$$

In a similar way, the measurement at t_k can be predicted by the state one-step estimation prediction $X_{k|k-1}^-$ and system measurement Equation (9) as follows:

$$Z_{k|k-1}^- = \mathbb{E}[H_k X_{k|k-1}^- + V_k] = H_k X_{k|k-1}^-. \tag{35}$$

In fact, there is difference between the measurement one-step prediction $Z_{k|k-1}^-$ and the actual measurement Z_k. The difference is denoted as measurement one-step prediction error:

$$\tilde{Z}_{k|k-1} = Z_k - Z_{k|k-1}^-. \tag{36}$$

Substituting the measurement Equations (9) and (35) into Equation (36) yields:

$$\tilde{Z}_{k|k-1} = Z_k - H_k X_{k|k-1}^- = H_k X_k + V_k - H_k X_{k|k-1}^- = H_k \tilde{X}_{k|k-1} + V_k. \tag{37}$$

In general, the measurement one-step prediction error $\tilde{Z}_{k|k-1}$ is called innovation in the classical Kalman filter theory, and it indicates the new information about the state estimate carried by the measurement one-step prediction error.

On the one hand, if the estimation of X_k only includes the state one-step prediction $X_{k|k-1}^-$ of the system state equation, the estimation accuracy will be low, as no information of the measurement equation has been used. On the other hand, according to Equation (37), the measurement one-step prediction error calculated using the system measurement equation contains the information of the state one-step prediction of $X_{k|k-1}^-$. Consequently, it is natural to consider all the state information that comes from the system state equation and the measurement equation, respectively, and correct the state one-step prediction mean $X_{k|k-1}^-$ with the measurement one-step prediction error $\tilde{Z}_{k|k-1}$. Thereby, the optimal estimation of X_k can be calculated by the combination of $X_{k|k-1}^-$ and $\tilde{Z}_{k|k-1}$ as follows:

$$\hat{X}_k = X_{k|k-1}^- + K_k \tilde{Z}_{k|k-1}, \tag{38}$$

where K_k is the undetermined correction factor matrix.

Substituting Equations (31) and (37) into Equation (38) obtains:

$$\hat{X}_k = X^-_{k|k-1} + K_k(Z_k - H_k X^-_{k|k-1}) = (I - K_k H_k) X^-_{k|k-1} + K_k Z_k \\ = (I - K_k H_k) \Phi_{k|k-1} \hat{X}_{k-1} + K_k Z_k. \quad (39)$$

From Equation (39), the current state estimation $\hat{X}_k$ is a linear combination of the last state estimation $\hat{X}_{k-1}$ and the current measurement Z_k, which considers the influence of the structural parameters $\Phi_{k|k-1}$ in the state equation and the structure parameters H_k in the measurement equation with different types of construction.

The state estimation error at the current time t_k is denoted as:

$$\tilde{X}_k = X_k - \hat{X}_k, \quad (40)$$

where X_k is the true values and $\hat{X}_k$ is the posterior estimation of X_k.

Substituting Equation (39) into Equation (40) obtains:

$$\tilde{X}_k = X_k - [X^-_{k|k-1} + K_k(Z_k - H_k X^-_{k|k-1})] = (X_k - X^-_{k|k-1}) - K_k(H_k X_k + V_k - H_k X^-_{k|k-1}) \\ = \tilde{X}_{k|k-1} - K_k(H_k \tilde{X}_{k|k-1} + V_k) = (I - K_k H_k) \tilde{X}_{k|k-1} - K_k V_k. \quad (41)$$

Then, the mean square error matrix of state estimation $\hat{X}_k$ is given by:

$$\Sigma_k = \mathbb{E}[\tilde{X}_k \tilde{X}_k^T] = \mathbb{E}\{[(I - K_k H_k) \tilde{X}_{k|k-1} - K_k V_k][(I - K_k H_k) \tilde{X}_{k|k-1} - K_k V_k]^T\} \\ = (I - K_k H_k) \mathbb{E}[\tilde{X}_{k|k-1} \tilde{X}^T_{k|k-1}](I - K_k H_k)^T + K_k \mathbb{E}[V_k V_k^T] K_k^T \\ = (I - K_k H_k) \Sigma_{k|k-1} (I - K_k H_k)^T + K_k R_k K_k^T. \quad (42)$$

Substituting Equation (34) into Equation (42) obtains:

$$\Sigma_k = (I - K_k H_k)[\Phi_{k|k-1} \Sigma_{k-1} \Phi^T_{k|k-1} + \Gamma_{k|k-1} Q_{k-1} \Gamma^T_{k|k-1}](I - K_k H_k)^T + K_k R_k K_k^T \\ = \Phi_{k|k-1} \Sigma_{k-1} \Phi^T_{k|k-1} + K_k H_k \Phi_{k|k-1} \Sigma_{k-1} \Phi^T_{k|k-1} H_k^T K_k^T - \Phi_{k|k-1} \Sigma_{k-1} \Phi^T_{k|k-1} H_k^T K_k^T \\ - K_k H_k \Phi_{k|k-1} \Sigma_{k-1} \Phi^T_{k|k-1} + \Gamma_{k|k-1} Q_{k-1} \Gamma^T_{k|k-1} - K_k H_k \Gamma_{k|k-1} Q_{k-1} \Gamma^T_{k|k-1} \\ - \Gamma_{k|k-1} Q_{k-1} \Gamma^T_{k|k-1} H_k^T K_k^T + K_k H_k \Gamma_{k|k-1} Q_{k-1} \Gamma^T_{k|k-1} H_k^T K_k^T + K_k R_k K_k^T. \quad (43)$$

We now use the approximation $\Phi_{k|k-1} \approx I + F(t_{k-1}) T_s$ as Equation (15). From Equation (22) with $\Gamma_{k|k-1} \approx G(t_{k-1})$, we have:

$$\Sigma_k = [I + F(t_{k-1}) T_s] \Sigma_{k-1} [I + F(t_{k-1}) T_s]^T + K_k H_k [I + F(t_{k-1}) T_s] \Sigma_{k-1} [I + F(t_{k-1}) T_s]^T H_k^T K_k^T \\ - [I + F(t_{k-1}) T_s] \Sigma_{k-1} [I + F(t_{k-1}) T_s]^T H_k^T K_k^T - K_k H_k [I + F(t_{k-1}) T_s] \Sigma_{k-1} [I + F(t_{k-1}) T_s]^T \\ + G(t_{k-1}) Q_{k-1} G^T(t_{k-1}) - K_k H_k G(t_{k-1}) Q_{k-1} G^T(t_{k-1}) - G(t_{k-1}) Q_{k-1} G^T(t_{k-1}) H_k^T K_k^T \\ + K_k H_k G(t_{k-1}) Q_{k-1} G^T(t_{k-1}) H_k^T K_k^T + K_k R_k K_k^T. \quad (44)$$

Note from Equation (24) that Q_k is of the order of T_s and from Equation (28) that $R_k = \frac{R(t_k)}{T_s}$; then, Equation (44) becomes:

$$\Sigma_k = [I + F(t_{k-1}) T_s] \Sigma_{k-1} [I + F(t_{k-1}) T_s]^T + K_k H_k [I + F(t_{k-1}) T_s] \Sigma_{k-1} [I + F(t_{k-1}) T_s]^T H_k^T K_k^T \\ - [I + F(t_{k-1}) T_s] \Sigma_{k-1} [I + F(t_{k-1}) T_s]^T H_k^T K_k^T - K_k H_k [I + F(t_{k-1}) T_s] \Sigma_{k-1} [I + F(t_{k-1}) T_s]^T \\ + G(t_{k-1}) Q(t_k) T_s G^T(t_{k-1}) - K_k H_k G(t_{k-1}) Q(t_k) T_s G^T(t_{k-1}) \\ - G(t_{k-1}) Q(t_k) T_s G^T(t_{k-1}) H_k^T K_k^T + K_k H_k G(t_{k-1}) Q(t_k) T_s G^T(t_{k-1}) H_k^T K_k^T + K_k \frac{R(t_k)}{T_s} K_k^T. \quad (45)$$

2.4. The Temporal Derivative of the Rényi Entropy and the Kalman Filter Gain

To obtain the continuous form of covariance matrix Σ, the limit will be taken. However, the relation between the undetermined correction factor matrix K_k and its continuous form still remains unknown. Therefore, we make the following assumption.

Assumption 1. K_k is of the order of T_s, that is:

$$K(t_k) = \frac{K_k}{T_s}. \tag{46}$$

From the conclusion, we can also derive this assumption conversely. We next draw the conclusion as one theorem under the assumption, as follows:

Theorem 1. The discrete form of the undetermined correction factor matrix is the same as the continuous form when the temporal derivative of Rényi entropy is minimized. This can be presented in a mathematical form as follows:

$$\{K_k = \Sigma_k H_k^T R_k, K = \Sigma H^T R^{-1} | K^* = \arg\min_K \dot{H}_R^{(\alpha)}(K)\}. \tag{47}$$

Proof of Theorem 1. We substitute the expression for K_k into Equation (45) and neglect higher-order terms in T_s; Equation (45) becomes:

$$\begin{aligned}
\Sigma_k &= [I + F(t_{k-1})T_s]\Sigma_{k-1}[I + F(t_{k-1})T_s]^T + T_s K(t_k) H_k \Sigma_{k-1}[I + F(t_{k-1})T_s]\Sigma_{k-1} \\
&\quad [I + F(t_{k-1})T_s]^T H_k^T T_s K^T(t_k) - [I + F(t_{k-1})T_s]\Sigma_{k-1}[I + F(t_{k-1})T_s]^T H_k^T T_s K^T(t_k) \\
&\quad - T_s K(t_k) H_k[I + F(t_{k-1})T_s]\Sigma_{k-1}[I + F(t_{k-1})T_s]^T + G(t_{k-1})Q(t_k)T_s G^T(t_{k-1}) \\
&\quad - T_s K(t_k) H_k G(t_{k-1})Q(t_k)T_s G^T(t_{k-1}) - G(t_{k-1})Q(t_k)T_s G^T(t_{k-1})H_k^T T_s K^T(t_k) \\
&\quad + T_s K(t_k) H_k G(t_{k-1})Q(t_k)T_s G^T(t_{k-1})H_k^T T_s K^T(t_k) + T_s K(t_k)\frac{R_k}{T_s}T_s K^T(t_k) \\
&= \Sigma_{k-1} + T_s F(t_{k-1})\Sigma_{k-1} + T_s \Sigma_{k-1} F^T(t_{k-1}) - \Sigma_{k-1} H_k^T T_s K(t_k)^T - T_s K(t_k) H_k \Sigma_{k-1} \\
&\quad + G(t_{k-1})Q(t_k)T_s G^T(t_{k-1}) + T_s K(t_k)\frac{R(t_k)}{T_s}T_s K^T(t_k).
\end{aligned} \tag{48}$$

Moving the first term of Equation (48) from right to left and dividing both sides by T_s to form the finite difference expression:

$$\begin{aligned}
\frac{\Sigma_k - \Sigma_{k-1}}{T_s} &= F(t_{k-1})\Sigma_{k-1} + \Sigma_{k-1}F^T(t_{k-1}) - \Sigma_{k-1}H_k^T K(t_k)^T - K(t_k)H_k \Sigma_{k-1} \\
&\quad + G(t_{k-1})Q(t_k)G^T(t_{k-1}) + K(t_k)R(t_k)K^T(t_k).
\end{aligned} \tag{49}$$

Finally, passing to the limit as $T_s \to 0$ and dropping of the subscripts lead to the matrix differential equation:

$$\dot{\Sigma} = F\Sigma + \Sigma F^T - \Sigma H^T K^T - KH\Sigma + GQG^T + KRK^T. \tag{50}$$

Σ is invertible, as it is a positive matrix. Multiplying Σ^{-1} with Equation (50), we can consider the temporal derivative of the Rényi entropy of the mean square error matrix Σ using Equation (2):

$$\begin{aligned}\dot{H}_R^{(\alpha)} &= \frac{1}{2}Tr\{\Sigma^{-1}\dot{\Sigma}\} \\ &= \frac{1}{2}Tr\{\Sigma^{-1}F\Sigma + F^T - H^T K^T - \Sigma^{-1}KH\Sigma + \Sigma^{-1}GQG^T + \Sigma^{-1}KRK^T\} \\ &= \frac{1}{2}Tr\{F + F^T - H^T K^T - KH + \Sigma^{-1}GQG^T + \Sigma^{-1}KRK^T\} \\ &= \frac{1}{2}Tr\{2F - 2KH + \Sigma^{-1}GQG^T + \Sigma^{-1}KRK^T\},\end{aligned} \quad (51)$$

where the invariance under the cyclic permutation property of the trace operator has been used to eliminate Σ^{-1} and Σ, as well as the truth that $Tr(F) = Tr(F^T)$ has been used to simplify the formula.

It is obvious that Equation (51) is a quadratic function of the undetermined correction factor matrix K. Thereby, there must be a minimum of $\dot{H}_R^{(\alpha)}(x)$ in a probabilistic sense. Taking the derivative of both sides of Equation (51) with respect to matrix K obtains:

$$\begin{aligned}\frac{\partial}{\partial K}\dot{H}_R^{(\alpha)} &= -2\frac{\partial Tr(KH)}{\partial K} + \frac{\partial Tr(\Sigma^{-1}KRK^T)}{\partial K} \\ &= -2H^T + \frac{Tr(\Sigma^{-1}KR(\partial K)^T)}{\partial K} + \frac{Tr(\Sigma^{-1}(\partial K)RK^T)}{\partial K} \\ &= -2H^T + \Sigma^{-1}KR + (RK^T\Sigma^{-1})^T.\end{aligned} \quad (52)$$

In addition, since Σ^{-1} and R_k are symmetric matrices, the result is:

$$\frac{\partial}{\partial K}\dot{H}_R^{(\alpha)} = -2H^T + 2\Sigma^{-1}KR. \quad (53)$$

R_k is invertible, as it is a positive matrix. According to the extreme value principle of the function, when the above are equal to zero, then we have:

$$K = \Sigma H^T R^{-1}. \quad (54)$$

So far, we have found the analytic solution to the undetermined correction factor matrix K, which is called the continuous-time Kalman filter gain in the classical Kalman filter. Then, the recursive formulations of the Kalman filter can be established through the Kalman filter gain K. Most importantly, this implies the connection between the temporal derivative of Rényi entropy and the classical Kalman filter: The temporal derivative of the Rényi entropy is minimized when the Kalman filter gain satisfies Equation (54).

Looking back to Assumption 1 and substituting Equation (28) into Equation (54), we obtain:

$$K(t_k) = \frac{K_k}{T_s} = K = \Sigma H^T R^{-1} = \Sigma_k H_k^T R_k(T_s) = \frac{\Sigma_k H_k^T R_k}{T_s}. \quad (55)$$

Therefore, the discrete-time Kalman filter gain can be expressed as follows:

$$K_k = \Sigma_k H_k^T R_k. \quad (56)$$

□

Remark 1. *The discrete-time Kalman filter gain has the same form as the continuous-time filter gain, as shown in the Equation (54). In principle, this is consistent with our intuition and proves the correctness and rationality of Assumption 1, in turn.*

Remark 2. *The Kalman filter gain is equivalent to the minimization of the temporal derivative of the Rényi entropy, although it has the same result as the original Kalman filter, which is deduced under the minimum mean square error criterion.*

Substituting Equation (54) into Equation (50), we have:

$$\begin{aligned}\dot{\Sigma} &= F\Sigma + \Sigma F^T - \Sigma H^T K^T - \Sigma H^T R^{-1} H\Sigma + GQG^T + \Sigma H^T R^{-1} R K^T \\ &= F\Sigma + \Sigma F^T - \Sigma H^T R^{-1} H\Sigma + GQG^T.\end{aligned} \quad (57)$$

This is a second-order nonlinear differential equation with respect to the mean square error matrix Σ, and it is commonly called the Riccati equation. This is the same result as that of the Bucy–Kalman filter [7].

If the system equation, Equation (3), and the measurement equation, Equation (4), form a linear time-invariant system with constant noise covariance, the mean square error matrix Σ may reach a steady-state value, and $\dot{\Sigma}$ may eventually reach zero. So, we have the continuous algebraic Riccati equation as follows:

$$\dot{\Sigma} = F\Sigma + \Sigma F^T - \Sigma H^T R^{-1} H\Sigma + GQG^T = 0. \quad (58)$$

As we can see, the time derivative of covariance at the steady state is zero; then, the temporal derivative of the Rényi entropy should also be zero:

$$\dot{H}_R^{(\alpha)} = 0. \quad (59)$$

This implies that when the system approaches a stable state, the Rényi entropy approaches a steady value so that the temporal derivative of the Rényi entropy is zero. This is reasonable when the steady system owns a constant Rényi entropy, as uncertainty is stable, which follows our intuitive understanding. Consequently, it is worth noting that whether the value of the Rényi entropy is stable or not can be a validated indicator of whether the system is approaching the steady state.

3. Simulations and Analysis

In this section, we give two experiments to show that when the nonlinear filter system approaches the steady state, the Rényi entropy of the system approaches stability. The first experiment is a numerical example of a falling body in noisy conditions, tracked by radar [30] using the UKF. The second experiment is a practical experiment of loosely coupled integration [29]. The simulations were carried out on MATLAB 2018a running on a computer with i5-5200U, 2.20 GHz CPU, and the graphs were plotted by MATLAB.

3.1. Falling Body Tracking

In the example of a falling body being tracked by radar, the body falls vertically. The radar is placed at a vertical distance L from the body, and the radar measures the distance y from the radar to the body. The state-space equation of the body is given by:

$$\begin{aligned}\dot{x}_1 &= x_2 \\ \dot{x}_2 &= d + g \\ \dot{x}_3 &= 0,\end{aligned} \quad (60)$$

where x_1 is the height, x_2 is the velocity, x_3 is the ballistic coefficient, $g = -9.81$ m/s^2 is the gravity acceleration, and d is the air drag, which could be approximated as:

$$d = \frac{\rho x_2^2}{2x_3} = \rho_0 \exp(-\frac{x_1}{k})\frac{x_2^2}{2x_3}, \quad (61)$$

where ρ is the air density with an initial value of $\rho_0 = 1.225$; $\rho_0 = 1.225$ and $k = 6705.6$ are constants.

The measurement equation is:

$$y = \sqrt{L^2 + x_1^2}. \quad (62)$$

It is worth noting that the drag and the square root cause severely nonlinearity in the state-space function and measurement function, respectively.

The discrete-time nonlinear system can be given by the Euler discretization method. Combining the additive process with Gaussian white noises for measurement, we can obtain:

$$\begin{aligned} x_1(n+1) &= x_1(n) + x_2(n) \cdot T + w_1(n) \\ x_2(n+1) &= x_2(n) + (d+g) \cdot T + w_2(n) \\ x_3(n+1) &= x_3(n) + w_3(n) \end{aligned} \tag{63}$$

$$y(n) = \sqrt{L^2 + x_1^2(n)} + v(n). \tag{64}$$

In the UKF numerical experiment, we set the sampling period to $T = 0.4$ s, the horizontal distance to $L = 100$ m, the maximum number of samples to $N = 100$, the process noise to $S_w = diag(10^5, 10^3, 10^2)$, the measurement noise to $S_v = 10^6$, and the initial state to $x = [10^5; -5000; 400]$. The results are shown as follows:

Figure 1 shows the evolution of covariance matrix Σ. Figures 2 and 3 show the Rényi entropy of covariance matrix Σ and its change in adjacent time, respectively. Notice that the uncertainty increases near the middle of the plots, which is coincident with the drag peak. However, the Rényi entropy fluctuates around 15; even the fourth element of Σ changes dramatically. Of course, the entropy changes are closely accompanied by the drag peak, which means the change of the entropy of covariance reflects the evolution of matrix Σ. Consequently, the Rényi entropy can be viewed as the indicator of whether the system is approaching the steady state or not.

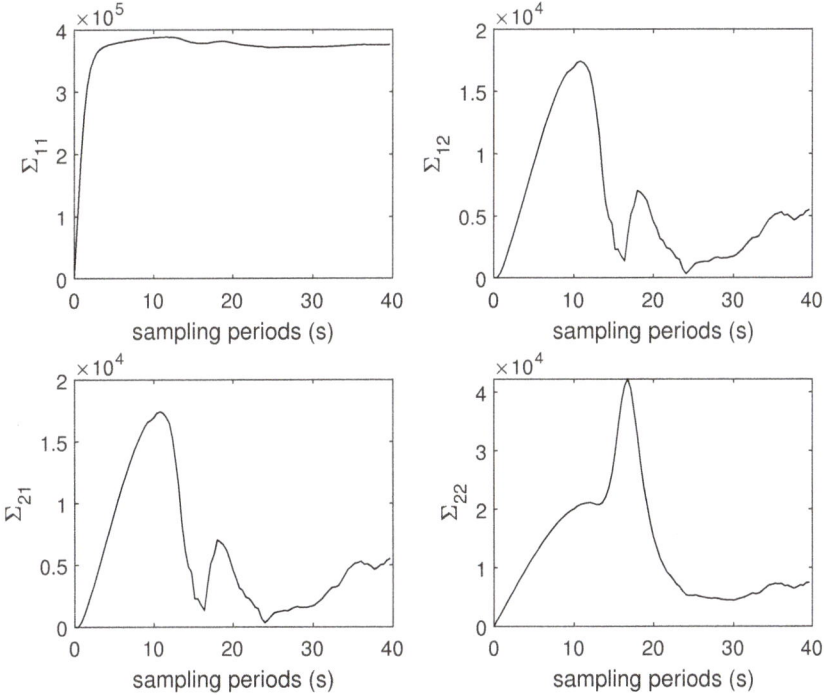

Figure 1. Evolution of matrix Σ.

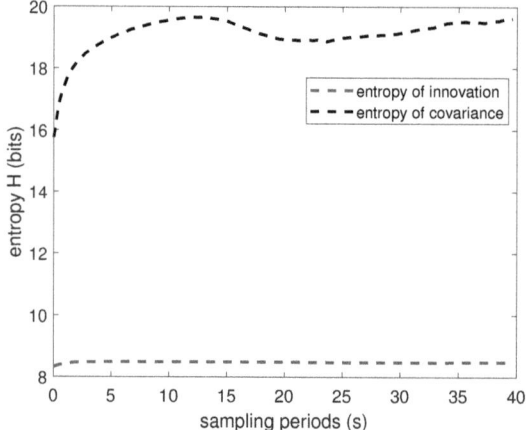

Figure 2. Simulation results for the entropy.

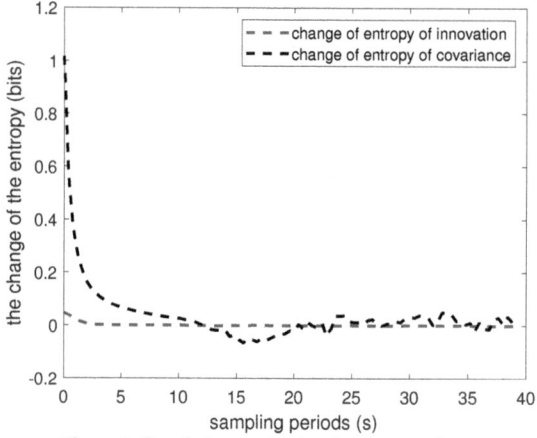

Figure 3. Simulation results for the change of entropy.

3.2. Practical Integrated Navigation

In the loosely integrated navigation system, the system state parameter x is composed of inertial navigation system (INS) error states in the North–East–Down (NED) local-level navigation frame, and can be expressed as follows:

$$x = [(\delta r^n)^T \quad (\delta v^n)^T \quad (\psi)^T \quad (b_g)^T \quad (b_a)^T]^T, \tag{65}$$

where δr^n, δv^n, and ψ represent the position error, the velocity error, and the attitude error, respectively; b_g and b_a are modeled as first-order Gauss–Markov processes, representing the gyroscope bias and the accelerometer bias, respectively.

The discrete-time state update equation is used to update state parameters as follows:

$$x_k = \Phi_{k|k-1} x_{k-1} + G_{k|k-1} w_{k-1}, \tag{66}$$

where $G_{k|k-1}$ is the system noise matrix, w_{k-1} is the system noise, and $\Phi_{k|k-1}$ is the state transition matrix from t_{k-1} to t_k; this is determined by the dynamic model of the state parameter.

In the loosely coupled integration, the measurement equation can be simply expressed as:

$$\delta z = H_k x_k + v_k, \qquad (67)$$

where v_k is the measurement noise, H_k is the measurement matrix, and z_k is the measurement vector calculated by subtracting the global navigation satellite system (GNSS) observation with the inertial navigation system (INS) mechanism.

The experiments reported in this section were carried out by processing the data from an unmanned ground vehicle test. The gyroscope random walk was set to $0.03 \text{ deg}/\sqrt{h}$ and the velocity random walk was set to $0.05 \text{ m/s}/\sqrt{h}$. The sampling rates of the inertial measurement unit (IMU) and the GNSS are 200 Hz and 1 Hz, respectively. The test lasts 48 min.

The position error curve, velocity error curve, and attitude error curve of the loosely coupled integration are shown in Figures 4–6. The root mean squares (RMSs) of the position errors in the north, east, and earth directions are 0.0057 m, 0.0024 m, and 0.0134 m, respectively. The RMS of the velocity errors in the north, east, and earth directions are 0.0023 m/s, 0.0021 m/s, and 0.0038 m/s, respectively. The RMSs of the attitude errors in the roll, pitch, and yaw directions are 0.0034 deg, 0.0030 deg, and 0.0178 deg, respectively.

The Rényi entropy of the covariance P is shown in Figure 7. As we can see, the Rényi entropy fluctuates around -100 once the filter converges, which is consistent with the conclusion from the entropy perspective.

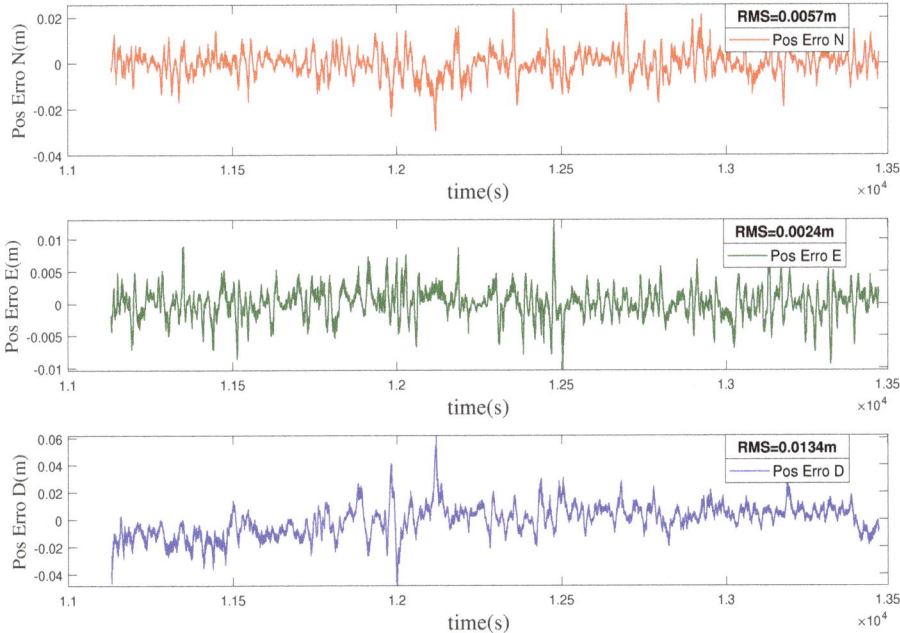

Figure 4. Position error of the loosely coupled integration.

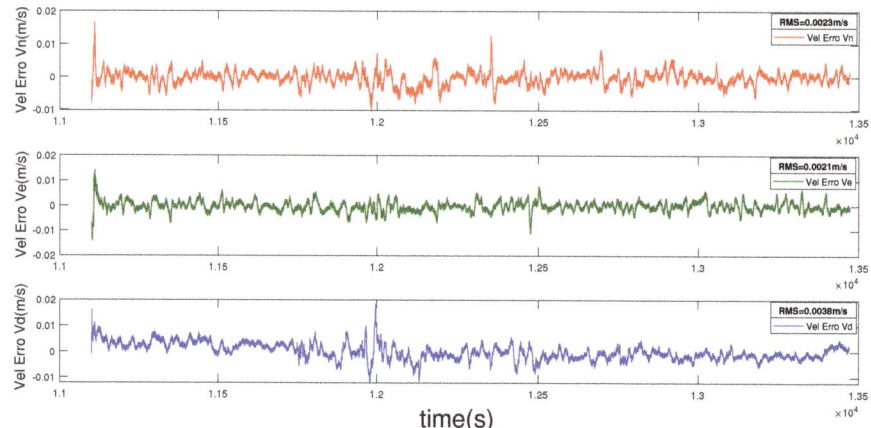

Figure 5. Velocity error of the loosely coupled integration.

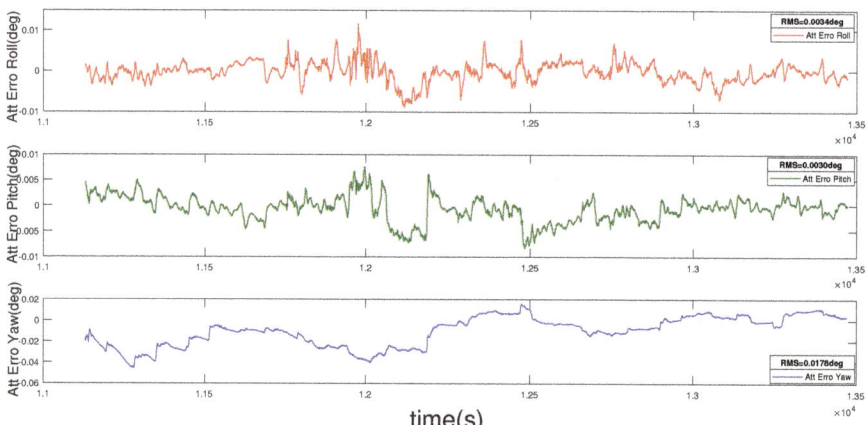

Figure 6. Attitude error of the loosely coupled integration.

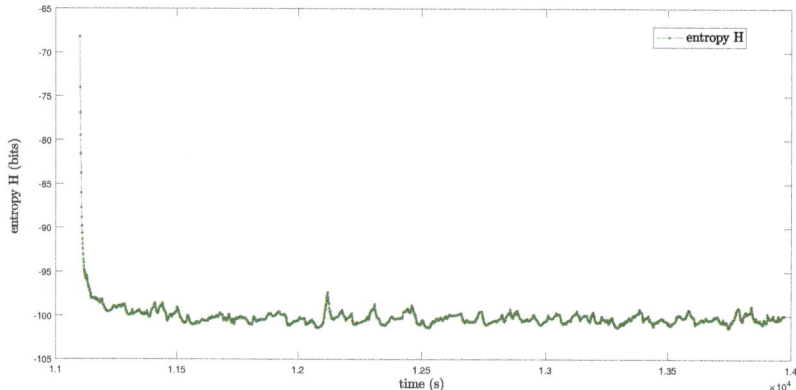

Figure 7. Rényi entropy of the covariance Σ.

4. Conclusions and Final Remarks

We have considered the original Kalman filter by taking the minimization of the temporal derivative of the Rényi entropy. In particular, we show that the temporal derivative of Rényi entropy is equal to zero when the Kalman filter system approaches the steady state, which means that the Rényi entropy approaches a stable value. Finally, simulation experiments and practical experiments show the Rényi entropy truly stays stable when the system becomes steady.

Future work includes calculating the Rényi entropy of the innovation term when the measurements and the noise are non-Gaussian [14] in order to evaluate the effectiveness of measurements and adjust the noise covariance matrix. Meanwhile, we can also calculate the Rényi entropy of the nonlinear dynamical equation to measure the nonlinearity in the propagation step.

Author Contributions: Conceptualization, Y.L. and C.G.; Funding acquisition, C.G. and J.L.; Investigation, Y.L.; Methodology, Y.L., C.G., and S.Y.; Project administration, J.L.; Resources, C.G.; Software, Y.L. and S.Y.; Supervision, J.L.; Validation, S.Y.; Visualization, S.Y.; Writing—original draft, Y.L.; Writing—review and editing, C.G., S.Y., and J.L. All authors have read and agreed to the published version of the manuscript.

Funding: This research was supported by a grant from the National Key Research and Development Program of China (2018YFB1305001).

Acknowledgments: In this section you can acknowledge any support given which is not covered by the author contribution or funding sections. This may include administrative and technical support, or donations in kind (e.g., materials used for experiments).

Conflicts of Interest: The authors declare no conflict of interest.

References

1. Shannon, C.E. A mathematical theory of communication. *Bell Syst. Tech. J.* **1948**, *27*, 379–423. [CrossRef]
2. Principe, J.C. *Information Theoretic Learning: Renyi's Entropy and Kernel Perspectives*; Springer Science & Business Media: Berlin, Germany, 2010.
3. He, R.; Hu, B.; Yuan, X.; Wang, L. *Robust Recognition via Information Theoretic Learning*; Springer International Publishing: Berlin, Germany, 2014.
4. Rényi, A. On measures of entropy and information. In Proceedings of the Fourth Berkeley Symposium on Mathematical Statistics and Probability, Berkeley, CA, USA, 20 June–30 July 1961.
5. Liang, X.S. Entropy evolution and uncertainty estimation with dynamical systems. *Entropy* **2014**, *16*, 3605–3634. [CrossRef]
6. Kullback, S.; Leibler, R.A. On information and sufficiency. *Ann. Math. Stat.* **1951**, *22*, 79–86. [CrossRef]
7. Kalman, R.E.; Bucy, R.S. New results in linear filtering and prediction theory. *J. Basic Eng.* **1961**, *83*, 95–108. [CrossRef]
8. DeMars, K.J. Nonlinear Orbit Uncertainty Prediction and Rectification for Space Situational Awareness. Ph.D. Thesis, The University of Texas at Austin, Austin, TX, USA, 2010.
9. DeMars, K.J.; Bishop, R.H.; Jah, M.K. Entropy-based approach for uncertainty propagation of nonlinear dynamical systems. *J. Guid. Control. Dyn.* **2013**, *36*, 1047–1057. [CrossRef]
10. Kim, H.; Liu, B.; Goh, C.Y.; Lee, S.; Myung, H. Robust vehicle localization using entropy-weighted particle filter-based data fusion of vertical and road intensity information for a large scale urban area. *IEEE Robot. Autom. Lett.* **2017**, *2*, 1518–1524. [CrossRef]
11. Zhang, J.; Du, L.; Ren, M.; Hou, G. Minimum error entropy filter for fault detection of networked control systems. *Entropy* **2012**, *14*, 505–516. [CrossRef]
12. Liu, Y.; Wang, H.; Hou, C. UKF based nonlinear filtering using minimum entropy criterion. *IEEE Trans. Signal Process.* **2013**, *61*, 4988–4999. [CrossRef]
13. Julier, S.; Uhlmann, J.; Durrant-Whyte, H.F. A new method for the nonlinear transformation of means and covariances in filters and estimators. *IEEE Trans. Autom. Control* **2000**, *45*, 477–482. [CrossRef]
14. Contreras-Reyes, J.E.; Cortés, D.D. Bounds on rényi and shannon entropies for finite mixtures of multivariate skew-normal distributions: Application to swordfish (xiphias gladius linnaeus). *Entropy* **2016**, *18*, 382. [CrossRef]

15. Ren, M.; Zhang, J.; Fang, F.; Hou, G.; Xu, J. Improved minimum entropy filtering for continuous nonlinear non-Gaussian systems using a generalized density evolution equation. *Entropy* **2013**, *15*, 2510–2523. [CrossRef]
16. Zhang, Q. Performance enhanced Kalman filter design for non-Gaussian stochastic systems with data-based minimum entropy optimisation. *AIMS Electron. Electr. Eng.* **2019**, *3*, 382. [CrossRef]
17. Chen, B.; Dang, L.; Gu, Y.; Zheng, N.; Príncipe, J.C. Minimum error entropy Kalman filter. *IEEE Trans. Syst. Man Cybern. Syst.* **2019**. [CrossRef]
18. Gultekin, S.; Paisley, J. Nonlinear Kalman filtering with divergence minimization. *IEEE Trans. Signal Process.* **2017**, *65*, 6319–6331. [CrossRef]
19. Darling, J.E.; DeMars, K.J. Minimization of the kullback–leibler divergence for nonlinear estimation. *J. Guid. Control Dyn.* **2017**, *40*, 1739–1748. [CrossRef]
20. Morelande, M.R.; Garcia-Fernandez, A.F. Analysis of Kalman filter approximations for nonlinear measurements. *IEEE Trans. Signal Process.* **2013**, *61*, 5477–5484. [CrossRef]
21. Raitoharju, M.; García-Fernández, Á.F.; Piché, R. Kullback–Leibler divergence approach to partitioned update Kalman filter. *Signal Process.* **2017**, *130*, 289–298. [CrossRef]
22. Hu, E.; Deng, Z.; Xu, Q.; Yin, L.; Liu, W. Relative entropy-based Kalman filter for seamless indoor/outdoor multi-source fusion positioning with INS/TC-OFDM/GNSS. *Clust. Comput.* **2019**, *22*, 8351–8361. [CrossRef]
23. Yu, W.; Peng, J.; Zhang, X.; Li, S.; Liu, W. An adaptive unscented particle filter algorithm through relative entropy for mobile robot self-localization. *Math. Probl. Eng.* **2013**. [CrossRef]
24. Arasaratnam, I.; Haykin, S. Cubature kalman filters. *IEEE Trans. Autom. Control* **2009**, *54*, 1254–1269. [CrossRef]
25. Kiani, M.; Barzegar, A.; Pourtakdoust, S.H. Entropy-based adaptive attitude estimation. *Acta Astronaut.* **2018**, *144*, 271–282. [CrossRef]
26. Giffin, A.; Urniezius, R. The Kalman filter revisited using maximum relative entropy. *Entropy* **2014**, *16*, 1047–1069. [CrossRef]
27. Chen, B.; Liu, X.; Zhao, H.; Principe, J.C. Maximum correntropy Kalman filter. *Automatica* **2017**, *76*, 70–77. [CrossRef]
28. Chen, B.; Xing, L.; Liang, J.; Zheng, N.; Principe, J.C. Steady-state mean-square error analysis for adaptive filtering under the maximum correntropy criterion. *IEEE Signal Process. Lett.* **2014**, *21*, 880–884.
29. Gongmin, Y.; Jun, W. *Lectures on Strapdown Inertial Navigation Algorithm and Integrated Navigation Principles*; Northwestern Polytechnical University Press: Xi'an, China, 2019.
30. Kumari, L.; Padma Raju, K. Application of Extended Kalman filter for a Free Falling body towards Earth. *IJACSA Ed.* **2011**, *2*, 4. [CrossRef]

© 2020 by the authors. Licensee MDPI, Basel, Switzerland. This article is an open access article distributed under the terms and conditions of the Creative Commons Attribution (CC BY) license (http://creativecommons.org/licenses/by/4.0/).

Article

Application of Cloud Model in Qualitative Forecasting for Stock Market Trends

Oday A. Hassen [1], Saad M. Darwish [2,*], Nur A. Abu [3] and Zaheera Z. Abidin [3]

1. Ministry of Education, Wasit Education Directorate, Kut 52001, Iraq; odayali@uowasit.edu.iq
2. Department of Information Technology, Institute of Graduate Studies and Research, Alexandria University, 163 Horreya Avenue, El–Shatby, Alexandria 21526, Egypt
3. Faculty of Information and Communication Technology, University Teknikal Malaysia Melaka, Melaka 76100, Malaysia; nura@utem.edu.my (N.A.A.); zaheera@utem.edu.my (Z.Z.A.)
* Correspondence: saad.darwish@alexu.edu.eg; Tel.: +20-122-263-2369

Received: 23 July 2020; Accepted: 26 August 2020; Published: 6 September 2020

Abstract: Forecasting stock prices plays an important role in setting a trading strategy or determining the appropriate timing for buying or selling a stock. The use of technical analysis for financial forecasting has been successfully employed by many researchers. The existing qualitative based methods developed based on fuzzy reasoning techniques cannot describe the data comprehensively, which has greatly limited the objectivity of fuzzy time series in uncertain data forecasting. Extended fuzzy sets (e.g., fuzzy probabilistic set) study the fuzziness of the membership grade to a concept. The cloud model, based on probability measure space, automatically produces random membership grades of a concept through a cloud generator. In this paper, a cloud model-based approach was proposed to confirm accurate stock based on Japanese candlestick. By incorporating probability statistics and fuzzy set theories, the cloud model can aid the required transformation between the qualitative concepts and quantitative data. The degree of certainty associated with candlestick patterns can be calculated through repeated assessments by employing the normal cloud model. The hybrid weighting method comprising the fuzzy time series, and Heikin–Ashi candlestick was employed for determining the weights of the indicators in the multi-criteria decision-making process. Fuzzy membership functions are constructed by the cloud model to deal effectively with uncertainty and vagueness of the stock historical data with the aim to predict the next open, high, low, and close prices for the stock. The experimental results prove the feasibility and high forecasting accuracy of the proposed model.

Keywords: cloud model; fuzzy time series; stock trend; Heikin–Ashi candlestick

1. Introduction

Forecasting stock prices is an attractive pursuit for investors and researchers who want to beat the stock market. The benefits of having a good estimation of the stock market behavior are well-known, minimizing the risk of investment and maximizing profits. Recently, the stock market has become an easily accessible investment tool, not only for strategic investors, but also for ordinary people. Over the years, investors and researchers have been interested in developing and testing models of stock price behavior. However, analyzing stock market movements and price behaviors is extremely challenging because of the market's dynamic, nonlinear, non–stationary, nonparametric, noisy, and chaotic nature [1]. Stock markets are affected by many highly interrelated uncertain factors that include economic, political, psychological, and company-specific variables. These uncertain factors are undesirable for the stock investor and make stock price prediction very difficult, but at the same time, they are also unavoidable whenever stock trading is preferred as an investment tool [1,2]. To invest in stocks and achieve high profits with low risks, investors have used technical and fundamental analysis as two major approaches in decision-making in financial markets [2].

Fundamental analysis studies all of the factors that have an impact on the stock price of the company in the future such as financial statements, management processes, industry, etc. It analyzes the intrinsic value of the firm to identify whether the stock is underpriced or overpriced. On the other hand, technical analysis uses past charts, patterns, and trends to forecast the price movements of the entity in the coming time [2,3]. The main weakness of fundamental analysis is that it is time-consuming as people cannot quickly locate and absorb the information needed to make thoughtful stock picks. People's judgments are subjective, as is their definition of fair value. The second drawback of a fundamental analysis is in relation to the efficient market hypothesis. Since all information about stocks is public knowledge—barring illegal insider information—stock prices reflect that knowledge.

A major advantage of technical analysis is its simple logic and application. It is seen in the fact that it ignores all economic, market, technological, and any other factors that may have an impact on the company and the industry and only focuses on the data on prices and the volume traded to estimate future prices. The second advantage of technical analysis is that it excludes the subjective aspects of certain companies such as the analyst's personal expectations [4]. However, technical analysis may get an investor trapped: when price movements are artificially created to lure an investor into the stock and once enough investors are entered, they start selling, and you may be trapped. Furthermore, it is too reliant on mathematics and patterns in the chart of the stock and ignores the underlying reasons or causes of price movements. As a result, the stock movements are too wild to handle or predict through technical analysis.

There exist two types of forecasting techniques to be implemented [5,6]: (a) qualitative forecasting models; and (b) quantitative forecasting models. The qualitative forecasting models are generally subjective in nature and are mostly based on the opinions and judgments of experts. Such types of methods are generally used when there is little or no past data available that can be used to base the forecast. Hence, the outcome of the forecast is based upon the knowledge of the experts regarding the problem. On the other hand, quantitative forecasting models make use of the data available to make predictions into the future. The model basically sums up the interesting patterns in the data and presents a statistical association between the past and current values of the variable. Management can use qualitative inputs in conjunction with quantitative forecasts and economic data to forecast sales trends. Qualitative forecasting is useful when there is ambiguous or inadequate data. The qualitative method of forecasting has certain disadvantages such as anchoring events and selective perception. Qualitative forecasts enable a manager to decrease some of this uncertainty to develop plans that are fairly accurate, but still inexact. However, the lack of precision in the development of a qualitative forecast versus a quantitative forecast ensures that no single qualitative technique produces an accurate forecast every time [2,4,7–10].

In nearly two decades, the fuzzy time series approach has been widely used for its superiorities in dealing with imprecise knowledge (like linguistic) variables in decision making. In the process of forecasting with fuzzy time series models, the fuzzy logical relationship is one of the most critical factors that influence the forecasting accuracy. Many studies seek to deploy neuro-fuzzy inference to the stock market in order to deal with probability. Fuzzy logic is known to be useful for decision-making where there is a great deal of uncertainty as well as vague phenomena, but lacks the learning capability; on the other hand, neural networks are useful in constructing an adaptive system that can learn from historical data, but are not able to process ambiguous rules and probabilistic datasets. It is tedious to develop fuzzy rules and membership functions and fuzzy outputs can be interpreted in a number of ways, making analysis difficult. In addition, it requires a lot of data and expertise to develop a fuzzy system.

Recently, a probabilistic fuzzy set was suggested for forecasting by introducing probability theory into a fuzzy set framework. It changes the secondary MF of type 2 fuzzy into the probability density function (PDF), so it is able to capture the random uncertainties in membership degree. It has the ability to capture uncertainties with fuzzy and random nature. However, the membership functions are difficult to obtain for existing fuzzy approaches of measurement uncertainty. In order to conquer this disadvantage, the cloud model was used to calculate the measurement uncertainty. A cloud is a new, easily visualized concept for uncertainty with well-defined semantics, mediating between the concept of a fuzzy set and that of a probability distribution [11–16]. A cloud model is an effective tool in transforming qualitative concepts and their quantitative expressions. The digital characteristics of cloud, expect value (*Ex*), entropy (*En*), and hyper–entropy (*He*), well integrate the fuzziness and randomness of linguistic concepts in a unified way. Cloud is combined with several cloud drops in which the shape of the cloud reflects the important characters of the quantity concept [17]. The essential difference between the cloud model and the fuzzy probability concept lies in the used method to calculate a random membership degree. Basically, with the three numerical characteristics, the cloud model can randomly generate a degree of membership of an element and implement the uncertain transformation between linguistic concepts and its quantitative instantiations.

Candlestick patterns provide a way to understand which buyer and seller groups currently control the price action. This information is visually represented in the form of different colors on these charts. Recently, several traders and investors have used the traditional Japanese candlestick chart pattern and analyzed the pattern visually for both quantitative and qualitative forecasting [6–10]. Heikin–Ashi candlesticks are an offshoot from Japanese candlesticks. Heikin–Ashi candlesticks use the open–close data from the prior period and the open–high–low–close data from the current period to create a combo candlestick. The resulting candlestick filters out some noise in an effort to better capture the trend.

1.1. Problem Statement

The price variation of the stock market is a non–linear dynamic system that deals with non–stationary and volatile data. This is the reason why its modeling is not a simple task. In fact, it is regarded as one of the most challenging modeling problems due to the fact that prices are stochastic. Hence, the best way to predict the stock price is to reduce the level of uncertainty by analyzing the movement of the stock price. The main motivation of our work was the successful prediction of stock future value that can yield enormous capital profits and can avoid potential market risk. Several classical approaches have been evolved based on linear time series models, but the patterns of the stock market are not linear. These approaches lead to inaccurate results, which may be susceptible to highly dynamic factors such as macroeconomic conditions and political events. Moreover, the existing qualitative based methods developed based on fuzzy reasoning techniques cannot describe the data comprehensively, which has greatly limited the objectivity of fuzzy time series in uncertain data forecasting. The most important disadvantage of the fuzzy time series approach is that it needs subjective decisions, especially in the fuzzification stage.

1.2. Contribution and Novelty

The objective of the work presented in this paper is to construct an accurate stock trend prediction model through utilizing a combination of the cloud model, Heikin–Ashi candlesticks, and fuzzy time series (FTS) in a unified model. The purpose of the cloud model is to add the randomness and uncertainty to the fuzziness linguistic definition of Heikin–Ashi candlesticks. FTS is utilized to abstract linguistic values from historical data, instead of numerical ones, to find internal relationship rules. Heikin–Ashi candlesticks were employed to give easier readability of the candle's features through the reduction of noise, eliminates the gaps between candles, and smoothens the movement of the market.

As far as the authors know, this is the first time that the cloud model has been used in forecasting stock market trends that is unlike the current methods that adopt a fuzzy probability approach for forecasting that requires an expert to define the extra parameters of the probabilistic fuzzy system such as output probability vector in probabilistic fuzzy rules and variance factor. These selected statistical parameters specify the degree of randomness. The cloud model not only focuses on the studies regarding the distribution of samples in the universe, but also try to generalize the point–based membership to a random variable on the interval [0, 1], which can give a brand new method to study the relationship between the randomness of samples and uncertainty of membership degree. More practically speaking, the degree with the aid of three numeric characteristics, by which the transformation between linguistic concepts and numeric values will become possible.

The outline of the remainder of this paper is as follows. Section 2 presents the background and summary of the state-of-the-art approaches. Section 3 describes the proposed model. The test results and discussion of the meaning are shown in Section 4. The conclusion of this work is given in Section 5.

2. Preliminaries and Literature Review

In this section, we summarize material that we need later that includes the cloud model, fuzzy time series, and Heikin–Ashi candlesticks. Finally, some state-of-the-art related works are discussed.

2.1. Cloud Model

The cloud model (CM) proposed by Li et al. [17] relies on probability statistics and traditional fuzzy theory [18,19]. The membership cloud model as shown in Figure 1 can mix the fuzziness and randomness to objectively describe the uncertainty of the complex system. This model makes it possible to obtain the range and the distribution of the quantitative data from qualitative information, which is described by linguistic value and effectively transits precise data into appropriate qualitative language value. The digital character of the cloud can be expressed by expected value (Ex), entropy (En), and hyper entropy (He). CM uses Ex to represent the qualitative concept and usually is the value of x corresponding to the cloud center. En represents the uncertainty measure of the qualitative concept. It measures the ambiguity of the quantitative numerical range. He symbols the uncertainty measure of entropy, namely the entropy of entropy, which reflects the dispersion degree of cloud, which appears in the size of the cloud's thickness [17–21].

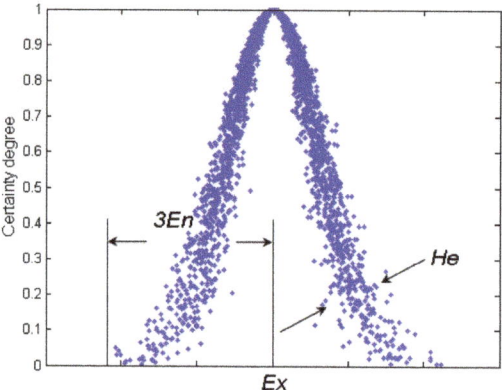

Figure 1. Cloud model.

The theoretical foundation of CM is the probability measure (i.e., the measure function in the sense of probability). On the basis of normal distribution and Gaussian membership function, CMs describe the vagueness of the membership degree of an element by a random variable defined in the universe. Being an uncertain transition way between a qualitative concept described by linguistic terms and its numerical representation, the cloud has depicted such abundant uncertainties in linguistic terms as randomness, fuzziness, and the relationship between them. CM can acquire the range and distributing law of the quantitative data from the qualitative information expressed in linguistic terms. CM has been successfully applied and gives better performance results in several fields such as intelligence control [11], data mining [19], and others. Figure 2 illustrates the types of cloud model (see [11,17] for more details).

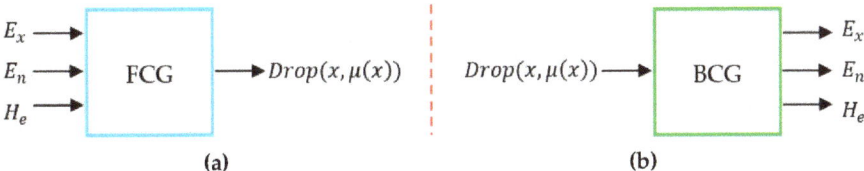

Figure 2. Two different types of cloud generators. (**a**) Forward cloud generator; (**b**) Backward cloud generator.

2.2. The Fuzzy Time Series Model

Fuzzy time series is another concept to solve forecasting problems in which the historical data are linguistic values. The fuzzy time series has recently received increasing attention because of its capability to deal with vague and incomplete data. There have been a variety of models developed to either improve forecasting accuracy or reduce computation overhead [22]. The fuzzy time series model uses a four–step framework to make forecasts, as shown in Figure 3: (1) define the universe of discourse and partition it into intervals; (2) determine the fuzzy sets on the universe of discourse and fuzzify the time series; (3) build the model of the existing fuzzy logic relationships in the fuzzified time series; and (4) make forecast and defuzzify the forecast values [23–25].

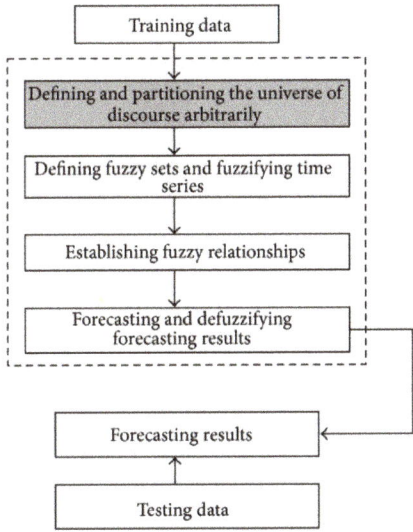

Figure 3. Processes of fuzzy time series forecasting.

Nevertheless, the forecasting performance can be significantly affected by the partition of the universe of discourse. Another issue is the consistency of the forecasting accuracy with the interval length. In general cases, better accuracy can be achieved with a shorter interval length. However, an effective forecasting model should adhere to the consistency principle. In accounting, consistency requires that a company's financial statements follow the same accounting principles, methods, practices, and procedures from one accounting period to the next. In general, the effect of some parameters in fuzzy time series such as population size, number of intervals, and order of fuzzy time series must be tested and analyzed [26,27].

2.3. Heikin–Ashi Candlestick Pattern

The current forecasting models do not contain the qualitative information that would help in predicting the future. Japanese candlesticks are a technical analysis tool that traders use to chart and analyze the price movement of securities. Japanese candlesticks provide more detailed and accurate information about price movements compared to bar charts. They provide a graphical representation of the supply and demand behind each time period's price action. Each candlestick includes a central portion that shows the distance between the open and the close of the security being traded, the area referred to as the body. The upper shadow is the price distance between the top of the body and the high for the trading period. The lower shadow is the price distance between the bottom of the body and the low for the trading period. The closing price of the security being traded determines whether the candlestick is bullish or bearish. The real body is usually white if the candlestick closes at a higher price than when it opened. In such a case, the closing price is located at the top of the real body and the opening price is located at the bottom. If the security being traded closed at a lower price than it opened for the time period, the body is usually filled up or black in color. The closing price is located at the bottom of the body and the opening price is located at the top. Modern candlesticks now replace the white and black colors of the body with more colors such as red, green, and blue. Traders can choose among the colors when using electronic trading platforms (see Figure 4) [6,7].

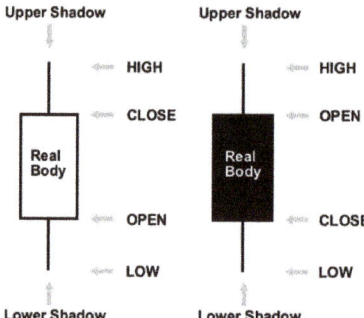

Figure 4. The dark candle and white candle.

Normal candlestick charts are composed of a series of open–high–low–close (OHLC) candles set apart by a time series. The Heikin–Ashi technique shares some characteristics with standard candlestick charts but uses a modified formula of close–open–high–low (COHL). There are a few differences to note between the two types of charts, and are demonstrated by the charts above. Heikin–Ashi has a smoother look as it essentially takes an average of the movement. There is a tendency with Heikin–Ashi for the candles to stay red during a downtrend and green during an uptrend, whereas normal candlesticks alternate colors, even if the price is moving dominantly in one direction. Since Heikin–Ashi takes an average, the current price on the candle may not match the price the market is actually trading at. For this reason, many charting platforms show two prices on the y-axis: one for the calculation of the Heikin–Ashi and another for the current price of the asset [7–10].

2.4. Related Work

Researchers that believe in the existence of patterns in a financial time series that make them predictable have centered their work mainly in two different approaches: statistical and artificial intelligence (AI). The statistical techniques most used in financial time series modeling are the autoregressive integrated moving average (ARIMA) and the smooth transition autoregressive (STAR) [2]. On the other hand, artificial intelligence provides sophisticated techniques to model time series and search for behavior patterns: genetic algorithms, fuzzy models, the adaptive neuro-fuzzy inference system (ANFIS), artificial neural networks (ANN), support vector machines (SVM), hidden Markov models, and expert systems, are some examples. Unlike statistical techniques, they are capable of obtaining adequate models for nonlinear and unstructured data. There exists a huge amount of literature that uses AI approaches for time series forecasting [2,4,8]. However, most of them are inaccurate: the computer programs are more effective in syntax analysis than semantic analysis. Furthermore, most of them follow the quantitative forecasting category; qualitative forecasting is useful when there is ambiguous or inadequate data. Most of the current studies were conducted from single time scale features of the stock market index, but it is also meaningful for studying from multiple time scale features [8]. With the development of deep learning, there are many methods based on deep learning used for stock forecasting and have drawn some essential conclusions [3].

In the literature, many studies have used an integrated neuro-fuzzy model to estimate the dynamics of the stock market using technical indicators [3]. This approach integrates the advantages in both the neural and fuzzy models to facilitate reliable intelligent stock value forecasting. However, most of these works did not consider the fractional deviation within a day. Another group of research work utilized hidden Markov models (HMMs) to predict the stock price based on the daily fractional change in the stock share value of intra-day high and low. To benefit from the correlation between the technical indicators and reduce the large dimensionality space, the principal component analysis (PCA) concept was deployed to select the most effective technical indicators among a large number of highly correlated variables. PCA linearly transforms the original large set of input variables into a smaller set of uncorrelated variables to reduce the large dimensionality space.

In addition, some researchers are currently using soft computing techniques (e.g., genetic algorithm) for selecting the most optimal subset of features among a large number of input features, and then selected features are given as input to the machine learning module (e.g., SVM Light software package). Technical analysis is carried out based on technical indicators from the stock to be predicted and also from other stocks that are highly correlated with it. However, the decision is carried out only based on the input feature variables of technical indicators. This leads to prediction errors due to the lack of precise domain knowledge and no consideration of various political and economic factors that affect the stock market other than the technical indicators [3,8].

Song and Chissom [13] suggested a forecasting model using fuzzy time series, which provided a theoretical framework to model a special dynamic process whose observations were linguistic values. The main difference between the traditional time series and fuzzy time series was that the observed values of the former were real numbers while the observed values of the latter were fuzzy sets or linguistic values. Chen et al. [16] presented a new method for forecasting university enrolment using fuzzy time series. Their method is more efficient than the suggested method by Song and Chissom due to the fact that their method used simplified arithmetic operation rather than the complicated MaxMin composition operation. Hwang [22] suggested a new method based on fuzzification to revise Song and Chissom's method. He used a different triangle fuzzification method to fuzzily crisp values. His method involved determining an interval of extension from both sides of crisp value in triangle membership function to get a variant degree of membership. The results obtained a better average forecasting error. In addition, the influences of factors and variables in a fuzzy time series model such as definition area, number and length of intervals, and the interval of extension in triangle membership function were discussed in detail. More techniques that used fuzzy time series for forecasting can be found in [23–27].

Nison [5] introduced the Japanese candlestick concepts to the Western world. Japanese candlestick patterns are believed to show both quantitative information like price, trend ... etc., and qualitative information like the psychology of the market. It considers not only the close values, but also the information on the body of the candlestick can offer an informative summary of the trading sessions [28] and some of its components are predictable [29]. Some researchers have combined technical patterns and candlestick information [30]. In the last decades, several researchers have used Japanese candlesticks in creative forecasting methods [31–36]. Lee et al. [31] suggested an expert system with IF–THEN rules to detect candlestick patterns, flag sell, and buy orders with good hit ratios in the Korean market. The authors in [32] displayed Japanese candlestick patterns using fuzzy linguistic variables and knowledge-based by fuzzing both the candle line and the candle lines relationship. In [33], a prediction model was suggested for the financial decision system based on fuzzy candlestick patterns. Lee [34] extended this work through creating and using personal candlestick pattern ontologies to allow different users to have their explanation of a candlestick pattern. Kamo et al. [8,35,36] suggested a model that combined neural networks, committee machines, and fuzzy logic to identify candlestick patterns and generate a market strength weight using fuzzy rules in [35], the type–1 fuzzy logic system in [36], and finally, the type–2 fuzzy logic system in [6].

Naranjo et al. [37] presented a model that used the K-nearest neighbors (KNN) algorithm to forecast the candlestick one day ahead using the fuzzy candlestick representation. Naranjo et al. [38] fuzzified the gap between candles and added it as an extended element in candlesticks patterns. However, Japanese candlestick has contradictory information due to the market's noise [38]. Recently, the Heikin–Ashi technique modifies the traditional candlestick chart and makes it easier to reduce the noise, eliminate the gaps between candles, and smoothen the movement of the market and let the traders focus on the main trend. The Heikin–Ashi graph is not only more readable than traditional candles, but is also a real trading system [10].

In general, most existing fuzzy time series forecasting models follow fuzzy rules according to the relationships between neighboring states without considering the inconsistency of fluctuations for a related period [38–40]. This paper proposes a new perspective to study the problem of prediction, in which inconsistency is quantified and regarded as a key characteristic of prediction rules by utilizing a combination of the cloud model, Heikin–Ashi candlesticks, and fuzzy time series (FTS) in a unified model that can represent both fluctuation trend and fluctuation consistency information.

3. Proposed Model

The purpose of the study is to predict and confirm accurate stock future trends due to a lack of insufficient levels of accuracy and certainty. However, there are many problems in previous studies. The main problems in data are uncertainty, noise, non-linearity, non-stationary, and dynamic process of stock prices in time series. In the prediction model, many models are used. The statistical method like the ARMA family is achieved with the trial and error basis iterations. Traders also have problems that include predicting the stock price every day, finding the reversal patterns of the stock price, the difficulty in model parameter tuning, and finally, the gap exists between prediction results and investment decision. Additionally, traditional candlestick patterns have problems such as the definition of the patterns itself being ambiguous and the largest number of patterns.

In order to deal with the above problems, the suggested prediction model uses both cloud model and Heikin–Ashi (HA) candlestick patterns. Figure 5 illustrates the main steps of the suggested model that include preparing historical data, HA candlestick processing, representing the HA candlestick using the cloud model, forecasting the next day price (open, high, low, close) using cloud–based time series prediction, formalizing the next day HA candlestick features, and finally, forecasting the trend and its strong patterns. The following subsection discusses each step in detail [9].

3.1. Step 1: Preparing the Historical Data

The publicly available stock market datasets contain historical data on the four price time series for several companies were collected from Yahoo (http://finance.yahoo.com). The dataset specifies the "opening price, lowest price, closing price, highest price, adjusted closing price, and volume" against each date. The data were divided into two parts: the training part and the testing part. The training part from the time series data was used for the formulation of the model while the testing part was used for the validation of the proposed model.

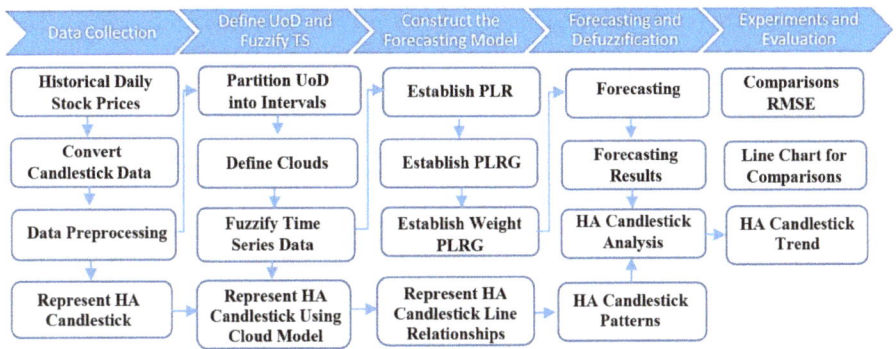

Figure 5. The procedure of the proposed forecasting model.

3.2. Step 2: Candlestick Data

The first stage in stock market forecasting is the selection of input variables. The two most common types of features that are widely used for predicting the stock market are fundamental indicators and technical indicators. The suggested model used technical indicators that are determined by employing candlestick patterns such as open price, close price, low price, and high price to try to find future stock prices [5,6]. A standard candlestick pattern is composed of one or more candlestick lines. However, the extended candlestick (Heikin–Ashi) patterns have one candlestick line. The HA candlestick uses the modified OHLC values as candlesticks that are calculated using [5]:

$$\left. \begin{array}{l} Ha_{Close} = \frac{(Open + High + Low + Close)}{4} \\ Ha_{Open} = \frac{(Ha_{Open}(Previous\ Bar) + Ha_{Close}(Previous\ Bar))}{2} \\ Ha_{High} = Max(High, Ha_{Open}, Ha_{Close}) \\ Ha_{Low} = Min(Low, Ha_{Open}, Ha_{Close}) \end{array} \right\} \quad (1)$$

Herein, each candlestick line has the following parameters: length of the upper shadow, length of the lower shadow, length of the body, color, open style, and close style. The open style and close style are formed by the relationship between a candlestick line and its previous candlestick line. The crisp value of the length of the upper shadow, length of the lower shadow, length of the body, and color play an important role in identifying a candlestick pattern and determining the efficiency of the candlestick pattern. The candlestick parameters are directly calculated using [9,10].

$$\left. \begin{array}{l} HaL_{Body} = \frac{Max(Ha_{open},\ Ha_{Close}) - Min(Ha_{open},\ Ha_{Close})}{Ha_{open}} \times 100 \\ HaL_{UpperShadow} = \frac{Ha_{High} - Max(Ha_{open},\ Ha_{Close})}{Open} \times 100 \\ HaL_{LowerShadow} = \frac{Min(Ha_{open},\ Ha_{Close}) - Ha_{Low}}{Ha_{open}} \times 100 \\ Ha_{Color} = Ha_{Close} - Ha_{open} \end{array} \right\} \quad (2)$$

where HaL indicates the length of the body, upper shadow, or lower shadow of the HA candlestick. The Ha_{COLOR} parameter represents the mean body color of the HA candlestick. Heikin–Ashi candlesticks are similar to conventional ones, but rather than using opens, closes, highs, and lows, they use average values for these four price metrics.

In stock market prediction, the quality of data is the main factor because the accuracy and the reliability of the prediction model depends upon the quality of data. Any unwanted anomalies in the dataset are known as noise. Outliers are the set of observations that do not obey the general behavior of the dataset. The presence of noise and outliers may result in poor prediction accuracy of forecasting models. The data must be prepared so that it covers the range of inputs for which the network is going to be used. Data pre-processing techniques attempt to reduce errors and remove outliers, hence improving the accuracy of prediction models. The purpose of HA charts is to filter noise and provide a clearer visual representation of the trend. Heikin–Ashi has a smoother look, as it is essentially taking an average of the movement [9,10].

3.3. Step 3: Cloud Model-Based Candlestick Representation

There is no crisp value to define the length of body and shadow in the HA candlestick; these variables are usually described as imprecise and vague. Herrin, to transform crisp candlestick parameters (HA quantitative values) to linguistic variables to define the candlestick (qualitative value), the cloud model was used. To achieve this goal, fuzzy HA candlestick pattern ontology was built that contains [4,8]:

- Candlestick Lines: Four fuzzy linguistic variables, equal, short, middle, and long, were defined to indicate the cloud model of the shadows and the body length. Figure 6 shows the membership function of the linguistic variables based on the cloud model, then used the maximum µ(x) to determine its linguistic variable. The ranges of body and shadow length were set to (0, p) to represent the percentage of the fluctuation of stock price. The parameter value of each fuzzy linguistic variable was set as stated in [8]. See [8] for more details regarding the rationale of using these values. These fuzzy linguistic variables are defined as:

$$Equal(x:a,b) = \begin{cases} 0 & x < a \\ \exp\left(-\frac{1}{2}\left(\frac{x-Ex}{En}\right)^2\right) & a \leq x \leq b \\ 0 & x > b \end{cases} \qquad (3)$$

$$Short/Middle(x:a,b,c,d) = \begin{cases} 0 & x < a \\ \exp\left(-\frac{1}{2}\left(\frac{x-Ex_1}{En_1}\right)^2\right) & a \leq x \leq b \\ 1 & b < x < c \\ \exp\left(-\frac{1}{2}\left(\frac{x-Ex_2}{En_2}\right)^2\right) & c \leq x \leq d \\ 0 & x > d \end{cases} \qquad (4)$$

$$Large(x:a,b) = \begin{cases} 0 & x < a \\ \exp\left(-\frac{1}{2}\left(\frac{x-Ex}{En}\right)^2\right) & a \leq x \leq b \\ 1 & x > b \end{cases} \qquad (5)$$

The body color $Body_{Color}$ is also an import feature of a candlestick line. It is defined by three terms Black, White, and Doji. A Doji term is defined to describe the situation where the open price equals the close price. In this case, the height of the body is 0, and the shape is represented by a horizontal bar. The definition of body color is defined as [10]:

$$\left. \begin{array}{l} If(Open - Close) > 0 \text{ Then } Body_{Color} = Black \\ If(Open - Close) < 0 \text{ Then } Body_{Color} = White \\ If(Open - Close) = 0 \text{ Then } Body_{Color} = Doji \end{array} \right\} \qquad (6)$$

- Candlestick Lines Relationships: This defines the place of the HA candlestick with the previous one to form open style and close style linguistic variables. In general, merging the description of the candlestick line and HA candlestick line relationship can create a HA candlestick pattern that is completely defined. Herein, five linguistic variables were defined to represent the relationship style (X style): low, equal low, equal, equal high, and high. Their membership function follows half bell cloud defined in Equation (7). Additionally, the parameter value of each fuzzy linguistic variable was set as stated in [8]. Figure 7 shows the membership function of the linguistic variable based on the cloud model:

$$X\_Style(x:a,b) = \begin{cases} 0 & x < a \\ exp\left(-\frac{1}{2}\left(\frac{x-Ex}{En}\right)^2\right) & a \leq x \leq b \\ 0 & x > b \end{cases} \quad (7)$$

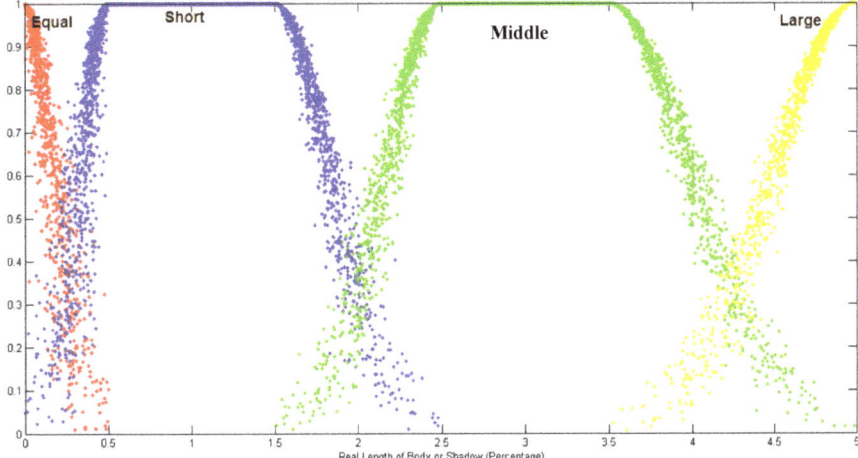

Figure 6. The membership function of the body and shadow length based on the cloud model.

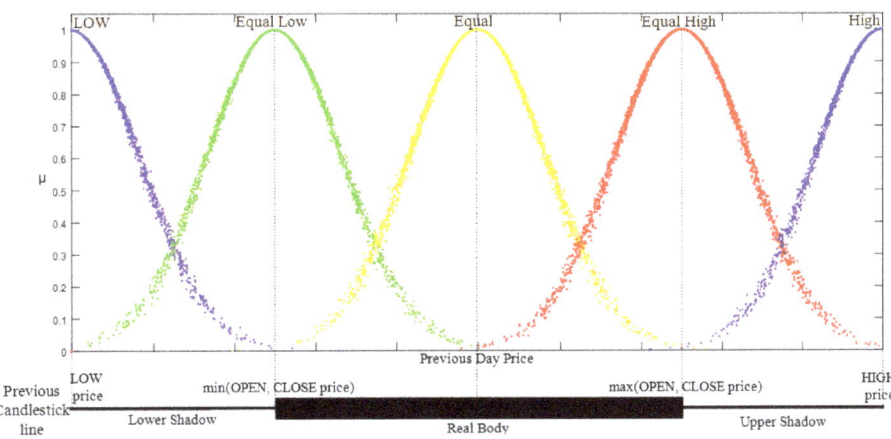

Figure 7. The membership function of the open and close styles based on the cloud model.

In our case, membership cloud function (forward normal cloud generator) converts the statistic results to fuzzy numbers, and constructs the one–to–many mapping model. The input of the forward normal cloud generator is three numerical characteristics of a linguistic term, (*Ex*, *En*, *He*), and the number of cloud drops to be generated, N, while the output is the quantitative positions of N cloud drops in the data space and the certain degree that each cloud drop can represent the linguistic term. The algorithm in detail is:

- Produce a normally distributed random number En' with mean En and standard deviation He;
- Produce a normally distributed random number x with mean Ex and standard deviation En';
- Calculate $Y = exp\left(-\frac{1}{2}\left(\frac{x-Ex}{En}\right)^2\right)$
- Drop (x,y) is a cloud drop in the universe of discourse; and
- Repeat step 1–4 until N cloud drops are generated.

Expectation value (Ex) at the center-of-gravity positions of cloud drops is the central value of distribution. Entropy (En) is the fuzzy measure of qualitative concept that describes the uncertainty and the randomness. The larger the entropy, the larger the acceptable interval of this qualitative concept, which represents that this conception is more fuzzy. Hyper entropy (He) is the uncertain measure of qualitative concept that describes the dispersion. The larger the hyper entropy, the thicker the shape of the cloud, which shows that this conception is more discrete [20,21].

– Forecast the next day price (open, high, low, close)

In the fuzzy candlestick pattern approach, the measured values are the open, close, high, and low price of trading targets in a specific time period. The features of the trading target price fluctuation are represented by the fuzzy candlestick pattern. The classification rules of fuzzy candlestick patterns can be determined by the investors or the computer system. In general, using a candlestick pattern approach for financial time series prediction consists of the following steps [21]:

- Partitioning the universe of discourse into intervals: In this case, after preparing the historical data and defining the range of the universe of discourse (UoD), open, high, low, and close prices should be established as a data price set for each one. Then, for each data price set, the variation percentage between two prices on time t and time $t + n$ is calculated $((Close_{t+n} - Close_t)/Close_t) \times 100$ to partition the universe of discourse dataset into intervals. Based on the variation, the minimum variation D_{min} and the maximum variation D_{max} are determined that define $U = [D_{min} - D_1, D_{max} + D_2]$, where D_1 and D_2 are suitable positive numbers.
- Classifying the historical data to its cloud: The next step determines the linguistic variables represented by clouds (see Figure 8) to describe the degree of variation between data of time t and time $t + n$ and defined it as a set of linguistic terms. Table 1 shows the digital characteristics of the cloud member function (Ex, En, He) for each linguistic term.
- Building the predictive logical relationships (PLR): The model builds the PLR to carry on the soft inference $A_{t-1} \rightarrow A_t$, where A_{t-1} and A_t are clouds representing linguistic concepts, by searching all clouds in time series with the pattern ($A_{t-1} \rightarrow A_t$).
- Building of predictive linguistic relationship groups (PLRG): In the training dataset, all PLRs with the same "current state" will be grouped into the same PLRG. If $A_1, A_2, \cdots, A_m$ is the "current state" of one PLR in the training dataset and there are r PLRs in the training dataset as $A_1 \rightarrow A_1$; $A_1 \rightarrow A_2$; ....; $A_1 \rightarrow A_m$, the r PLRs can be grouped into the same PLRG, as $A_1 \rightarrow A_1, A_2, ..., A_m$. Then, assign the weight elements for each PLRG. Assume A_i has n_1 relationships with A_1, n_2 relationships with A_2, and so on. The weight values (w) can be assigned as w_i = (number of recurrence of A_i)/(total number of PLRs).
- Calculating the predicted value via defuzzification: Then the model forecasts the next day (open, high, low, close) prices through defuzzification and calculates the predicted value at time t P(t) by following the rule:

  ✔ Rule 1: If there is only one PLR in the PLRG, ($A_1 \rightarrow A_i$) then,

$$P(t) = \frac{Ex_i + S(t-1)}{2} \qquad (8)$$

✔ Rule 2: If there is r PLR in the PLRG, ($A_1 \to A_1, A_2, \ldots, A_p$) then,

$$P(t) = \frac{1}{2}\left(\left(\frac{(n_1 \times Ex_1) + (n_2 \times Ex_2) + \ldots + (n_p \times Ex_p)}{n_1 + n_1 + \ldots + n_p}\right) + S(t-1)\right) \quad (9)$$

✔ Rule 3: If there is no PLR in the PLRG, ($A_1 \to \#$) where the symbol "#" denotes an unknown value; then apply Equation (8). Ex_i is the expectation of the Gaussian cloud C_i corresponding to A_i, n_i is the number of A_i appearing in the PLRG, $1 \le i \le r$, and $S(t-1)$ denotes the observed value at time $t-1$.

- Transforming the forecasting results (open, high, low, and close) to the next HA candlestick. through the following rules [9]:

✔ **Rule 1: If** $Body_{Color}$ is White and HaL_{Body} is Long **Then**, UP Trend.
✔ **Rule 2: If** $Body_{Color}$ is Black and HaL_{Body} is Long **Then**, Down Trend.
✔ **Rule 3: If** $Body_{Color}$ is White and HaL_{Body} is Long and $HaL_{LowerShadow}$ is Equal **Then**, Strong UP Trend.
✔ **Rule 4: If** $Body_{Color}$ is Black and HaL_{Body} is Long and $HaL_{UpperShadow}$ is Equal **Then**, Strong Down Trend.
✔ **Rule 5: If** (HaL_{Body} is Equal) and ($HaL_{UpperShadow}$ & $HaL_{LowerShadow}$) is Long **Then** Change of Trend.
✔ **Rule 6: If** (HaL_{Body} is Short) and ($HaL_{UpperShadow}$ & $HaL_{LowerShadow}$) is not Equal **Then**, Consolidation Trend.
✔ **Rule 7: If** (HaL_{Body} is Short or Equal) and (Ha_{Open}_Style and Ha_{Close}_Style) is (Low_Style or EqualLow_Style) and $HaL_{UpperShadow}$ is Equal **Then** Weaker Trend.

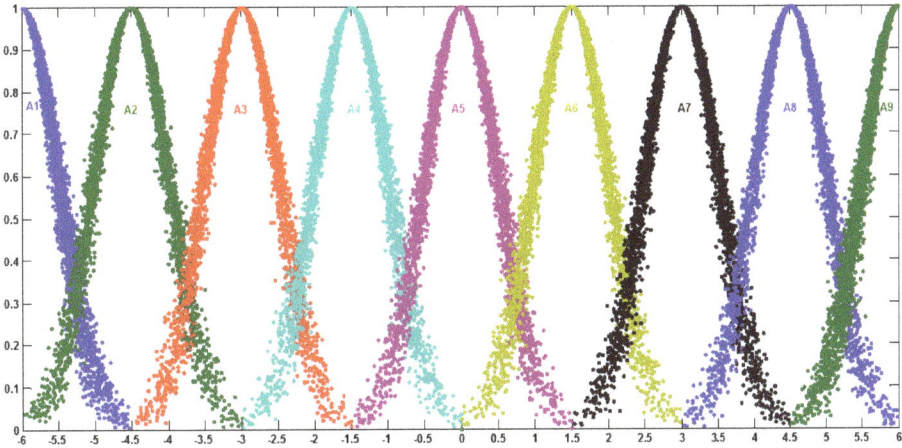

Figure 8. The clouds of the linguistic terms.

Table 1. The digital characteristics of cloud member function for each linguistic term.

Price Variation		[−6, −4.5]	[−6, −3]	[−4.5, −1.5]	[−3, 0]	[−1.5, 1.5]	[0, 3]	[1.5, 4.5]	[3, 6]	[4.5, 6]
Linguistic Terms		A1 Extreme Decrease	A2 Large Decrease	A3 Normal Decrease	A4 Small Decrease	A5 No Change	A6 Small Increase	A7 Normal Increase	A8 Large Increase	A9 Extreme Increase
CG	Ex	−6	−4.5	−3	−1.5	0	1.5	3	4.5	6
	En	0.5	0.5	0.5	0.5	0.5	0.5	0.5	0.5	0.5
	He	0.05	0.05	0.05	0.05	0.05	0.05	0.05	0.05	0.05

4. Experimental Results

In order to test the efficiency and validity of the proposed model, the model was implemented in MATLAB language. The prototype verification technique was built in a modular fashion and has been implemented and tested in a Dell™ Inspiron™ N5110 Laptop machine, Dell computer Corporation, Texas, which had the following features: Intel(R) Core(TM) i5–2410M CPU@ 2.30GHz, and 4.00 GB of RAM, 64–bit Windows 7. A dataset composed of real-time stocks series of the NYSE (New York Stock Exchange) was used in the experimentation. The dataset had 13 time series of NYSE companies, each one with the four prices (open, high, low, and close). Time series were downloaded from the Yahoo finance website (http://finance.yahoo.com), Table 2 shows the companies' names, symbol, and starting date and ending date for the selected dataset. The dataset was divided into 2/3 for training and the other 1/3 for testing.

Table 2. Selected time series datasets.

Company	Symbol	from	to
Boeing Company	BA	02/01/1962	27/06/2018
Bank of America	BAC	03/01/2000	12/12/2014
DuPont	DD	03/01/2000	12/12/2014
Ford Motor Co.	F	03/01/2000	12/12/2014
General Electric	GE	03/01/2000	12/12/2014
Hewlett–Packard	HPQ	03/01/2000	12/12/2014
Microsoft	MSFT	03/01/2000	12/12/2014
Monsanto	MON	18/10/2000	12/12/2014
Toyota Motor	TM	03/01/2000	12/12/2014
Wells Fargo	WFC	01/06/1972	27/06/2018
Yahoo	YHOO	03/01/2005	12/12/2014
Exxon Mobil	XOM	02/01/1970	21/05/2018
Walt Disney	DIS	02/01/1962	27/06/2018

In the proposed forecasting model, the parameters were set as follows: the ranges of body (p) and shadow length were set to (0, 14) to represent the percentage of the fluctuation of stock price because the varying percentages of the stock prices are limited to 14 percent in the Taiwanese stock market, for example. It should be noted that although we limited the fluctuation of body and shadow length to 14 percent, in other applications, the designer can change the range of the fluctuation length to any number [4]. The four parameters (a–d) of the function to describe the linguistic variables SHORT and MIDDLE were (0, 0.5, 1.5, 2.5) and (1.5, 2.5, 3.5, 5). The parameters (a, b) that were used to model the EQUAL fuzzy set were equal to (0, 0.5). Regarding the two parameters D_1 and D_2, which are used to determine the UOD, we can set $D_1 = 0{:}17$ and $D_2 = 0{:}34$, so the UoD can be represented as [6,8]. Finally, the number of drops in the cloud model used to build the membership function is usually equal to the number of samples in the dataset to describe the data efficiently. The mean squared error (MSE) and mean absolute percentage error (MAPE) that are used by academicians and practitioners [4,21] were used to evaluate the accuracy of the proposed method. Tables 3–6 show the output of applying each model step for the Yahoo dataset.

$$MSE = \frac{\sum_{i=1}^{n}(Forcasted\ Value - Actual\ Value)^2}{n} \qquad (10)$$

$$MAPE = \frac{1}{n}\sum_{i=1}^{n}\left|\frac{(Actual\ Value)_i - (Forcasted\ Value)_i}{(Actual\ Value)_i}\right| \qquad (11)$$

Table 3. Heiken–Ashi candlestick patterns derived from Yahoo training data.

Date	Open	High	Low	Close	HA Open	HA High	HA Low	HA Close	HA Body	HA Upper Shadow	HA Lower Shadow	HA Color	HA Open Style	HA Close Style
10/01/2005	36.00	36.76	35.51	36.32	36.16	36.76	35.51	36.15	EQUAL	SHORT	SHORT	BLACK	HIGH	HIGH
11/01/2005	36.31	36.58	35.39	35.66	36.15	36.58	35.39	35.99	SHORT	SHORT	SHORT	BLACK	EQUAL_HIGH	EQUAL_HIGH
12/01/2005	35.88	36.18	34.80	36.14	36.07	36.18	34.80	35.75	SHORT	SHORT	MIDDLE	BLACK	EQUAL_HIGH	EQUAL_HIGH
13/01/2005	36.12	36.32	35.26	35.33	35.91	36.32	35.26	35.76	SHORT	SHORT	SHORT	BLACK	EQUAL_HIGH	EQUAL_HIGH
14/01/2005	35.86	36.70	35.83	36.70	35.83	36.70	35.83	36.27	SHORT	SHORT	EQUAL	WHITE	EQUAL_HIGH	HIGH
⋮	⋮	⋮	⋮	⋮	⋮	⋮	⋮	⋮	⋮	⋮	⋮	⋮	⋮	⋮
28/11/2014	51.87	52.00	51.64	51.74	51.73	52.00	51.64	51.81	EQUAL	SHORT	EQUAL	WHITE	EQUAL_HIGH	EQUAL_HIGH
01/12/2014	51.43	51.43	49.66	50.10	51.77	51.77	49.66	50.66	MIDDLE	EQUAL	SHORT	BLACK	EQUAL_HIGH	LOW
02/12/2014	50.27	51.12	50.01	50.67	51.21	51.21	50.01	50.52	SHORT	EQUAL	SHORT	BLACK	EQUAL_HIGH	EQUAL_HIGH
03/12/2014	50.71	50.97	50.20	50.28	50.87	50.97	50.20	50.54	SHORT	EQUAL	SHORT	BLACK	EQUAL_HIGH	EQUAL_HIGH
04/12/2014	50.19	50.67	49.90	50.41	50.70	50.70	49.90	50.29	SHORT	EQUAL	SHORT	BLACK	EQUAL_HIGH	EQUAL_HIGH
05/12/2014	51.03	51.25	50.51	50.99	50.50	51.25	50.50	50.95	SHORT	SHORT	EQUAL	WHITE	EQUAL_HIGH	HIGH
08/12/2014	50.52	50.90	49.22	49.62	50.72	50.90	49.22	50.07	SHORT	SHORT	SHORT	BLACK	LOW	LOW
09/12/2014	48.75	50.53	48.29	50.51	50.39	50.53	48.29	49.52	SHORT	MIDDLE	MIDDLE	BLACK	EQUAL_HIGH	EQUAL_HIGH
10/12/2014	50.33	50.69	49.19	49.21	49.96	50.69	49.19	49.86	EQUAL	SHORT	SHORT	BLACK	EQUAL_HIGH	EQUAL_HIGH

Table 4. Yahoo dataset, one day variations, and its cloud.

Date	Open	High	Low	Close	One Day Variations Open	Cloud	One Day Variations High	Cloud	One Day Variations Low	Cloud	One Day Variations Close	Cloud
	O	H	L	C	O		H		L		C	
03/01/2005	38.36	38.9	37.65	38.18	0.00		0.00		0.00		0.00	
04/01/2005	38.45	38.54	36.46	36.58	0.23	A5	−0.93	A4	−3.16	A3	−4.19	A2
05/01/2005	36.69	36.98	36.06	36.13	−4.58	A2	−4.05	A2	−1.10	A4	−1.23	A4
06/01/2005	36.32	36.5	35.21	35.43	−1.01	A4	−1.30	A4	−2.36	A4	−1.94	A4
07/01/2005	35.99	36.46	35.41	35.96	−0.91	A4	−0.11	A4	0.57	A5	1.50	A6
10/01/2005	36.00	36.76	35.51	36.32	0.03	A4	0.82	A6	0.28	A5	1.00	A6
⋮	⋮	⋮	⋮	⋮	⋮	⋮	⋮	⋮	⋮	⋮	⋮	⋮
10/12/2014	50.33	50.69	49.19	49.21	3.24	A7	0.32	A5	1.86	A6	−2.57	A4
11/12/2014	49.54	50.58	49.43	49.94	−1.57	A4	−0.22	A4	0.49	A5	1.48	A6

Table 5. The PLR results.

Date	Open PLR	High PLR	Low PLR	Close PLR
03/01/2005				
04/01/2005	$A_5 \to A2$	$A4 \to A2$	$A3 \to A4$	$A2 \to A4$
05/01/2005	$A2 \to A4$	$A2 \to A4$	$A4 \to A4$	$A4 \to A4$
06/01/2005	$A4 \to A4$	$A4 \to A4$	$A4 \to A5$	$A4 \to A6$
07/01/2005	$A4 \to A4$	$A4 \to A6$	$A5 \to A5$	$A6 \to A6$
....	....	....	....	....
....	....	....	....	....
09/12/2014	$A2 \to A7$	$A4 \to A5$	$A4 \to A6$	$A6 \to A4$
10/12/2014	$A7 \to A4$	$A5 \to A4$	$A6 \to A5$	$A4 \to A6$
11/12/2014	$A4 \to A4$	$A4 \to A6$	$A5 \to A4$	$A6 \to A5$

Table 6. PLRG for close PLR.

	Close	To									Total Count
		A1	A2	A3	A4	A5	A6	A7	A8	A9	
From	A1	0	2	0	14	1	2	2	0	3	24
	A2	2	7	3	21	4	16	6	6	5	70
	A3	0	1	1	15	3	7	2	2	1	32
	A4	6	30	15	370	79	170	67	29	17	783
	A5	4	3	3	100	22	28	11	7	1	179
	A6	5	10	3	152	42	64	32	13	6	327
	A7	3	9	3	68	19	21	4	6	3	136
	A8	0	3	2	25	8	15	10	7	3	73
	A9	4	5	2	18	1	4	2	3	5	44
											1668

The suggested model was verified with respect to the RMS on both the training and testing data. The predicted prices of the model were found to be correct and close to the actual prices. There was a clear difference between the MSE values for the training and testing data, showing that the model was overfitting the training data as the error on the training dataset was minimized. The reason for this is that the model was not as generalized and was specialized to the structure in the training dataset. Using cross validation represents one possible way to handle overfitting, and using multiple runs of cross validation is better again. The model RMS is summarized in Table 7.

Table 7. Average MSE of the suggested model for all dataset.

MSE	Open	High	Low	Close
Training Data	0.09	0.19	0.16	0.20
Testing Data	0.03	0.07	0.07	0.07

Table 8 shows the comparison results between our two versions of the suggested model: the first one uses open, high, low, and close price as the initial price in the cloud FTS model (Cloud FTS) and the second method uses HaOpen, HaHigh, HaLow, and HaClose prices as the initial price in the cloud FTS model (HA Cloud FTS), and other two standard Song fuzzy time series (FTS) [13,14] and Yu weighted fuzzy time series (WFTS) models [23]. In Song's studies, the fuzzy relationships were treated as if they were equally important, which might not have properly reflected the importance of each individual fuzzy relationship in forecasting. In Yu's study, it is recommended that different weights be assigned to various fuzzy relationships. From Table 8, the MSE of the forecasting results of the proposed model was smaller than that of the other methods for all datasets. That is, the proposed model could obtain a higher forecasting accuracy rate for forecasting stock prices than the Song FTS and Yu WFTS models. In general, the MSE values changed according to the nature of each dataset. It can be noted from the table that the Wells Fargo dataset yielded the best results in terms of RMS for both the training and testing data. In general, the Wells Fargo dataset is a small dataset (2,313 row and 12 column) that is probably linearly separable, so it produced high accuracy. This is a bit difficult to accomplish with larger data, so the algorithm produced lower accuracy.

Table 8. MSE Comparison for CLOSE price prediction between HA Cloud FTS, Cloud FTS, Yu WFTS and Song.

Company	MSE	HA Cloud FTS		Cloud FTS		Yu WFTS [23]		Song FTS [14]	
		Train	Test	Train	Test	Train	Test	Train	Test
Boeing Company	BA	0.048	0.672	0.078	0.960	5.290	3.460	5.954	3.725
Bank of America	BAC	0.941	0.023	1.124	0.029	6.503	2.592	2.756	0.960
DuPont	DD	0.270	0.116	0.397	0.152	5.336	2.496	14.516	7.076
Ford Motor Co.	F	0.168	0.020	0.203	0.026	5.905	2.690	4.080	1.588
General Electric	GE	3.204	0.023	3.423	0.036	8.526	2.403	9.425	2.074
Hewlett–Packard	HPQ	1.392	0.096	1.769	0.130	7.182	2.756	6.605	2.372
Microsoft	MSFT	0.740	0.048	0.922	0.068	5.905	2.403	7.129	2.372
Monsanto	MON	1.904	0.314	2.528	0.476	8.009	3.028	6.052	1.588
Toyota Motor	TM	1.166	0.449	1.369	0.504	6.300	2.856	19.272	9.303
Wells Fargo	WFC	0.023	0.102	0.040	0.144	4.928	2.624	3.133	1.638
Yahoo	YHOO	0.203	0.073	0.250	0.090	5.664	2.624	6.052	2.496
Exxon Mobil	XOM	0.040	0.221	0.068	0.314	4.580	2.560	6.656	3.572
Walt Disney	DIS	0.023	0.130	0.036	0.194	5.198	2.723	4.580	2.250
AVERAGE		0.779	0.176	0.939	0.240	6.102	2.709	7.400	3.155

One possible explanation of these results is that, compared with standard models that use FTS only, utilizing FTS with the cloud model helps to automatically produces random membership grades of a concept through a cloud generator. In this way, the membership functions are built based on the characteristics of the data instead of traditional fuzzy–based forecasting methods that depend on the expert. From the point of view of the importance of using HA candlesticks with the cloud model for forecasting, utilizing the HA candlesticks showed significant features that could identify market turning points and also the direction of the trend that helps improve prediction accuracy.

The last set of experiments was fulfilled to validate the efficiency of the suggested model compared to state-of-the-art models listed in Figure 9 using the Taiwan Capitalization Weighted Stock Index (TAIEX). The data used for comparison were obtained from a website https://www.twse.com.tw/ that provided the stock prices prevailing at the NASDAQ stock quotes. As shown in Figure 9, the proposed model can perform effective prediction where the predicted stock price closely resembles the actual price in the stock market. The MSE of the suggested model was 665.40 compared with 1254.90, 4530.45, and 4698.78 for the other methods, respectively. Clearly, the suggested model had a smaller MSE than the previous methods. One of the reasons for this result is due to the merging between the cloud model and HA candlesticks, which makes it possible to account for the vagueness and uncertainty of the pattern features based on data characteristics.

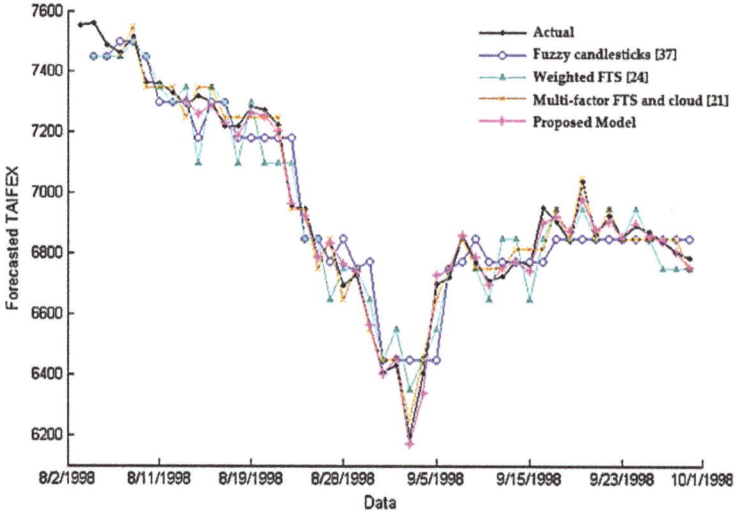

Figure 9. Comparison of the forecasting values of different methods.

5. Conclusions

In recent years, mathematical and computational models from artificial intelligence have been used for forecasting. Knowing about future values and the stock market trend has attracted a lot of attention by researchers, investors, financial experts, and brokers. This work analyzed stock trading due to its high non-linear, uncertain, and dynamic data over time. Therefore, this paper presented a Japanese candlestick-based cloud model for stock price prediction that minimizes the investor risk while investing money in the stock market. The proposed work presented an enhanced fuzzy time series forecasting model based on the cloud model and Heikin–Ashi Japanese candlestick to predict and confirm the accurate stock trends. The objective of this model was to handle qualitative forecasting and not quantitative only. The experimental result showed that using HA Cloud FTS and Cloud FTS had a lower average than the other methods used in the literature. This low average proves the high accuracy of the proposed model. HA Cloud FTS provided a MSE = 0.779 for the training data and 0.176 for the test data and Cloud FTS gave a MSE of 0.939 for the training data and 0.240 for the test data; these results mean that the HA Cloud FTS method, which uses HaOpen, HaHigh, HaLow, HaClose prices as the initial price, has a significant improvement in stock market trend prediction. Future work includes embedding Neutrosophic logic to enhance qualitative forecasting.

Author Contributions: Conceptualization, S.M.D. and O.A.H.; Methodology, S.M.D., O.A.H. and N.A.A.; Software, O.A.H., N.A.A. and Z.Z.A.; Validation, S.M.D., N.A.A. and O.A.H.; Formal analysis, S.M.D., N.A.A. and O.A.H.; Investigation, S.M.D., N.A.A. and O.A.H.; Resources, O.A.H., N.A.A. and Z.Z.A.; Data curation, O.A.H., N.A.A. and Z.Z.A.; Writing—original draft preparation, S.M.D., N.A.A. and O.A.H.; Writing—review and editing, S.M.D., O.A.H. and N.A.A.; Visualization, O.A.H., N.A.A. and Z.Z.A.; Supervision, S.M.D. and N.A.A. All authors have read and agreed to the published version of the manuscript.

Funding: This research received no external funding.

Conflicts of Interest: The authors declare no conflict of interest.

References

1. Gunduz, H.; Yaslan, Y.; Cataltepe, Z. Intraday prediction of Borsa Istanbul using convolutional neural networks and feature correlations. *Knowl. Based Syst.* **2017**, *137*, 138–148. [CrossRef]
2. Abbad, J.; Fardousi, B.; Abbad, M. Advantages of using technical analysis to predict future prices on the Amman stock exchange. *Int. J. Bus. Manag.* **2014**, *9*, 1–16. [CrossRef]

3. Sharma, A.; Bhuriya, D.; Singh, U. Survey of stock market prediction using machine learning approach. *Int. Conf. Electron. Commun. Aerosp. Technol.* **2017**, 506–509. [CrossRef]
4. Hon, C.; Leon, L.; Alan, L. Pattern discovery of fuzzy time series for financial prediction. *IEEE Trans. Knowl. Data Eng.* **2006**, *18*, 613–625. [CrossRef]
5. Nison, S. *Japanese Candlestick Charting Techniques*; New York Institute of Finance: New York, NY, USA, 1991; ISBN 0–13–931650–7.
6. Kamo, T. Integrated computational intelligence and Japanese candlestick method for short–term financial forecasting. Ph.D. Thesis, Missouri University of Science and Technology, Rolla, MO, USA, 2011.
7. Chandrinos, S.; Lagaros, N. Construction of currency portfolios by means of an optimized investment strategy. *Oper. Res. Perspect.* **2018**, *5*, 32–44. [CrossRef]
8. Lorenzo, R. *Basic Technical Analysis of Financial Markets: A Modern Approach, Perspectives in Business Culture*; Springer–Italia: Milan, Italy, 2013. [CrossRef]
9. Valcu, D. Using the Heikin–Ashi technique. *Tech. Anal. Stock. Com–Modities Mag. Ed.* **2004**, *22*, 16–29.
10. Lorenzo, R. *Heikin Ashi: How to Make Money by Fast Trading. Perspectives in Business Culture*; Springer: Berlin/Heidelberg, Germany, 2012; pp. 165–169. [CrossRef]
11. Zhang, F.; Fan, Y.; Shen, C.; Li, D. Intelligent control based on membership cloud generators. *Acta Aeronaut. Astronaut. Sin. Ser. A B* **1999**, *20*, 89–92.
12. Li, D. Uncertainty in knowledge representation. *J. Eng. Sci.* **2000**, *2*, 73–79.
13. Song, Q.; Chissom, B. Fuzzy time series and its models. *Fuzzy Sets Syst.* **1993**, *54*, 269–277. [CrossRef]
14. Song, Q.; Chissom, B. Forecasting enrollments with fuzzy time series—part 1. *Fuzzy Sets Syst.* **1993**, *54*, 1–9. [CrossRef]
15. Song, Q.; Chissom, B. Forecasting enrollments with fuzzy time series—part 2. *Fuzzy Sets Syst.* **1994**, *62*, 1–8. [CrossRef]
16. Chen, S. Forecasting enrollments based on fuzzy time series. *Fuzzy Sets Syst.* **1996**, *81*, 311–319. [CrossRef]
17. Li, D.; Meng, H.; Shi, X. Membership clouds and membership cloud generators. *J. Comput. Res. Dev.* **1995**, *32*, 15–20.
18. Li, D. Knowledge representation in KDD based on linguistic atoms. *J. Comput. Sci. Technol.* **1997**, *12*, 481–496. [CrossRef]
19. Wang, S.; Li, D.; Shi, W.; Li, D.; Wang, X. Cloud model–based spatial data mining. *Int. J. Geogr. Inf. Sci.* **2003**, *9*, 60–70. [CrossRef]
20. Deng, W.; Wang, G.; Zhang, X. A novel hybrid water quality time series prediction method based on cloud model and fuzzy forecasting. *Chemom. Intell. Lab. Syst.* **2015**, *149*, 39–49. [CrossRef]
21. Dong, L.; Wang, P.; Yan, F. Damage forecasting based on multi–factor fuzzy time series and cloud model. *J. Intell. Manuf.* **2016**, *30*, 521–538. [CrossRef]
22. Hwang, J.; Chen, S.; Lee, C. Handling forecasting problems using fuzzy time series. *Fuzzy Sets Syst.* **1998**, *100*, 217–228. [CrossRef]
23. Yu, H. Weighted fuzzy time series models for TAIEX forecasting. *Physics A* **2004**, *349*, 609–624. [CrossRef]
24. Cheng, C.; Chen, T.; Chiang, C. Trend–weighted fuzzy time series model for TAIEX forecasting. *Lect. Notes Comput. Sci.* **2006**, *4234*, 469–477. [CrossRef]
25. Cheng, C.; Chen, T.L.; Teoh, H.; Chiang, C. Fuzzy time series based on adaptive expectation model for TAIEX forecasting. *Expert Syst. Appl.* **2008**, *34*, 1126–1132. [CrossRef]
26. Huarng, K. Effective lengths of intervals to improve forecasting in fuzzy time series. *Fuzzy Sets Syst.* **2001**, *123*, 387–394. [CrossRef]
27. Xihao, S.; Yimin, L. Average–based fuzzy time series models for forecasting Shanghai compound index. *World J. Model. Simul.* **2008**, *4*, 104–111.
28. Fiess, N.; MacDonald, R. Towards the fundamentals of technical analysis: Analyzing the information content of high, low and close prices. *Econ. Model.* **2002**, *19*, 353–374. [CrossRef]
29. Arroyo, J. Forecasting candlesticks time series with locally weighted learning methods. In *Classification as a Tool for Research*; Springer: Berlin/Heidelberg, Germany, 2010; pp. 603–611.
30. Cervelló–Royo, R.; Guijarro, F.; Michniuk, K. Stock market trading rule based on pattern recognition and technical analysis: Forecasting the DJIA index with intraday data. *Expert Syst. Appl.* **2015**, *42*, 5963–5975. [CrossRef]

31. Lee, K.H.; Jo, G.S. Expert system for predicting stock market timing using a candlestick chart. *Expert Syst. Appl.* **1999**, *16*, 357–364. [CrossRef]
32. Chiung–Hon, L.; WenSung, C.; Alan, L. An implementation of knowledge based pattern recognition for financial prediction. In Proceedings of the 2004 IEEE Conference on Cybernetics and Intelligent Systems, Singapore, 1–3 December 2004; pp. 218–223. [CrossRef]
33. Chiung–Hon, L.; Lee, W.; Liu, C. Candlestick Tutor: An intelligent tool for investment knowledge learning and sharing. In Proceedings of the Fifth IEEE International Conference on Advanced Learning Technologies (ICALT'05), Kaohsiung, Taiwan, 5–8 July 2005; pp. 38–240. [CrossRef]
34. Lee, C. Modeling personalized fuzzy candlestick patterns for investment decision making. In Proceedings of the 2009 Asia–Pacific Conference in Information Processing, Shenzhen, China, 18–19 July 2009; pp. 286–289. [CrossRef]
35. Kamo, T.; Dagli, C. Fuzzy Logic–Based Japanese Candlestick Pattern Recognition and Financial Forecast. In *Intelligent Engineering Systems through Artificial Neural Networks*; Cihan, H.D., Ed.; ASME Press: New York, NY, USA, 2007; Volume 17, pp. 267–272. [CrossRef]
36. Kamo, T.; Dagli, C. Hybrid approach to the Japanese candlestick method for financial forecasting. *Expert Syst. Appl.* **2009**, *36*, 5023–5030. [CrossRef]
37. Naranjo, R.; Santos, M. Fuzzy candlesticks forecasting using pattern recognition for stock markets. *Adv. Intell. Syst. Comput.* **2016**, *527*, 323–333.
38. Naranjo, R.; Arroyo, J.; Santos, M. Fuzzy modeling of stock trading with fuzzy candlesticks. *Expert Syst. Appl.* **2018**, *93*, 15–27. [CrossRef]
39. Meesad, P.; Thi, M. Forecasting and Trading Stock Using Technical Analysis and Neural fuzzy Network. Master's Thesis, King Mongkut's University of Technology North Bangkok, Bangkok, Thailand, 2006.
40. Govindasamy, V.; Thambidurai, P. Probabilistic fuzzy logic based stock price prediction. *Int. J. Comput. Appl.* **2013**, *71*, 28–32. [CrossRef]

© 2020 by the authors. Licensee MDPI, Basel, Switzerland. This article is an open access article distributed under the terms and conditions of the Creative Commons Attribution (CC BY) license (http://creativecommons.org/licenses/by/4.0/).

Article

A Novel Comprehensive Evaluation Method for Estimating the Bank Profile Shape and Dimensions of Stable Channels Using the Maximum Entropy Principle

Hossein Bonakdari [1], Azadeh Gholami [2], Amir Mosavi [3,4,5,*], Amin Kazemian-Kale-Kale [2], Isa Ebtehaj [1] and Amir Hossein Azimi [6]

1. Department of Soils and Agri-Food Engineering, Université Laval, Québec, QC G1V0A6, Canada; Hossein.bonakdari@fsaa.ulaval.ca (H.B.); isa.ebtehaj.1@ulaval.ca (I.E.)
2. Environmental Research Centre, Department of Civil Engineering, Razi University, Kermanshah 6714414971, Iran; Gholamiazadeh1@gmail.com (A.G.); Aminkazemi_akkk@yahoo.com (A.K.-K.-K.)
3. Faculty of Civil Engineering, Technische Universität Dresden, 01069 Dresden, Germany
4. Institute of Research and Development, Duy Tan University, Da Nang 550000, Vietnam
5. School of Economics and Business, Norwegian University of Life Sciences, 1430 Ås, Norway
6. Department of Civil Engineering, Lakehead University, 955 Oliver Rd, Thunder Bay, ON P7B 5E1, Canada; azimi@lakeheadu.ca
* Correspondence: amir.mosavi@mailbox.tu-dresden.de

Received: 31 July 2020; Accepted: 23 October 2020; Published: 26 October 2020

Abstract: This paper presents an extensive and practical study of the estimation of stable channel bank shape and dimensions using the maximum entropy principle. The transverse slope (S_t) distribution of threshold channel bank cross-sections satisfies the properties of the probability space. The entropy of S_t is subject to two constraint conditions, and the principle of maximum entropy must be applied to find the least biased probability distribution. Accordingly, the Lagrange multiplier (λ) as a critical parameter in the entropy equation is calculated numerically based on the maximum entropy principle. The main goal of the present paper is the investigation of the hydraulic parameters influence governing the mean transverse slope ($\overline{S_t}$) value comprehensively using a Gene Expression Programming (GEP) by knowing the initial information (discharge (Q) and mean sediment size (d_{50})) related to the intended problem. An explicit and simple equation of the $\overline{S_t}$ of banks and the geometric and hydraulic parameters of flow is introduced based on the GEP in combination with the previous shape profile equation related to previous researchers. Therefore, a reliable numerical hybrid model is designed, namely Entropy-based Design Model of Threshold Channels (EDMTC) based on entropy theory combined with the evolutionary algorithm of the GEP model, for estimating the bank profile shape and also dimensions of threshold channels. A wide range of laboratory and field data are utilized to verify the proposed EDMTC. The results demonstrate that the used Shannon entropy model is accurate with a lower average value of Mean Absolute Relative Error (MARE) equal to 0.317 than a previous model proposed by Cao and Knight (1997) (MARE = 0.98) in estimating the bank profile shape of threshold channels based on entropy for the first time. Furthermore, the EDMTC proposed in this paper has acceptable accuracy in predicting the shape profile and consequently, the dimensions of threshold channel banks with a wide range of laboratory and field data when only the channel hydraulic characteristics (e.g., Q and d_{50}) are known. Thus, EDMTC can be used in threshold channel design and implementation applications in cases when the channel characteristics are unknown. Furthermore, the uncertainty analysis of the EDMTC supports the model's high reliability with a Width of Uncertainty Bound (WUB) of ±0.03 and standard deviation (S_d) of 0.24.

Keywords: water resources; channel; mathematical entropy model; bank profile shape; gene expression programming (GEP); entropy; genetic programming; artificial intelligence; data science; big data

1. Introduction

The sections and dimensions of rivers and alluvial channels change due to the constant interactions between water and sediments. River and channel plans and cross-sections undergo dimensional changes until equilibrium or stable state is attained. After equilibrium, the average dimensions of a stable cross-section do not change over time; in fact, the rate of sedimentation and erosion in a channel cross-section is theoretically in equilibrium [1–3]. In this case, the particles on the bed and at the channel banks are in dynamic balance. In channels with coarse particles, the movement of sediments at any location in the channel contradicts the term "channel stability" [4,5]. In this type of channel, it is not possible for sediments to move without changing the channel dimensions and width [6]. Moreover, the channel dimensions and width of water surface are only preserved (channel stability) in a state when the sediment particles on the channel bed move slightly and at the banks are in the threshold of motion [7]. In such case, one problem related to river morphology is with predicting the erosion process of river banks and profile shape formation until stable sections are achieved [8,9].

The S_t is distributed between zero value on the channel bed and the maximum S_t value (S_t^+) at the free water surface at the water margin. S_t distribution is related to the lateral distance (x) from the channel bed ($x = 0$) to the water margin. At the water margin, x is named L which is equal to the half-width of the free water surface ($B/2$) ($L = B/2$). Therefore, it is worth using the entropy concept in the study of the S_t of bank profiles because the entropy concept is based on the probability principle and its relation to a channel's geometric parameters. Furthermore, since the S_t^+ value at the free water surface is equal to μ (submerged static coefficient of Coulomb friction), the S_t of the banks is affected by the hydraulic parameters of the channel cross-sections too (including flow and sediment characteristics). The $\overline{S_t}$ value in channels is due to the homogeneity of S_t^+ values as a result of these conditions. Because the $\overline{S_t}$ value is not specified for channels (and also there is no specified relation for computing it), a uniform distribution of the transverse bank slope is assumed to obtain the $\overline{S_t}$ value from the ratio of the maximum flow depth at the channel centerline (h_c) to the corresponding lateral distance of this depth from the central channel axis (L). Therefore, if the channel dimension values are not specified, the $\overline{S_t}$ value cannot be obtained. Therefore, a novel relationship would have existed to estimated $\overline{S_t}$ values based on available datasets (not only channel dimensions).

Furthermore, with the obtained entropy equation it is possible to accurately predict the S_t of the banks depending on the correct values of the Lagrange multipliers contained in the equation. Therefore, if the entropy equation can predict the transverse bank slope correctly, multiplier λ should be closely related to the hydraulic and geometric parameters of the banks, which has not been investigated so far except the recent study of authors. Gholami et al. [10] analyzed the sensitivity of λ multiplier to different hydraulic and geometric parameters. They referred to considerable impact of the maximum slope of the bank profile and the dimensionless lateral distance of the river banks on λ variations. Therefore, by investigating the relationship between the entropy parameters and the hydraulic and geometric parameters of a channel, it is possible to achieve a simpler equation for the transverse bank slope distribution and thus, the bank profile shape. Based on Gholami et al.'s [10] study results, a simple relation is presented based on the maximum entropy principle to compute entropy parameter using maximum and mean values of S_t. In the Consequently, the fraction obtained with the $\overline{S_t}$ to S_t^+ ratio (δ) is evaluated and a relationship between the δ ratio and the entropy parameter ($K = \lambda\mu$) is presented. Moreover, a regression model based on GEP is used to create a relationship between the $\overline{S_t}$ of the banks and the geometric and hydraulic parameters of the flow (when the channel dimensions are unknown and only the hydraulic characteristics (e.g., Q and d_{50}) are available). This relationship is combined with Vigilar and Diplas' [11] polynomial equation to present an equation for estimating the stable free surface width based on the relationship between δ and K. The EDMTC proposed in this paper is used together with the bank profile shape equation to obtain the channel bank dimensions.

2. Literature Review

So far, many studies have been carried out to examine channel dimensions in dynamic equilibrium state [12–19]. However, few studies have examined the bank profile shape of threshold channels or the static equilibrium of channels. Parker [6] did extensive research in this field and justified the stable channel paradox with the nonuniform shear stress distribution on the channel bed and banks due to the longitudinal transformation of the lateral flow momentum. Parker's model estimated the bank profile shape as a cosine curve. Later, Ikeda [20] conducted extensive laboratory studies to investigate the shape of stable channel banks. Ikeda then employed a mathematical model based on Parker's idea and presented an exponential equation for bank profile shapes. Ikeda [20] pointed out that the most influential parameters in determining the shape of stable channels are the Q and d_{50}. Diplas [21] used an analytical model with their experimental data and proposed a special case of Ikeda's [20] equation as an exponential function for a bank profile shape. Pizzuto [22] examined the stability criterion using an analytical solution of the widening process at the free water surface. Pizzuto [22] considered the shear stress redistribution due to lateral diffusion and reported an exponential function for a bank profile after channel widening stops. Diplas and Vigilar [23] presented a numerical model to assess the difference between the shape of threshold channels and a previous conventional shape (cosine) for banks. They stated that with particles that do not move along the banks, the transverse slope of the banks should be milder, in which case a wider and deeper channel would form. Hence, they introduced a fifth-degree polynomial profile shape of stable channel banks. Vigilar and Diplas [11,24,25] provided graphs for use to predict the dimensions and profile shapes of stable channel banks with a third-degree polynomial equation. This equation can accurately predict the bank profile shape, because it is in accordance with the results obtained with the equations of several other researchers who have used various other methods [26,27]. Babaeyan [7] did an extensive laboratory study and according to their observational data introduced a hyperbolic bank profile shape. Cao and Knight [28] were the first to examine the shape of bank profiles using the entropy concept. By applying the shape equation obtained with the maximum entropy principle, they reported a parabolic equation. In solving their entropy equation, the Lagrange multiplier (λ) contained within were tested numerically. The equation was validated according to Chow's [29] definition of natural rivers considering a value of zero for λ. Cao and Knight [28] emphasized the need to further consider the physical concept of multiplier λ. Following Cao and Knight's [28] brief study, no other study has been based on the entropy concept to predict the S_t and hence the bank profile shape of stable channels. Gholami et al. [30–34] assessed the ability of different artificial intelligence (AI) methods in the estimation of bank profile shapes of threshold channels. They referred to high efficiency in these methods in estimation and the necessity of further researches about on forming stable shape of bank profiles.

Due to the significance of the entropy concept, many studies have addressed entropy in examining different variables [35–38]. In hydraulic science, Chiu [39] was the first to examine the flow velocity distribution using entropy. Later, other considerations were applied to evaluate the mean and maximum velocity ratio, shear stress and sediment concentration distributions in the cross sections of channels [40–53]. In the field of application of entropy concepts in determining S_t of stable channels, recently, Gholami et al. [54,55] assessed the ability of Tsallis and Shannon entropy concepts in estimation of S_t of stable channels banks. They extensively assessed the variation of different entropy parameters and their signs in obtained entropy-based equations. However, they presented no reports about the significant effects of relations of maximum and mean values of S_t with entropy parameters and the other hydraulic and geometric conditions.

3. Materials and Methods

3.1. Maximum Entropy Principle in Estimating the Transverse Slope of Stable Banks

Cao and Knight [28] evaluated the S_t of banks in threshold state using the principle of maximum entropy for the first time. In the following, Gholami et al. [54,56] modified the application of maximum entropy principle used by Cao and Knight [28]. Cao and Knight [28] employed the Shannon entropy [56] in the form of Equation (1) and presented Equation (2) considering the S_t of stable banks as a random variable and the principle of maximum entropy [57,58] associated with the two constraint conditions of continuity and momentum in Equations (3) and (4) [59].

$$H(S_t) = - \int p(S_t) \ln p(S_t) dS_t, \quad (1)$$

where $p(S_t)$ is the Probability Density Function (PDF) of the S_t of the banks, and H is the amount of entropy.

$$S_t = \frac{1}{\lambda} \ln\left[1 + (e^{\lambda \mu} - 1)\frac{x}{L}\right] \quad (2)$$

$$\int_0^\mu p(S_t) dS_t = 1, \quad (3)$$

$$\int_0^\mu S_t p(S_t) dS_t = \overline{S_t}, \quad (4)$$

where x is the lateral distance of points on the banks from the channel centerline and λ is the Lagrange multiplier. Figure 1 represents a symmetrical bank cross section of stable alluvial channels. In stable channels, S_t of the banks changes monotonically from the centerline of the channel bed ($x = 0$ and $y = 0$) that is zero ($S_t = 0$) to the S_t^+ value at the free water surface at the water margin ($x = L = B/2$ and $y = h_c$), which is equal to μ (the submerged static coefficient of Coulomb friction).

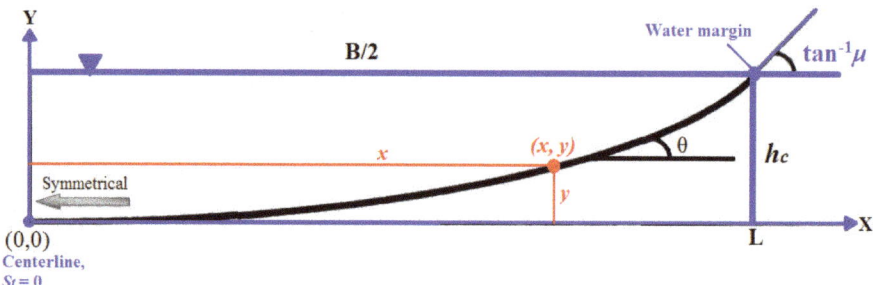

Figure 1. Symmetrical cross section of alluvial threshold channels and its characteristics.

Cao and Knight [28] carried out numerical testing and considered a specified range for λ (1, 5, 10, 50, 100). They stated that when λ tends toward zero, the cross-sectional bank shape is a parabolic curve. Consequently, this multiplier was deleted from their equation. The following equation was presented with numerical justification for bank profile shape estimation:

$$y* = \left(\frac{\mu^2}{4}\right) x*^2, \quad (5)$$

where $x* = x/h_c$ is the dimensionless lateral distance from the channel centerline and $y* = y/h_c$ is the dimensionless vertical boundary level. The Lagrange multiplier is a key component of the maximum entropy principle. In the following, Gholami et al. [54] presented an equation based on the maximum entropy principle to caculate λ numerically [54] which is explained in summary in the following. Accordingly, by using the Lagrange Multiplier Method (LMM) and variable calculation technique [39,60,61], the equation below is obtained for $p(S_t)$:

$$p(S_t) = \exp(\lambda_1 + \lambda S_t - 1). \quad (6)$$

Equation (6) is used with the first constraint (Equation (3)) to obtain the following equation:

$$e^{\lambda_1 - 1} = \lambda(e^{\lambda \mu} - 1)^{-1}, \tag{7}$$

where λ_1 is Lagrange multipliuer and equal to: $\lambda_1 = \ln[\lambda/(e^{\lambda \mu} - 1)] + 1$.

Furthermore, by replacing Equations (6) and (7) in the second constraint condition (Equation (4)), the following equation is obtained to calculate λ:

$$\overline{S_t} = \frac{\mu e^{\lambda \mu}}{(e^{\lambda \mu} - 1)} - \frac{1}{\lambda}. \tag{8}$$

On the other hand, by dividing the sides of Equation (8) by μ, the following equation is obtained:

$$\frac{\overline{S_t}}{\mu} = \delta = \frac{e^K}{(e^K - 1)} - \frac{1}{K} \tag{9}$$

where K is a dimensionless parameter known as the entropy parameter used to measure the uniformity of the probability and distribution of the S_t, which is equal to $K = \lambda \mu$, and δ is the ratio of $\overline{S_t}$ to S_t^+ (=μ). In the present study, when the values of h_c, L, and S_t^+ (=μ) are known, the $\overline{S_t}$ value along the banks is obtained by assuming the uniform distribution of S_t as equal to the h_c/L ratio. Therefore, λ is obtained by numerically solving Equation (8). Then, the S_t distribution of stable banks can be computed according to Equation (2). Moreover, physical justifications of λ multiplier and the effect of different hydraulic and geometric parameters on it is investigated in Gholami et al. [10]. On the other hand, the S_t at each point on the channel banks is formulated as $S_t = dy/dx$, where y is the vertical boundary level of the points. By integrating this, the bank profile shape equation for threshold channels becomes Equation (10), where the integral constant (C) is obtained by applying the boundary condition at the channel centerline (x and $y = 0$).

$$y = \frac{1}{\lambda} \left[\left(x + \frac{L}{e^{\lambda \mu} - 1} \right) \ln \left(1 + (e^{\lambda \mu} - 1) \frac{x}{L} \right) - x \right]. \tag{10}$$

This is introduced as the bank profile shape equation based on developed entropy model which is extended in Gholami et al. [54] in details. If the channel dimensions (B and h_c) are not specified, it is not possible to estimate λ and hence, the S_t and y values. Therefore, in this paper, the next section presents a numerical model for when the channel dimensions are not specified and only Q and d_{50} are known from the problem condition.

3.2. Calculating μ

The μ value can be calculated as $\mu = \tan \varphi$, where φ is the angle of sediment reposition. Furthermore, since the value of μ changes with the sand size and roughness [5,62], the following relationship between the φ and sediment size (d_{50}) can be utilized in the current study to compute φ in uniform sediments [10,27,54]:

$$\varphi = \left[\begin{array}{l} 0.302(\log d_{50})^5 + 0.126(\log d_{50})^4 - 1.811(\log d_{50})^3 \\ -0.57(\log d_{50})^2 + 5.952(\log d_{50}) + 37.52 \end{array} \right] \tag{11}$$

where φ is in degree and d_{50} should be inserted in centimeters.

3.3. Entropy-Based Design Model of Threshold Channels (EDMTC)

As stated in the previous section, by assuming a uniform distribution for S_t value, the $\overline{S_t}$ value can be obtained by the h_c/L when the values of h_c, and L (=$B/2$) are known. Accordingly, if the h_c and B

values are not known, it is not possible to calculate $\overline{S}_t$. In this section, an explicit relationship will be provided to calculate the $\overline{S}_t$ value for the cases that the channel dimensions (h_c, B) are not available.

In this way, using several series of available observational data with different hydraulic conditions, the Q, d_{50} and μ values are determined and a relationship for the $\overline{S}_t$ value based on these parameters is applied to calculate the $\overline{S}_t$ value for any other data where the channel dimensions are not specified. Accordingly, considering Q, d_{50} and μ parameters as input parameters and $\overline{S}_t$ as output parameter based on a numerical GEP model (Figure 2) [32,63,64] provide a relationship for predicting $\overline{S}_t$ in the form of Equation (12):

$$\overline{S}_t = G1 + G2 + G3,$$
$$G1 = e\left\{-\left\{[\mu^2 - 2\mu + \ln(\mu + 4.433)] + \left[\exp(-(Q+\mu)^2)\right] + \exp\left[-((0.936 + d_{50})^2)\right]^2\right\}\right\},$$
$$G2 = e\left\{-\{[(17.693 - 1.565Q) + (1/d_{50})] + [\mu + 1.565 - \mu Q]\}^2\right\},$$
$$G3 = e\left\{-\{[(1.112Qd_{50} - \ln(6.5Q))/\mu] + \mu\}^2\right\}.$$
(12)

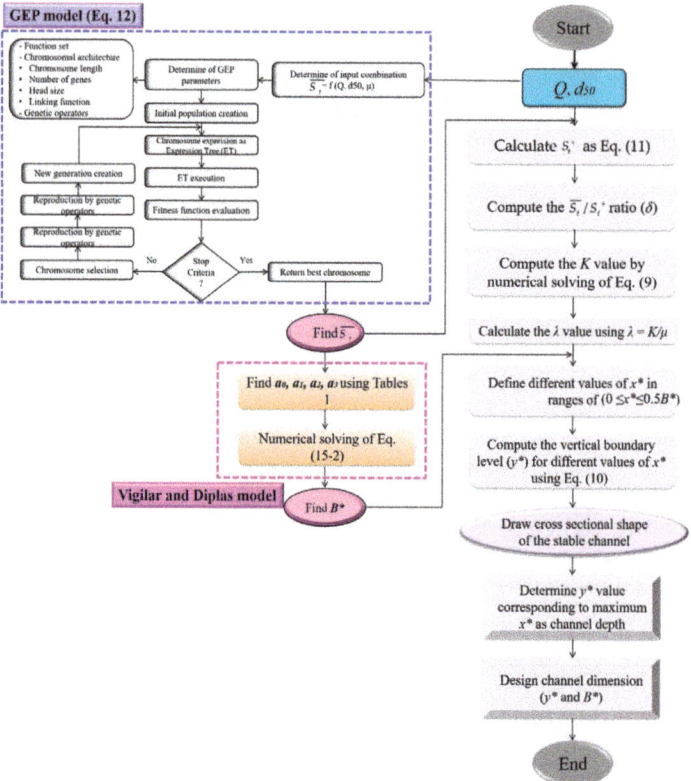

Figure 2. Flowchart of the proposed Entropy-based Design Model of Threshold Channels (EDMTC) computational procedure for designing the dimensions and shape of threshold channels in the present study.

In fact, with input parameters Q, d_{50} and μ (=S_t^+) the value of $\overline{S}_t$ is calculated using Equation (12). Now by knowing the $\overline{S}_t$ value for any channel whose stability dimensions are not specified, in addition to bank profile shape, the width and depth of the channel after stability can be determined. To do this,

$\overline{S}_t$ can be calculated by using the equations presented by former researchers who have applied analytical and theoretical frameworks to derive the relationships. As stated, the polynomial shape proposed by some researchers is an acceptable shape than the previous classic cosine, parabolic, and exponential forms [23]. Therefore, in the present study, the polynomial function provided by Vigilar and Diplas [11] is used to estimate the bank profile shape of stable channels as follows [11]:

$$y* = 1 - a_3 x*^3 - a_2 x*^2 - a_1 x* - a_0. \tag{13}$$

Coefficients a_0, a_1, a_2 and a_3 depend on the values of δ^*_{cr} and μ, which are obtained from Table 1 for each given dataset [11]. δ^*_{cr} is the dimensionless critical stress depth ($\delta^*_{cr} = \delta_{cr}/h_c$) in critical condition of sediments in the bank profile. In this case, the shear stress depth (δ') is $\delta' = \tau/\rho g S$, where τ is the shear stress along the channel and S is the longitudinal slope of the water surface. The value of δ^*_{cr} can be obtained according to the ($\mu - \delta^*_{cr}$) figures related to Vigilar and Diplas [11].

Table 1. Coefficients in the bank profile shape equation related to Vigilar and Diplas [11] (Equation (13)) for different values of μ and δ^*_{cr} [11].

a_0	a_1	a_2	a_3	δ^*_{cr}
		$\mu = 0.4$		
1.0001	−0.0135	−0.0411	0	0.93
1.0004	−0.0236	−0.0412	0	0.935
1.0008	−0.0307	−0.0412	0	0.94
1.0009	−0.0342	−0.0413	0	0.945
		$\mu = 0.55$		
1.0003	−0.018	−0.0503	−0.0029	0.9
1.0006	−0.0299	−0.0527	−0.0027	0.905
1.0008	−0.0366	−0.0547	−0.0025	0.91
1.001	−0.0416	−0.0565	−0.0022	0.915
1.0011	−0.0463	−0.0586	−0.0019	0.921
		$\mu = 0.65$		
1.0006	−0.0278	−0.0543	−0.006	0.885
1.001	−0.0444	−0.06	−0.0054	0.895
1.0013	−0.0529	−0.0647	−0.0048	0.905
1.0041	−0.0556	−0.0665	−0.0045	0.909
		$\mu = 0.76$		
1.0009	−0.0365	−0.0544	−0.0105	0.87
1.0014	−0.0531	−0.061	−0.0101	0.88
1.0017	−0.0621	−0.0662	−0.0095	0.89
1.0018	−0.0662	−0.0701	−0.009	0.897
		$\mu = 0.84$		
1.0011	−0.0418	−0.0516	−0.0146	0.86
1.0016	−0.0594	−0.059	−0.0143	0.87
1.002	−0.0697	−0.0634	−0.0141	0.88
1.0021	−0.0742	−0.0708	−0.013	0.89
		$\mu = 1.0$		
1.0016	−0.0571	−0.0466	−0.0233	0.845
1.0022	−0.0738	−0.0531	−0.0237	0.855
1.0025	−0.0828	−0.0589	−0.0236	0.865
1.0028	−0.0884	−0.0656	−0.023	0.875
1.0028	−0.0892	−0.0683	−0.0226	0.878

Now, the derivative of the above function (Equation (13)) versus $dx*$ yields the transverse slope function at different points in the channel as follows:

$$S_t = \frac{dy*}{dx*} = -3a_3 x*^2 - 2a_2 x* - a_1. \tag{14}$$

Now, according to the mean value theorem in integral, the mean slope value of the bank profiles ($\overline{S_t}$) is computed based on the mean value theorem for definite integrals for $y*$ distribution (Equation (13)) along the transverse interval in range of ($0 \leq x* \leq 0.5B*$) according to the following Equations (15a–c):

$$\overline{S_t} = \frac{1}{0.5B*} \int_0^{0.5B*} y*(x)dx, \qquad (15a)$$

$$\overline{S_t} = \frac{2}{B*}\left[1 - a_3\frac{B*^3}{8} - a_2\frac{B*^2}{4} - a_1\frac{B*}{2} - a_0\right], \qquad (15b)$$

$$\overline{S_t} = -a_3\frac{B*^2}{4} - a_2\frac{B*}{2} - a_1 - \frac{2}{B*}(a_0 - 1). \qquad (15c)$$

Therefore, by obtaining $\overline{S_t}$ value using Equation (12), $B*$ value of the free water surface of bank profile is obtained with Equation (15b). In fact, with input parameters Q, d_{50} and μ ($=S_t^+$) the value of $\overline{S_t}$ is calculated using Equation (12). Then, Equation (15c) is used to obtain the value of $B*$ based on obtained $\overline{S_t}$ values according to Equation (12). Accordingly, in this study, the EDMTC (Figure 2) is presented to predict the dimensions and shape of bank profiles using the entropy principle. The value of $x*$ (lateral distance from the channel axis) is selected for a specific range of arbitrary $x*$ values at a distance of $0 \leq x*_i \leq 0.5B*\,(=L)$. The values of $y*$ obtained by the entropy facilitate plotting the bank shape profiles against different x_i. Figure 2 shows the flowchart of the GEP model and model developed in the present study (EDMTC) to predict the shape and dimensions of threshold channels.

3.4. Experimental Data

The observational data series used in the present study were collected in previous investigations by Mikhailova et al. [65], Ikeda [20], Diplas [19], Babaeyan [7], Macky [66], Hassanzadeh et al. [67], and Khodashenas [68]. The hydraulic and geometric conditions of the data vary, with different ranges of Q and d_{50} values in the channel as well as geometric conditions of the laboratory flumes used with each data series. Furthermore, several tests were carried out for different discharge rates with each data series, and the channels had different conditions until reaching equilibrium state. In each observational data series, in addition to the channel dimensions (B and h_c) the coordinate data of the points in stable bank profiles (x, y) were extracted for some discharge values as well. Moreover, all experiments were done in laboratory flumes with different aspect ratios ($B/h_c = \alpha$) in the range (4–30). In each test, the sediment sizes selected were somewhat course, so the corresponding proportional discharge in the channels would cause no movement of sediment particles in the channels. Hence, the stresses on the walls and channel bed were respectively less and more than the critical stress until threshold channel conditions would govern. Table 2 summarizes the hydraulic and geometric conditions for the data used.

Table 2. Summary of experimental characteristics for the data used in the present study.

Researchers	Runs. No.	No. of Series	d_{50} [mm]	Discharge (Q) [L/s]	Water Surface Half-Width (B/2) [cm]	Central Water Depth (h_c) [cm]
Mikhailova et al. [65]	2	S1	0.2	65	112	10.4
		S2	0.2	69	132.5	14.4
Ikeda [20]	1	S3	1.3	16.28	24.8	3.54
Diplas [21]	1	S4	1.9	12.526	33	3.85
Babaeyan [7]	1	S5	1	2.5	52.6	2.63
Macky [66] (Field data)	1	S6	3.42	64.3	127	3.7
Hassanzadeh et al. [67]	2	S7	1.2	11.09	32	8.6
		S8	1.6	20.07	40.6	10.9
Khodashenas [68]	4	S9	0.53	6.2	21.7	8
		S10	0.53	2.57	16	6.3
		S11	0.53	2.18	17	6.12
		S12	0.53	1.157	9.5	3.7

3.5. Used Data in Modeling

As stated in the previous section, in this paper, 12 numbers of observed runs (S1–S12) (according to Table 1) with different hydraulic and geometry characteristics are selected for training and testing the EDMTC model. The hydraulic and geometric conditions of the data series are varied, so that the range of Q and d_{50} values in the channel, as well as the geometric conditions of the laboratory flumes used in each data series, are different. Furthermore, in each seven available observational data series (Mikhailova et al. 1980; Ikeda 1981; Diplas 1990; Babaeyan 1996; Macky 1999; Hassanzadeh et al. 2014; and Khodashenas 2016), there are several runs related to them according below with different discharges, therefore, the stable channel shape formed on banks in each observed run is different.

- Ikeda (1981) → one run as **S3** (8 samples)
- Diplas (1990) → one run as **S4** (25 samples)
- Babaeyan (1996) → one run as **S5** (8 samples)
- Macky (1999) → one run as **S6** (101 samples)
- Hassanzadeh et al. (2014) → two runs as **S7** (33 samples) and **S8** (38 samples)
- and Khodashenas (2016) → four runs as **S9** (44 samples), **S10** (33 samples), **S11** (57 samples) and **S12** (20 samples)

In fact, in this paper, external-validation is performed. External validation means that among 12 numbers of data series (totally 367 sample numbers), some data series are used for training and some data series are selected for testing the models. Accordingly, in this paper, 10 data series of S1, S2, S3, S7, S8, S9, S10, S11, and S12 (65% of all samples: 233 samples) are used for training the EDMTC model and three data series of S4, S5, and S6 (35% of all samples: 134 samples) related to Diplas' (1990), Babaeyan's (1996) and Macky's (1999) data series are selected for testing the EDMTC model. This kind of validation is acceptable, because the proposed EDMTC model is trained and tested based on data series with different hydraulic and geometry characteristics.

3.6. Evaluation of Model Efficiency

In order to evaluate the methods presented in this study, several statistical indices are used: The determination coefficient (R^2), Root Mean Squared Error (RMSE), Mean Absolute Relative Error (MARE), Mean Absolute Error (MAE), and Bias. These evaluation criteria are defined by Equations (16)–(20):

$$R^2 = 1 - \frac{\sum\limits_{i=1}^{n}(y_i - x_i)^2}{\sum\limits_{i=1}^{n}(y_i - \bar{y})^2}, \qquad (16)$$

$$RMSE = \sqrt{\frac{1}{n}\sum_{i=1}^{n}(x_i - y_i)^2}, \qquad (17)$$

$$MARE = \frac{1}{n}\sum_{i=1}^{n}\left(\frac{|x_i - y_i|}{x_i}\right), \qquad (18)$$

$$MAE = \frac{1}{n}\sum_{i=1}^{n}|x_i - y_i|, \qquad (19)$$

$$Bias = \frac{1}{n}\sum_{i=1}^{n}(x_i - y_i), \qquad (20)$$

where y_i and x_i denote the estimated and observed values, $\bar{y}$ represents the mean modeled values and n is the sample size. The closer the R^2 coefficient is to the unit value (1), the higher the agreement there is between the observed and predicted values. The closer the results of MARE, RMSE, Bias, and MAE indices are to zero, the higher the estimation accuracy is as well. Positive and negative Bias values imply model over and underestimation, respectively [69–71]. Therefore, computing several evaluation criteria can better reveal the model performance [72,73].

4. Results

In the first section, the ability of entropy model is evaluated to predict bank profile shapes. In the second section, the EDMTC proposed in this study is examined in detail. At the end, the uncertainty of the proposed EDMTC is examined using different uncertainty indexes.

4.1. Entropy Model in Predicting Bank Profile Shapes

In Figure 3, the vertical boundary level of stable channel banks is estimated by the developed entropy model based on the maximum entropy principle which is proposed in Gholami et al. [54] for the first time. The λ value is obtained by numerical solution of Equation (8). Accordingly, for each data series (each bank profile shape), one λ value is obtained by numerically solving Equation (8). In Equation (8), $\overline{S_t}$ value is calculated by assuming uniform distribution of S_t, according to ratio of h_c/L. Using obtained λ value, the y value is computed based on entropy method by solving Equation (10). The y^* distribution obtained by Equation (10) corresponding each x^* value is drawn for each data series in Figure 3. Moreover, the results of Cao and Knight's [28] model (CKM) (according to Equation (5)) are extracted and their proposed bank profile shape is drawn in Figure 3 to evaluate the entropy model performance. Table 3 contains the different error indices for entropy model and CKM. Figure 3 indicates that entropy model exhibits acceptable conformity with the corresponding observational data series in predicting the vertical boundary level and hence, estimates the bank profile shape with low error values. According to all data series, entropy model is able to estimate the governing bank profile shape trend with lower MARE and RMSE values equal to 0.317 and 0.08 better than CKM with 0.981 and 0.363 values respectively. Figure 3 also shows that for two data series, i.e., S1 and S2 (Mikhailova et al.'s [65] data), CKM has high error values in y^* estimation and high accuracy in the area near the free water surface, where high MARE values in the 2–4 range are observed for these data series. However, the proposed entropy model is able to detect the bank profile shape trend with lower error values (MARE = 0.2 and 0.8 for S1 and S2 datasets respectively) than CKM with 1.95 and 3.95 MARE values, which represents the significant superiority of entropy model. This process is repeated for the S2 and S3 data series. Although CKM exhibits acceptable performance, entropy model is more accurate with lower error values and coincides closely with the observed values (especially in the area near the surface). For the S6 field data series, although both models do not perform well (with close Bias values of −0.31 and 0.45 for entropy model and CKM respectively), entropy model again performs with lower error (MARE = 0.58) than CKM (MARE = 1.03). Furthermore, the high MARE index value for CKM is representative of its inability to estimate low y^* values (in the vicinity of channel bed), a problem that is solved by entropy model significantly. Furthermore, the RMSE values of CKM and entropy model which is equal to 0.5 and 0.38 respectively approved the inefficiency of CKM in estimating low y^* levels. With data series S7, the improvement of entropy model over CKM by about 60% and 85% in the MARE and RMSE values respectively is observed clearly in Figure 3, as entropy model highly conforms to the observational data with R^2 values of 0.98. With Khodashenas' [68] data (S9–S12), the higher efficiency of entropy model over CKM is evident with lower MARE and RMSE values in entropy model than CKM. Furthermore, entropy model is able to estimate the water surface widening with high y^* values well with low values of RMSE and Bias values close to 0. The negative and positive Bias value represents the underestimation and overestimation of the models respectively. As it can be seen in the Bias values, the CKM in most of the datasets have positive Bias values and overestimates the y^* values in comparison with the corresponding observed values. It can thus be said that the entropy model proposed in the present study based on the maximum entropy principle is more accurate in the estimating the bank profile shape of stable channels than CKM, which suggests a parabolic curve (Equation (5)) for channel banks. A notable point in this paper is the significant physical effect of λ values on the accurate estimation of the intended variables, which is negligible with CKM. The λ values obtained by entropy model in this study are gathered in Table 3, where it can be seen that this multiplier is in a specified range of −2 to 2 with almost all data series (except with 1–2 data series). Furthermore, the λ values are the same for different runs of one experiment.

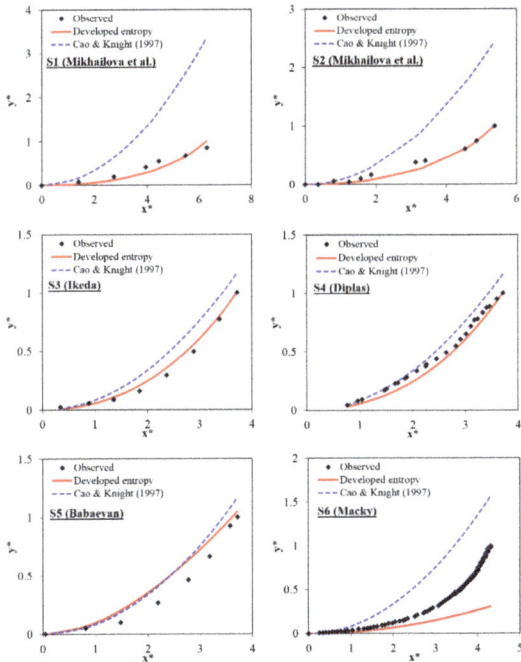

Figure 3. *Cont.*

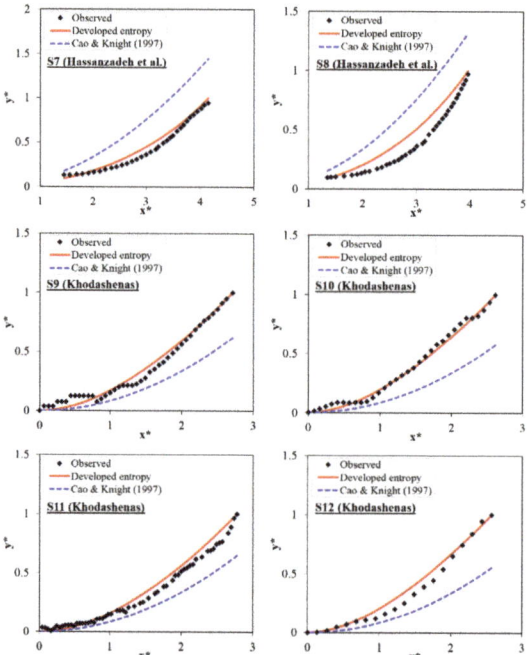

Figure 3. Bank profile shape predicted by developed entropy model and Cao and Knight's [28] model (CKM) for different observational data series (S1–S12).

Table 3. Assessment of the efficiency of developed entropy model (DEM) and CKM compared with different observational data series according to different error indices and λ values related to DEM in this paper.

Data Series	MARE		RMSE		Bias		R^2		λ
	DEM	CKM	DEM	CKM	DEM	CKM	DEM	CKM	DEM
S1	0.254	1.95	0.103	1.31	−0.04	0.99	0.93	0.981	−5.56
S2	0.86	3.95	0.057	0.7	−0.036	0.47	0.98	0.988	−4.26
S3	0.228	0.47	0.037	0.141	0.022	0.116	0.99	0.981	−1.62
S4	0.15	0.11	0.053	0.08	−0.05	0.064	0.99	0.997	−1.75
S5	0.43	0.42	0.1	0.135	−0.08	0.114	0.99	0.988	2.11
S6	0.58	1.03	0.38	0.5	−0.31	0.45	0.96	0.957	1.5
S7	0.147	0.86	0.056	0.37	0.045	0.35	0.98	0.966	−2.46
S8	0.315	0.99	0.109	0.34	0.098	0.32	0.97	0.95	−2.2
S9	0.26	0.50	0.044	0.184	0.008	−0.148	0.98	0.989	1.72
S10	0.18	0.56	0.028	0.24	−0.01	−0.192	0.99	0.987	2.2
S11	0.23	0.46	0.05	0.14	0.03	−0.108	0.99	0.996	1.4
S12	0.17	0.47	0.05	0.22	0.03	−0.16	0.985	0.996	2.4
Averaged	0.317	0.981	0.08	0.363	−0.02	0.189	0.978	0.981	-

4.2. Presenting the Entropy-Based Design Model of Threshold Channels (EDMTC)

In previous sections, the entropy model was evaluated for its prediction ability of bank profile shapes in case the depth and width of the free water surface in the channel are determined. In this study, EDMTC based on the relationship between the entropy parameter and the S_t of channel banks to predict the channel dimensions as well as the bank profile shape is presented and explained in detail in

Section 3.2 and Figure 2. The proposed EDMTC is evaluated in the first subsequent section and the model's uncertainty is examined in the second part.

Evaluation of EDMTC Performance

Figure 4 displays scatter plots of the EDMTC proposed in this study for several observational data series. The left side of the figure contains the regression plots of the y^* values predicted by EDMTC compared to the corresponding observational values. The right side of the figure shows the cross-sectional profile shapes predicted by EDMTC compared with the profile shapes obtained with observational values. Table 4 lists the error indices of EDMTC compared to the corresponding observational values. The scatter plots indicate that EDMTC can very accurately predict the vertical elevation of stable channel banks, as most data is compressed around the trend line and slight scattering is observed for some of the datasets. In Figure 4, the trend line is mapped to the data and the resulting equation is $y = ax + b$. Closer a and b values to 1 and 0, respectively, represent acceptable model prediction performance. According to the trend line, for all datasets the predicted values are concentrated around this line and the values of a, b are close to 1, 0, respectively. This indicates the high efficiency of the proposed EDMTC in predicting the vertical elevation of channel banks. Moreover, the R^2 index value in this figure is higher than 0.95 for all observational data series, indicating the high EDMTC prediction accuracy. The value of this index is very close to 1 for some of the observational data [20,21,68], signifying very high model conformity to the corresponding observational values. Furthermore, according to the diagrams on the right side of Figure 4, the EDMTC is able to accurately estimate the bank profile shape trend for all data series. Although some differences between the values y^* predicted by the model and the observational values are seen, it is notable that EDMTC is able to model the vertical bank elevation (from the channel center on the bed to the free water surface margins) and the water surface widening near the water surface levels similar to the corresponding observational values. The error index values in Table 4 are also validated accordingly. This table shows that the MARE values for all datasets are 0.3–0.5, which is close to 0. This index indicates the accuracy of the proposed EDMTC in predicting the vertical elevation of banks as well as the free water surface width in stable channels. An important point is that the proposed EDMTC predicts the profile shape trend successfully and can therefore be used to design the width and depth (dimensions) of stable channels when only flow inputs such as Q, d_{50} and μ are known. The high accuracy of this model is confirmed, and achieving such a model with the least parameters to predict the dimensions and cross-sectional bank shapes formed in stable channels is of considerable importance. Also, EDMTC not only considers the geometric conditions of the channel cross sections but also involves the hydraulic conditions of the problem (by using Vigilar and Diplas' [11] equation), which is one of the notable features of this model. Based on most observational data series, the estimated channel width is very similar to the observational values (in some cases it is slightly less). For example, for the EDMTC profile predictions based on the observational data from Diplas [21], Babaeyan [7], and Hassanzadeh et al. [67], the water surface width is estimated very close to the observed values. Furthermore, for most observational datasets, the proposed model estimates greater values for the vertical elevation of the water surface, although the estimated profile trend fits the observational values perfectly. The partial error values of EDMTC that are mostly seen in the areas near the channel bed and the free water surface with some of the datasets can be considered measurement errors of the observational data [74]. For some data, e.g., Hassanzadeh et al. [67] and Khodashenas [68] this error is seen at the channel bed. Additionally, Figure 4 shows that EDMTC based on Khodashenas' [68] data estimates lower y^* than the actual values, which results in a negative Bias and an absolute error increase of 14% in MAE value according Table 4 (MAE represents the absolute magnitude of the difference between observational values and the model). It is worth noting that the EDMTC can estimate a more logical shape than the profile derived from the corresponding observational values, which has a uniform distribution from the bed to the water surface. With the rest of the data series, EDMTC estimates roughly higher partial values equal to the observational values for y^*, as the RMSE error value is about 0.9–0.13, which is acceptable. Therefore, EDMTC with low average error values (MARE = 0.55 and MAE = 0.19) is generally highly accurate in predicting bank profiles and stable channel dimensions.

Table 4. Evaluation of the EDMTC proposed in the present study in estimating the dimensions of stable channels in comparison with several available observational data series.

Dataset	R^2	MARE	RMSE	MAE	Bias
Ikeda [20] (S3)	0.995	0.357	0.098	0.078	0.064
Diplas [21] (S4)	0.991	0.186	0.132	0.097	0.094
Babaeyan [7] (One set) (S5)	0.961	0.400	0.124	0.095	−0.095
Macky [66] (S6)	0.942	0.568	0.556	0.381	0.380
Hassanzadeh et al. [67] (S7)	0.986	1.164	0.456	0.436	0.436
Hassanzadeh et al. [67] (S8)	0.981	1.146	0.380	0.364	0.364
Khodashenas [68] (S9)	0.992	0.426	0.127	0.109	−0.109
Khodashenas [68] (S10)	0.979	0.473	0.169	0.143	−0.143
Khodashenas [68] (S11)	0.994	0.361	0.096	0.076	−0.076
Khodashenas [68] (S12)	0.995	0.475	0.193	0.147	−0.147
Average	0.9816	0.5556	0.2331	0.1926	0.0768

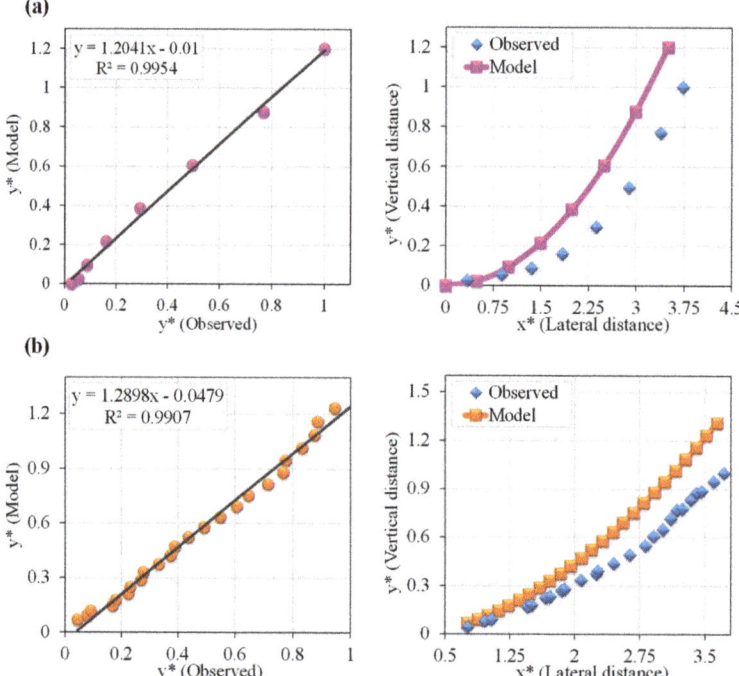

Figure 4. *Cont.*

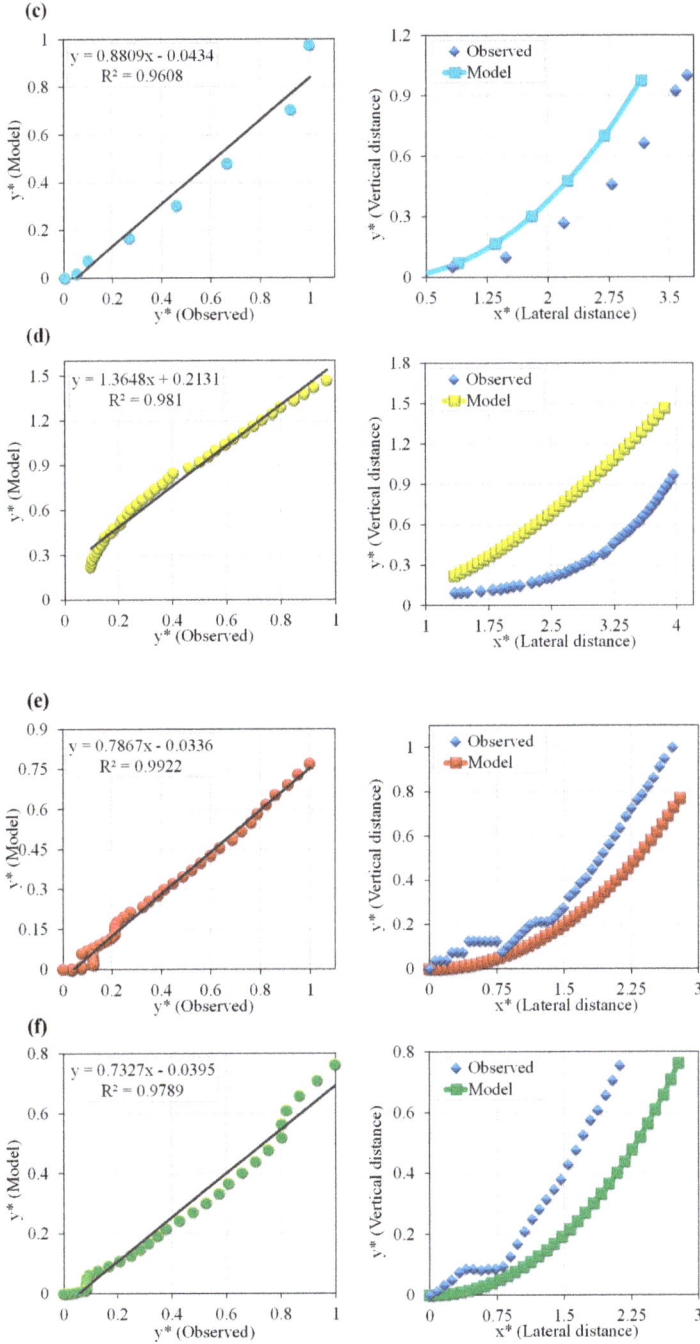

Figure 4. Comparison of values predicted for the vertical boundary level of stable channels by the EDMTC proposed in the present study using scatter plots (left side) and cross-sectional profile shapes (right side) for different observational data: (**a**) Ikeda [20]-S3, (**b**) Babaeyan [7]-S5, (**c**) Diplas [21]-S4, (**d**) Hassanzadeh et al. [67]-S7, (**e**) Khodashenas [68]-S9, and (**f**) Khodashenas [68]-S12.

4.3. Uncertainty Analysis of the Proposed EDMTC and GEP Model

In this section, the uncertainty of EDMTC in predicting the bank profile shape based on entropy model ans also GEP model in predicting $\overline{S_t}$ of bank (according Equation (12)) is examined and the uncertainty indices are shown in Table 5. With the Uncertainty Wilson Score Method (UWSM) [10], ref. [19,75–79], the error of the $\overline{S_t}$ predicted by the GEP model and the y^* values predicted by EDMTC is calculated and compared with the corresponding observation values. The error between estimated and observed values (e_i) and the corresponding the Mean Prediction Error (MPE or $\bar{e}$) and standard deviation (S_d) for error values calculated for data is obtained as Equations (21)–(23):

$$e_i = x_i - y_i, \tag{21}$$

$$MPE = \bar{e} = \frac{1}{n}\sum_{i=1}^{n} e_i, \tag{22}$$

$$S_d = \sqrt{\sum_{i=1}^{n}\left(\frac{(e_i - \bar{e})^2}{n-1}\right)}, \tag{23}$$

where n is the sample size. With these indices, the WUB are calculated as Equation (24):

$$WUB = \frac{1}{n^{0.5}}(I_{lt}\, S_d), \tag{24}$$

where I_{lt} is the left-tailed inverse of the error distribution that represent the probability of error distrubution associated with the numebr of degree of freedom with which to characterize the distribution [76,80]. In the present paper, the probability of 0.05 error (95% Confidence Bound (CB)) with degree of freedom equals to $n-1$ is considered in I_{lt}-value calculation [80]. Moreover, CB is the 95% quantile of the I_{lt} distribution with 1 degree of freedom. In the following, CB can be defined. In this range, the WUB represents the upper and lower uncertainty bounds of CB respectively as Upper Bound (UB) and Lower Bound (LB). UB and LB can be calculated by $\bar{e} \pm WUB$. Moreover, the CB represents the mean value of error. Furthermore, $\bar{d}_x$ represents the average width of CB which is calculated as Equation (25). The lower average width of the CB associated with the lower values of S_d and WUB provides the high certainty of model.

$$\bar{d}_x = \frac{1}{n}\sum_{i=1}^{n}(UB - LB) = \frac{1}{n}\sum_{i=1}^{n}\bar{e} \pm WUB, \tag{25}$$

Table 5. Uncertainty analysis for the Gene Expression Programming (GEP) model in $\overline{S_t}$ prediction according to Equation (12) and EDMTC.

Model	Datasets	Sample Number	S_d	MPE	WUB	$\bar{d}_x$	CB
EDMTC	Ikeda [20] (S3)	8	0.08	−0.064	±0.07	0.065	−0.13 to 0.00
	Diplas [21] (S4)	25	0.09	−0.094	±0.04	0.09	−0.13 to −0.05
	Babaeyan [7] (S5)	8	0.08	0.095	±0.075	0.095	+0.02 to +0.17
	Khodashenas [68] (S9)	44	0.07	0.109	±0.02	0.11	+0.09 to +0.13
	Khodashenas [68] (S10)	33	0.09	0.143	±0.035	0.145	+0.11 to +0.18
	Khodashenas [68] (S12)	20	0.13	0.147	±0.06	0.15	+0.09 to +0.21
	All datasets	266	0.33	−0.14	±0.04	0.14	−0.18 to −0.10
GEP, Equation (12)	All datasets	20	0.02	−0.009	±0.01	±0.01	−0.02 to 0.00

The ideal certainty analysis is achieved when most of the estimated values are bracketed within the CB and also the narrowest width is achieved.

Table 5 shows the MPE, CB, $\bar{d}_x$, and WUB for predicting the S_t^- using the GEP model as well as the values of these indices for the EDMTC. Figure 5 displays the CB calculated using MPE for several observational data series (S3, S4, S5, S9, S10, and S12). In EDMTC, according Table 5, for all datasets, the low values of $\bar{d}_x$ (0.14), WUB (±0.04) and the low value of MPE (−0.14) represent the low uncertainty and high precision of proposed EDMTC in predicting y^* values. It is clear that for almost each observational data series, 95% of predicted and observed values are within the CB range beside the narrow WUB. This represents the acceptable accuracy of the proposed models in predicting the vertical boundary elevation of stable channel profiles. According Table 5, in S3 [20] and S5 [7] data, almost all of the y^* values predicted by EDMTC model are located within the one side of CB. Because, in these series of data, the more underestimation and overestimation performance of the EDMTC causes the almost high values of WUB. Morover, CB is calculated based on mean error values, therefore, the higher and lower predicted y^* values than observed values are located in one side of CB. For the rest of the data, as more than 95% of the data are within this bound. According Table 5, the WUB in all test is low for EDMTC and for GEP model the WUB is 0.01. The low WUB and associate with the low $\bar{d}_x$ values provides a high certainty and precision of EDMTC for S3 [20] and S4 [21], and S5 [7]. While in S12 [68] the low values of WUB is associated with high S_d values. The low values of S_d and WUB in GEP model represents the high precision (low MPE value) and certainty of model simultaneously. Therefore, according to the explanations and results presented, it can be said that the proposed EDMTC and GEP has great certainty and their ability to predict the dimensions and stable bank profiles with high accuracy is assured. Therefore, the models proposed in this study can be used to predict channel dimensions in cases when there is little channel information given. Besides, the proposed model is capable of predicting the profile shape of stable channel banks when observational data for the bank profile shape is not available.

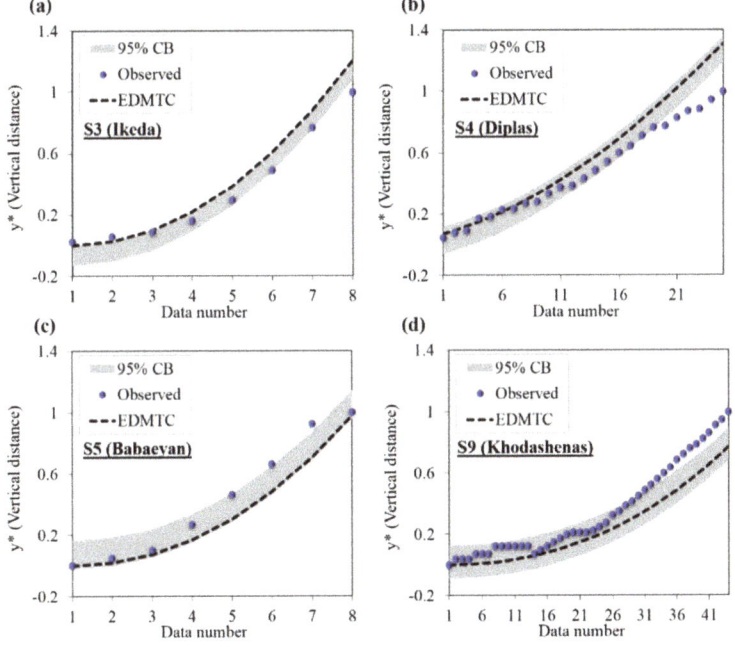

Figure 5. Cont.

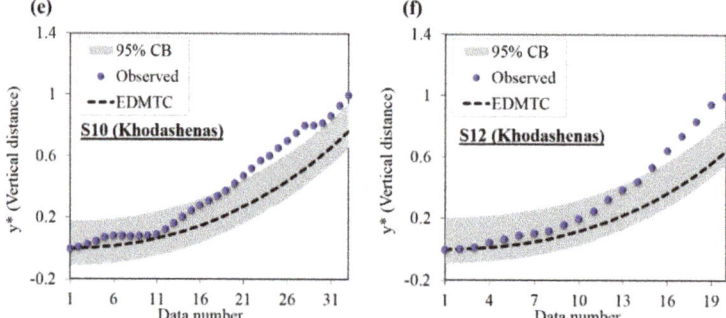

Figure 5. CB (95%) ranges for the observational values and values predicted by EDMTC for the vertical boundary elevation of stable channels based on different datasets of (**a**) Ikeda [20] (S3), (**b**) Diplas [21] (S4), (**c**) Babaeyan [7]-S5, (**d**) Khodashenas [68]-S9, (**e**) Khodashenas [68]-S10, and (**f**) Khodashenas [68]-S11.

Finally, the proposed EDMTC can be used to determine the maximum value of y^* as the maximum dimensionless depth at the channel center and the predicted free surface width. In this case, the channel dimensions can be obtained using the proposed model.

5. Conclusions

In the present study, the maximum entropy principle was employed to provide an equation to calculate the Lagrange multipliers. Accordingly, an equation was developed to predict the bank profile shape of threshold channels. The relation between (δ) ratio with the entropy parameter (K) and the hydraulic and geometric characteristics of channels was evaluated. Next, the EDMTC computational model for estimating the shape of banks profiles and the channel dimensions (B and h_c) was designed based on the maximum entropy principle in combination with the GEP regression model for cases when only the Q and d_{50} are known as problem conditions. The results indicate that the entropy model is capable of predicting the bank profile shape trend with acceptable error values (MARE = 0.317, RMSE = 0.09) according to the experimental data in comparison with the Cao and Knight's [28] model (MARE = 0.317, RMSE = 0.09). Therefore, the λ multiplier has a significant role in determining the transverse slope and consequently the vertical elevation of banks, and the physical meaning of λ is associated with the hydraulic parameters governing the problem. The EDMTC proposed in this study with R^2 greater than 0.95 and MAE in the 0.076–0.436 range for different observational data series is able to predict the bank profile shape trend as well as the free water surface level in threshold channels. In addition, the uncertainty analysis of EDMTC demonstrated that more than 95% of predicted and observed data are within the CB with low WUB, and the model reliability is largely assured. The EDMTC computational model presented in this paper can be used widely to predict stable channel profiles when the given problem information only includes the Q and d_{50}. This study was developed on Shannon entropy concept, it is suggested to improve the obtained results with other generalized entropies. It is further recommended that other equations provided by different researchers be used to estimate the free surface width of channels. Regression and AI models based on more field data also ought to be used to estimate the mean transverse slope of banks as well as other entropy model types to examine the accuracy of the model presented in this study.

Author Contributions: Conceptualization, H.B. and A.G.; methodology, H.B., A.G., and I.E.; software, A.G. and I.E.; validation, H.B., A.G., and A.M.; formal analysis, H.B., A.G., and A.K.-K.-K.; investigation, H.B. and A.G.; resources, H.B., A.G., and A.H.A.; data curation, H.B., A.G., and I.E.; writing—original draft preparation, H.B. and A.G.; writing—review and editing, H.B., A.G., A.K.-K.-K., A.H.A., and A.M.; visualization, H.B. and A.G.; supervision, H.B.; project administration, H.B.; funding acquisition, A.M. All authors have read and agreed to the published version of the manuscript.

Funding: This research received no funding.

Acknowledgments: This research in part is supported by the Hungarian State and the European Union under the EFOP-3.6.2-16-2017-00016 project. Support of European Union, the new Szechenyi plan, European Social Fund and the Alexander von Humboldt Foundation are also acknowledged.

Conflicts of Interest: The authors declare no conflict of interest.

References

1. Julien, P.Y.; Wargadalam, J. Alluvial Channel Geometry: Theory and Applications. *J. Hydraul. Eng.* **1995**, *121*, 312–325. [CrossRef]
2. Parker, G. Self-formed straight rivers with equilibrium banks and mobile bed. Part 1. The sand-silt river. *J. Fluid Mech.* **1978**, *89*, 109–125. [CrossRef]
3. Wolman, M.G.; Brush, L.M. *Factors Controlling the Size and Shape of Stream Channels in Coarse Noncohesive Sands*; US Government Printing Office: Washington, DC, USA, 1961.
4. Glover, R.E.; Florey, Q.L. *Stable Channel Profiles*; Lab. Rep. 325Hydraul; U.S. Bureau of Reclamation: Washington, DC, USA, 1951.
5. Lane, E.W. Progress report on studies on the design of stable channels by the Bureau of Reclamation. *Proc. Am. Soc. Civ. Eng. ASCE* **1953**, *79*, 1–31.
6. Parker, G. Self-formed straight rivers with equilibrium banks and mobile bed, Part 2. The gravel river. *J. Fluid Mech.* **1978**, *89*, 127–146. [CrossRef]
7. Babaeyan-Koopaei, K. A Study of Straight Stable Channels and Their Interactions with Bridge Structures. Ph.D. Thesis, University of Newcastle Upon Tyne, Newcastle Upon Tyne, UK, 1996.
8. Hey, R.D.; Heritage, G.L. Dimensional and dimensionless regime equations for gravel-bed rivers. In *International Conference on River Regime*; Hydraulics Research Limited; Wiley: Wallingford, UK, 1988; pp. 1–8.
9. Lawrence, S.D. *Fluvial Hydraulics*; Oxford University Press: Oxford, UK, 2009; pp. 92–111.
10. Gholami, A.; Bonakdari, H.; Mohammadian, M.; Zaji, A.H.; Gharabaghi, B. Assessment of geomorphological bank evolution of the alluvial threshold rivers based on entropy concept parameters. *Hydrol. Sci. J.* **2019**, *64*, 856–872. [CrossRef]
11. Vigilar, G.; Diplas, P. Stable channels with mobile bed: Model verification and graphical solution. *J. Hydraul. Eng. Asce* **1998**, *124*, 1097–1108. [CrossRef]
12. Afzalimehr, H.; Abdolhosseini, M.; Singh, V.P. Hydraulic geometry relations for stable channel design. *J. Hydrol. Eng.* **2010**, *15*, 859–864. [CrossRef]
13. Gholami, A.; Bonakdari, H.; Ebtehaj, I.; Shaghaghi, S.; Khoshbin, F. Developing an expert group method of data handling system for predicting the geometry of a stable channel with a gravel bed. *Earth Surf. Process. Landf.* **2017**, *42*, 1460–1471. [CrossRef]
14. Hey, R.D.; Thorne, C.R. Stable channels with mobile gravel beds. *J. Hydraul. Eng.* **1986**, *112*, 671–689. [CrossRef]
15. Bonakdari, H.; Gholami, A.; Gharabaghi, B. Modelling Stable Alluvial River Profiles Using Back Propagation-Based Multilayer Neural Networks. In Proceedings of the Intelligent Computing-Proceedings of the Computing Conference, London, UK, 16–17 July 2019; Springer: Cham, Switzerland, 2019; pp. 607–624.
16. Métivier, F.; Lajeunesse, E.; Devauchelle, O. Laboratory rivers: Lacey's law, threshold theory, and channel stability. *Earth Surf. Dyn.* **2017**, *5*, 187. [CrossRef]
17. Bonakdari, H.; Gholami, A.; Sattar, A.M.; Gharabaghi, B. Development of robust evolutionary polynomial regression network in the estimation of stable alluvial channel dimensions. *Geomorphology* **2020**, *350*, 106895. [CrossRef]
18. Gholami, A.; Bonakdari, H.; Ebtehaj, I.; Khodashenas, S.R. Reliability and sensitivity analysis of robust learning machine in prediction of bank profile morphology of threshold sand rivers. *Measurement* **2020**, *153*, 107411. [CrossRef]
19. Singh, V.P. *Entropy Theory in Hydraulic Engineering: An Introduction*; American Society of Civil Engineers: Reston, VA, USA, 2014.
20. Ikeda, S. Self-formed straight channels in sandy beds. *J. Hydraul. Div. Asce* **1981**, *107*, 389–406.
21. Diplas, P. Characteristics of self-formed straight channels. *J. Hydraul. Eng. ASCE* **1990**, *116*, 707–728. [CrossRef]
22. Pizzuto, J.E. Numerical simulation of gravel river widening. *Water Resour. Res.* **1990**, *26*, 1971–1980. [CrossRef]
23. Diplas, P.; Vigilar, G. Hydraulic geometry of threshold channels. *J. Hydraul. Eng. Asce* **1992**, *118*, 597–614. [CrossRef]

24. Vigilar, G.; Diplas, P. Design of a threshold channel. In *Hydraulic Engineering: Saving a Threatened Resource—In Search of Solutions*; ASCE: Reston, VA, USA, 1992; pp. 729–734.
25. Vigilar, G.; Diplas, P. Stable channels with mobile bed: Formulation and numerical solution. *J. Hydraul. Eng. Asce* **1997**, *123*, 189–199. [CrossRef]
26. Dey, S. Bank profile of threshold channels: A simplified approach. *J. Irrig. Drain. Eng. Asce* **2001**, *127*, 184–187. [CrossRef]
27. Yu, G.; Knight, D.W. Geometry of self-formed straight threshold channels in uniform material. In *Water Maritime and Energy, Proceedings of the Institute of Civil Engineering, London, UK*; Institute of Civil Engineering: London, UK, 14 September 1998; Volume 130, pp. 31–41.
28. Cao, S.; Knight, D.W. Entropy-based design approach of threshold alluvial channels. *J. Hydraul. Res.* **1997**, *35*, 505–524. [CrossRef]
29. Chow, V.D. *Open Channel Hydraulics*; McGraw-Hill: New York, NY, USA, 1959; pp. 20–21.
30. Gholami, A.; Bonakdari, H.; Ebtehaj, I.; Mohammadian, M.; Gharabaghi, B.; Khodashenas, S.R. Uncertainty Analysis of Intelligent Model of Hybrid Genetic Algorithm and Particle Swarm Optimization with ANFIS to Predict Threshold Bank Profile Shape Based on Digital Laser Approach Sensing. *Measurement* **2018**, *121*, 294–303. [CrossRef]
31. Gholami, A.; Bonakdari, H.; Ebtehaj, I.; Gharabaghi, B.; Khodashenas, S.R.; Talesh, S.H.A.; Jamali, A.A. Methodological approach of predicting threshold channel bank profile by multi-objective evolutionary optimization of ANFIS. *Eng. Geol.* **2018**, *239*, 298–309. [CrossRef]
32. Gholami, A.; Bonakdari, H.; Ebtehaj, I.; Talesh, S.H.A.; Khodashenas, S.R.; Jamali, A. Analyzing bank profile shape of alluvial stable channels using robust optimization and evolutionary ANFIS methods. *Appl. Water Sci.* **2019**, *9*, 40. [CrossRef]
33. Gholami, A.; Bonakdari, H.; Zeynoddin, M.; Ebtehaj, I.; Gharabaghi, B.; Khodashenas, S.R. Reliable method of determining stable threshold channel shape using experimental and gene expression programming techniques. *Neural Comput. Appl.* **2019**, *31*, 5799–5817. [CrossRef]
34. Gholami, A.; Bonakdari, H.; Samui, P.; Mohammadian, M.; Gharabaghi, B. Predicting stable alluvial channel profiles using emotional artificial neural networks. *Appl. Soft Comput.* **2019**, *78*, 420–437. [CrossRef]
35. Deng, Z.Q.; Singh, V.P. Mechanism and conditions for change in channel pattern. *J. Hydraul. Res.* **1999**, *37*, 465–478. [CrossRef]
36. Liang, J.H.; Ghidaoui, M.S.; Deng, J.Q.; Gray, W.G. A Boltzmann-based finite volume algorithm for surface water flows on cells of arbitrary shapes. *J. Hydraul. Res.* **2007**, *45*, 147–164. [CrossRef]
37. Eskov, V.M.; Eskov, V.V.; Vochmina, Y.V.; Gorbunov, D.V.; Ilyashenko, L.K. Shannon entropy in the research on stationary regimes and the evolution of complexity. *Mosc. Univ. Phys. Bull.* **2017**, *72*, 309–317. [CrossRef]
38. Zhao, J.; Wang, D.; Yang, H.; Sivapalan, M. Unifying catchment water balance models for different time scales through the maximum entropy production principle. *Water Resour. Res.* **2016**, *52*, 7503–7512. [CrossRef]
39. Chiu, C.L. Entropy and probability concepts in hydraulics. *J. Hydraul. Eng.* **1987**, *113*, 583–599. [CrossRef]
40. Araujo, J.C.D.; Chaudhry, F.H. Experimental evaluation of 2-D entropy model for open-channel flow. *J. Hydraul. Eng.* **1998**, *124*, 1064–1067. [CrossRef]
41. Bonakdari, H. Establishment of relationship between mean and maximum velocities in narrow sewers. *J. Environ. Manag.* **2012**, *113*, 474–480. [CrossRef]
42. Bonakdari, H.; Sheikh, Z.; Tooshmalani, M. Comparison between Shannon and Tsallis entropies for prediction of shear stress distribution in open channels. *Stoch. Environ. Res. Risk Assess.* **2015**, *29*, 1–11. [CrossRef]
43. Chiu, C.L.; Said, C.A.A. Maximum and mean velocities and entropy in open-channel flow. *J. Hydraul. Eng.* **1995**, *121*, 26–35. [CrossRef]
44. Chiu, C.L.; Chen, Y.C. An efficient method of discharge estimation based on probability concept. *J. Hydraul. Res.* **2003**, *41*, 589–596. [CrossRef]
45. Cui, H.; Singh, V.P. Application of minimum relative entropy theory for streamflow forecasting. *Stoch. Environ. Res. Risk Assess.* **2017**, *31*, 587–608. [CrossRef]
46. Kazemian-Kale-Kale, A.; Bonakdari, H.; Gholami, A.; Khozani, Z.S.; Akhtari, A.A.; Gharabaghi, B. Uncertainty analysis of shear stress estimation in circular channels by Tsallis entropy. *Phys. A Stat. Mech. Its Appl.* **2018**, *510*, 558–576. [CrossRef]

47. Gholami, A.; Bonakdari, H.; Zaji, A.H.; Akhtari, A.A. Simulation of open channel bend characteristics using computational fluid dynamics and artificial neural networks. *Eng. Appl. Comput. Fluid Mech.* **2015**, *9*, 355–369. [CrossRef]
48. Marini, G.; De Martino, G.; Fontana, N.; Fiorentino, M.; Singh, V.P. Entropy approach for 2D velocity distribution in open-channel flow. *J. Hydraul. Res.* **2011**, *49*, 784–790. [CrossRef]
49. Moramarco, T.; Saltalippi, C.; Singh, P. Estimation of mean velocity in natural channels based on Chiu's velocity distribution equation. *J. Hydrol. Eng.* **2004**, *9*, 42–50. [CrossRef]
50. Singh, V.P.; Luo, H. Entropy theory for distribution of one-dimensional velocity in open channels. *J. Hydrol. Eng.* **2011**, *16*, 725–735. [CrossRef]
51. Singh, V.P.; Cui, H. Suspended sediment concentration distribution using Tsallis entropy. *Phys. A Stat. Mech. Its Appl.* **2014**, *414*, 31–42. [CrossRef]
52. Sterling, M.; Knight, D.W. An attempt at using the entropy approach to predict the transverse distribution of boundary shear stress in open channel flow. *Stochastic Environ. Res. Risk Assess.* **2002**, *16*, 127–142. [CrossRef]
53. Choo, Y.M.; Yun, G.S.; Choo, T.H.; Kwon, Y.B.; Sim, S.Y. Study of shear stress in laminar pipe flow using entropy concept. *Environ. Earth Sci.* **2017**, *76*, 616. [CrossRef]
54. Gholami, A.; Bonakdari, H.; Mohammadian, M. Enhanced formulation of the probability principle based on maximum entropy to design the bank profile of channels in geomorphic threshold. *Stoch. Environ. Res. Risk Assess.* **2019**, *33*, 1013–1034. [CrossRef]
55. Gholami, A.; Bonakdari, H.; Mohammadian, A. A method based on the Tsallis entropy for characterizing threshold channel bank profiles. *Phys. A Stat. Mech. Its Appl.* **2019**, *526*, 121089. [CrossRef]
56. Shannon, C.E. A mathematical theory of communication. *Bell Syst. Tech. J.* **1948**, *27*, 623–656. [CrossRef]
57. Jaynes, E.T. Information theory and statistical mechanics I. *Phys. Rev.* **1957**, *106*, 620–630. [CrossRef]
58. Jaynes, E.T. Information theory and statistical mechanics II. *Phys. Rev.* **1957**, *108*, 171–190. [CrossRef]
59. Barbe, D.E.; Cruise, J.F.; Singh, V.P. Solution of three constraint entropy-based velocity distribution. *J. Hydraul. Eng.* **1991**, *117*, 1389–1396. [CrossRef]
60. Cao, S.; Chang, H.H. Entropy as a probability concept in energy-gradient distribution. In Proceedings of the National Conference Hydraulic Engineering, Colorado Springs, CO, USA, 8–12 August 1988; ASCE: New York, NY, USA, 1988; pp. 1013–1018.
61. Pipes, L.A. *Applied Mathematics for Engineering and Physicists*; McGraw-Hill: London, UK, 1970.
62. Van Burkalow, A. Angle of repose and angle of sliding friction: An experimental study. *Geol. Soc. Am. Bull.* **1945**, *56*, 669–707. [CrossRef]
63. Ebtehaj, I.; Sattar, A.; Bonakdari, H.; Zaji, A.H. Prediction of scour depth around bridge piers using self-adaptive extreme learning machine. *J. Hydroinform.* **2017**, *19*, 207–224. [CrossRef]
64. Gholami, A.; Bonakdari, H.; Zaji, A.H.; Akhtari, A.A.; Khodashenas, S.R. Predicting the Velocity Field in a 90° Open Channel Bend Using a Gene Expression Programming Model. *Flow Meas. Instrum.* **2015**, *46*, 189–192. [CrossRef]
65. Mikhailova, N.A.; Shevchenko, O.B.; Selyametov, M.M. Laboratory of Investigation of the formation of stable channels. *Hydro Tech. Constr.* **1980**, *14*, 714–722. [CrossRef]
66. Macky, G.H. Large flume experiments on the stable straight gravel bed channel. *Water Resour. Res.* **1999**, *35*, 2601–2603. [CrossRef]
67. Hassanzadeh, Y.; Majdzadeh, T.M.R.; Imanshoar, F.; Jafari, A. Validation of river bank profiles in sand-bed rivers. *J. Civ. Environ. Eng.* **2014**, *43*, 59–68.
68. Khodashenas, S.R. Threshold gravel channels bank profile: A comparison among 13 models. *Int. J. River Basin Manag.* **2016**, *14*, 337–344. [CrossRef]
69. Gholami, A.; Bonakdari, H.; Zaji, A.H.; Michelson, D.G.; Akhtari, A.A. Improving the performance of multi-layer perceptron and radial basis function models with a decision tree model to predict flow variables in a sharp 90 bend. *Appl. Soft Comput.* **2016**, *48*, 563–583. [CrossRef]
70. Gholami, A.; Bonakdari, H.; Akhtari, A.A. Developing finite volume method (FVM) in numerical simulation of flow pattern in 60 open channel bend. *J. Appl. Res. Water Wastewater* **2016**, *3*, 193–200.
71. Gholami, A.; Bonakdari, H.; Ebtehaj, I.; Akhtari, A.A. Design of an adaptive neuro-fuzzy computing technique for predicting flow variables in a 90° sharp bend. *J. Hydroinform.* **2017**, *19*, 572–585. [CrossRef]

72. Gholami, A.; Bonakdari, H.; Zaji, A.H.; Fenjan, S.A.; Akhtari, A.A. New radial basis function network method based on decision trees to predict flow variables in a curved channel. *Neural Comput. Appl.* **2018**, *30*, 2771–2785. [CrossRef]
73. Harman, C.; Stewardson, M.; DeRose, R. Variability and uncertainty in reach bankfull hydraulic geometry. *J. Hydrol.* **2008**, *351*, 13–25. [CrossRef]
74. Ebtehaj, I.; Bonakdari, H. No-deposition sediment transport in sewers using of gene expression programming. *Soft Comput. Civ. Eng.* **2017**, *1*, 29–53.
75. Newcombe, R.G. Two-sided confidence intervals for the single proportion: Comparison of seven methods. *Stat. Med.* **1998**, *17*, 857–872. [CrossRef]
76. Gholami, A.; Bonakdari, H.; Zaji, A.H.; Akhtari, A.A. A comparison of artificial intelligence-based classification techniques in predicting flow variables in sharp curved channels. *Eng. Comput.* **2020**, *36*, 295–324. [CrossRef]
77. Gholami, A.; Bonakdari, H.; Zaji, A.H.; Akhtari, A.A. An efficient classified radial basis neural network for prediction of flow variables in sharp open-channel bends. *Appl. Water Sci.* **2019**, *9*, 145. [CrossRef]
78. Gholami, A.; Akhtari, A.A.; Minatour, Y.; Bonakdari, H.; Javadi, A.A. Experimental and numerical study on velocity fields and water surface profile in a strongly-curved 90 open channel bend. *Eng. Appl. Comput. Fluid Mech.* **2014**, *8*, 447–461. [CrossRef]
79. Berry, G.; Armitage, P. Mid-P confidence intervals: A brief review. *J. R. Stat. Soc. Ser. D (Stat.)* **1995**, *44*, 417–423. [CrossRef]
80. Cox, D.R.; Hinkley, D.V. *Theoretical Statistics*; Chapman and Hall: London, UK, 1974.

Publisher's Note: MDPI stays neutral with regard to jurisdictional claims in published maps and institutional affiliations.

© 2020 by the authors. Licensee MDPI, Basel, Switzerland. This article is an open access article distributed under the terms and conditions of the Creative Commons Attribution (CC BY) license (http://creativecommons.org/licenses/by/4.0/).

MDPI
St. Alban-Anlage 66
4052 Basel
Switzerland
Tel. +41 61 683 77 34
Fax +41 61 302 89 18
www.mdpi.com

Entropy Editorial Office
E-mail: entropy@mdpi.com
www.mdpi.com/journal/entropy

www.ingramcontent.com/pod-product-compliance
Lightning Source LLC
LaVergne TN
LVHW070446100526
838202LV00014B/1673